Computer Aided Machine Drawing Practice

AUTHOR

Dr. Kanak Kalita
Associate Professor,
Dept. of Mechanical Engineering (ME),Vel Tech Rangarajan
Dr.Sagunthala R&D Institute of Science and Technology,
Chennai, Tamil Nadu

REVIEWER

Dr. Anil Kumar Agarwal
Director
Technical Education Rajasthan

KHANNA BOOK PUBLISHING CO. (P) Ltd.
4C/4344, Ansari Road, Darya Ganj,
New Delhi, India

All India Council for Technical Education
Nelson Mandela Marg, Vasant Kunj,
New Delhi, India

Dear Readers,

To prevent the piracy, this book is secured with HIGH SECURITY HOLOGRAM on the front title cover. In case you don't find the hologram on the front cover title, please write us to at contact@khannabooks.com or whatsapp us at +91-99109 09320 and avail special gift voucher for yourself.

Specimen of Hologram on front Cover title:

Moreover, there is a SPECIAL DISCOUNT COUPON for you with EVERY HOLOGRAM.
How to avail this SPECIAL DISCOUNT:
Step 1: Scratch the hologram
Step 2: Under the scratch area, your "coupon code" is available
Step 3: Logon to www.khannabooks.com
Step 4: Use your "coupon code" in the shopping cart and get your copy at a special discount
Step 5: Enjoy your reading!

ISBN: 978-93-55383-57-0
Book Code: DIP452EN

Computer Aided Machine Drawing Practice

by Dr. Kanak Kalita

First Edition: 2025

Originally Published by:
All India Council for Technical Education, Ministry of Education, GOI

Exclusively Published & Distributed by:
Khanna Book Publishing Co. (P) Ltd.
4C/4344, Ansari Road, Daryaganj,
New Delhi (INDIA)

Visit us at: **www.khannabooks.com**
Write us at: **contact@khannabooks.com**

To view complete list of books, Please scan the QR Code:

KPH

Printed in Bharat.

प्रो. टी. जी. सीताराम
अध्यक्ष
Prof. T. G. Sitharam
Chairman

सत्यमेव जयते

आज़ादी का अमृत महोत्सव

अखिल भारतीय तकनीकी शिक्षा परिषद्
(भारत सरकार का एक सांविधिक निकाय)
(शिक्षा मंत्रालय, भारत सरकार)
नेल्सन मंडेला मार्ग, वसंत कुंज, नई दिल्ली–110070
दूरभाष : 011–26131498
ई–मेल : chairman@aicte-india.org

ALL INDIA COUNCIL FOR TECHNICAL EDUCATION
(A STATUTORY BODY OF THE GOVT. OF INDIA)
(Ministry of Education, Govt. of India)
Nelson Mandela Marg, Vasant Kunj, New Delhi-110070
Phone : 011-26131498
E-mail : chairman@aicte-india.org

FOREWORD

Engineers are the backbone of any modern society. They are the ones responsible for the marvels as well as the improved quality of life across the world. Engineers have driven humanity towards greater heights in a more evolved and unprecedented manner.

The All India Council for Technical Education (AICTE), have spared no efforts towards the strengthening of the technical education in the country. AICTE is always committed towards promoting quality Technical Education to make India a modern developed nation emphasizing on the overall welfare of mankind.

An array of initiatives has been taken by AICTE in last decade which have been accelerated now by the National Education Policy (NEP) 2020. The implementation of NEP under the visionary leadership of Hon'ble Prime Minister of India envisages the provision for education in regional languages to all, thereby ensuring that every graduate becomes competent enough and is in a position to contribute towards the national growth and development through innovation & entrepreneurship.

One of the spheres where AICTE had been relentlessly working since past couple of years is providing high quality original technical contents at Under Graduate & Diploma level prepared and translated by eminent educators in various Indian languages to its aspirants. For students pursuing 2nd year of their Engineering education, AICTE has identified 87 books, which shall be translated into 12 Indian languages - Hindi, Tamil, Gujarati, Odia, Bengali, Kannada, Urdu, Punjabi, Telugu, Marathi, Assamese & Malayalam. In addition to the English medium, books in different Indian Languages are going to support the students to understand the concepts in their respective mother tongue.

On behalf of AICTE, I express sincere gratitude to all distinguished authors, reviewers and translators from the renowned institutions of high repute for their admirable contribution in a record span of time.

AICTE is confident that these outcomes based original contents shall help aspirants to master the subject with comprehension and greater ease.

T.G. Sitharam

(Prof. T. G. Sitharam)

ACKNOWLEDGEMENT

I am deeply grateful to the authorities of AICTE, with special acknowledgment to Prof. T. G. Sitharam, Chairman; Dr. Abhay Jere, Vice-Chairman; Prof. Rajive Kumar, Member-Secretary; and Dr. Ramesh Unnikrishnan, Advisor-II and Dr. Sunil Luthra, Director, Training and Learning Bureau. Their collective vision and planning were instrumental in the publication of the book on ***Computer Aided Machine Drawing Practice***. Sincere appreciation is extended to Prof. Anil Kumar Agarwal, Director Technical Education Rajasthan. As the reviewer of this book, Prof. Agarwal has made invaluable contributions, ensuring the content is not only student-friendly but also aesthetically pleasing and well-organized.

I extend my heartfelt thanks to Founder President and Chancellor Col. Prof. Vel. Dr. R. Rangarajan, Foundress President Dr. Sagunthala Rangarajan and Chairperson Managing Trustee Mrs. Rangarajan Mahalakshmi Kishore. Their unwavering institutional support was crucial in the successful completion of this book. My gratitude also goes to Prof. Dr. S. Salivahanan, Vice-Chancellor and Prof. Dr. E. Kannan, Registrar, for their motivational and continuous encouragement throughout the process. I express my deepest thanks to Dr P Chandrakumar, Dean Research & Development, for his motivation.

Special acknowledgment goes to my friend Vikas Kumar and my wife Swarna, both of whom provided invaluable support with the illustrations in this book.

I wish to thank my parents, whose nurturing and support have been my pillars through the significant events of my life. Lastly, I extend my profound thanks to my son Aatherv, for his understanding and patience, excusing my work during weekends and family time.

This book is an outcome of various suggestions of AICTE members, experts and authors who shared their opinion and thought to further develop the engineering education in our country. Acknowledgements are due to the contributors and different workers in this field whose published books, review articles, papers, photographs, footnotes, references and other valuable information enriched me at the time of writing the book.

Dr. Kanak Kalita

PREFACE

The advent of Computer-Aided Design (CAD) has revolutionized the field of engineering drawing, providing tools that allow for efficient and precise creation, modification and documentation of design. "Computer Aided Machine Drawing Practice" is a comprehensive guide meticulously crafted to bridge the gap between theoretical principles and their practical applications in the realm of machine drawing.

The book commences with an introduction to CAD software, providing a conceptual framework with a special emphasis on AutoCAD. It elucidates the benefits of utilizing AutoCAD, the latest features in AutoCAD 2022 and the system requirements for running the software effectively.

As readers delve deeper, they will navigate through the components of the AutoCAD software window, each of which is explained in detail to facilitate a smooth user experience. From understanding the basic functionalities to mastering the viewport, each section is designed to equip the readers with the knowledge and skills required to operate AutoCAD with confidence and ease.

The book also provides detailed drawings and explanations of various mechanical components, including cotter joints, knuckle joints, flanged couplings, universal couplings, plummer blocks, footstep bearings, eccentrics, machine vices, connecting rods, stuffing boxes and screw jacks. Each component is explained with detailed drawings.

"Computer Aided Machine Drawing Practice" is not just a book; it is a practical tool designed to empower engineering students with the knowledge and skills required to excel in the field of machine drawing using CAD software. Through its systematic approach, detailed explanations and practical examples, this book aims to make the process of learning AutoCAD not only easy but also enjoyable.

I hope that this book serves as a valuable resource for students, educators, and professionals in the field of engineering drawing and CAD. Your feedback is invaluable to me and I welcome any suggestions and comments that can help improve future editions of this book. Happy drawing!

Dr. Kanak Kalita

OUTCOME BASED EDUCATION

For the implementation of an outcome based education the first requirement is to develop an outcome based curriculum and incorporate an outcome based assessment in the education system. By going through outcome based assessments, evaluators will be able to evaluate whether the students have achieved the outlined standard, specific and measurable outcomes. With the proper incorporation of outcome based education there will be a definite commitment to achieve a minimum standard for all learners without giving up at any level. At the end of the programme running with the aid of outcome based education, a student will be able to arrive at the following outcomes:

Programme Outcomes (POs) are statements that describe what students are expected to know and be able to do upon graduating from the program. These relate to the skills, knowledge, analytical ability attitude and behaviour that students acquire through the program. The POs essentially indicate what the students can do from subject-wise knowledge acquired by them during the program. As such, POs define the professional profile of an engineering diploma graduate.

National Board of Accreditation (NBA) has defined the following seven POs for an Engineering diploma graduate:

PO1. Basic and Discipline specific knowledge: Apply knowledge of basic mathematics, science and engineering fundamentals and engineering specialization to solve the engineering problems.

PO2. Problem analysis: Identify and analyses well-defined engineering problems using codified standard methods.

PO3. Design/ development of solutions: Design solutions for well-defined technical problems and assist with the design of systems components or processes to meet specified needs.

PO4. Engineering Tools, Experimentation and Testing: Apply modern engineering tools and appropriate technique to conduct standard tests and measurements.

PO5. Engineering practices for society, sustainability and environment: Apply appropriate technology in context of society, sustainability, environment and ethical practices.

PO6. Project Management: Use engineering management principles individually, as a team member or a leader to manage projects and effectively communicate about well-defined engineering activities.

PO7. Life-long learning: Ability to analyse individual needs and engage in updating in the context of technological changes.

COURSE OUTCOMES

After By the end of the course the students are expected to learn:By the end of the course the students are expected to learn:

CO-1: Understand the representation of materials used in machine drawing.

CO-2: Draw the development of surfaces for sheet metal working applications.

CO-3: Draw the machine elements including keys, couplings, cotters, riveted, bolted and welded joints.

CO-4: Construct an assembly drawing using part drawings of machine components.

CO-5: Represent tolerances and the levels of surface finish of machine elements.

Mapping of Course Outcomes with Programme Outcomes to be done according to the matrix given below:

Course Outcomes	EXPECTED MAPPING WITH PROGRAMME OUTCOMES (1- Weak Correlation; 2- Medium correlation; 3- Strong Correlation)						
	PO-1	PO-2	PO-3	PO-4	PO-5	PO-6	PO-7
CO-1	3	3	3	3	1	1	3
CO-2	3	2	2	2	1	1	3
CO-3	3	2	2	2	1	1	3
CO-4	3	2	2	3	1	1	3
CO-5	3	3	3	3	1	1	3

GUIDELINES FOR TEACHERS

To implement Outcome Based Education (OBE) knowledge level and skill set of the students should be enhanced. Teachers should take a major responsibility for the proper implementation of OBE. Some of the responsibilities (not limited to) for the teachers in OBE system may be as follows:

- Within reasonable constraint, they should manoeuvre time to the best advantage of all students.
- They should assess the students only upon certain defined criterion without considering any other potential ineligibility to discriminate them.
- They should try to grow the learning abilities of the students to a certain level before they leave the institute.
- They should try to ensure that all the students are equipped with the quality knowledge as well as competence after they finish their education.
- They should always encourage the students to develop their ultimate performance capabilities.
- They should facilitate and encourage group work and team work to consolidate newer approach.
- They should follow Blooms taxonomy in every part of the assessment.

Bloom's Taxonomy

Level	Teacher should Check	Student should be able to	Possible Mode of Assessment
Creating	Students ability to create	Design or Create	Mini project
Evaluating	Students ability to Justify	Argue or Defend	Assignment
Analysing	Students ability to distinguish	Differentiate or Distinguish	Project/Lab Methodology
Applying	Students ability to use information	Operate or Demonstrate	Technical Presentation/ Demonstration
Understanding	Students ability to explain the ideas	Explain or Classify	Presentation/Seminar
Remembering	Students ability to recall (or remember)	Define or Recall	Quiz

GUIDELINES FOR STUDENTS

Students should take equal responsibility for implementing the OBE. Some of the responsibilities (not limited to) for the students in OBE system are as follows:

- Students should be well aware of each UO before the start of a unit in each and every course.
- Students should be well aware of each CO before the start of the course.
- Students should be well aware of each PO before the start of the programme.
- Students should think critically and reasonably with proper reflection and action.
- Learning of the students should be connected and integrated with practical and real life consequences.
- Students should be well aware of their competency at every level of OBE.

LIST OF FIGURES

Unit :1 Introduction to CAD Software

Fig. 1.1: AutoCAD Application window 6
Fig. 1.2: Ribbon Toolbar 6
Fig. 1.3: Quick Access Toolbar 7
Fig. 1.4: Draw Toolbar 7
Fig. 1.5: Modify Toolbar 8
Fig. 1.6: UCS Icon 8
Fig. 1.7: Status Bar 9
Fig. 1.8: Steps to open a new file 9
Fig. 1.9: Steps to save a file 10
Fig. 1.10: Steps to open an existing drawing file 10
Fig. 1.11: Steps to quitting a session 11
Fig. 1.12: Example of setting units 12
Fig. 1.13: Example of setting limits 12
Fig. 1.14: Example of toggling grid on and off 13
Fig. 1.15: Example of setting Snap and adjusting its properties 13
Fig. 1.16: Dynamic input 13
Fig. 1.17: Example of orthomode 14
Fig. 1.18: Example of polar tracking 14
Fig. 1.19: Snap tracking 15
Fig. 1.20: Object snap options 15
Fig. 1.21: Isodraft mode 15
Fig. 1.22: Three different isometric planes 16
Fig. 1.23: Lineweight display 16
Fig. 1.24: Transparency display 18
Fig. 1.25: Example of selection cycling 18
Fig. 1.26: Dynamic UCS 18
Fig. 1.27: Graphic Configures 19
Fig. 1.28: Undo and Redo command 19
Fig. 1.29: Example of application of absolute coordinates 19
Fig. 1.30: Example of application of relative coordinates 20
Fig. 1.31: Example of application of polar coordinates 21
Fig. 1.32: Example of application of cartesian coordinates 21

Unit : 2 Drawing Aid and Editing Commands in AutoCAD

Fig. 2.1: Point command 30
Fig. 2.2: Example of Point command 30
Fig. 2.3: Line command 31

Fig. 2.4: Example of Line command 32
Fig. 2.5: Ray command 32
Fig. 2.6: Example of Ray command 33
Fig. 2.7: Construction Line command 34
Fig. 2.8: Example of Construction Line command 34
Fig. 2.9: Circle command 35
Fig. 2.10: Different options for Circle command 36
Fig. 2.11: Example of center-radius Circle command 36
Fig. 2.12: Example of center-diameter Circle command 37
Fig. 2.13: Example of 2-points Circle command 37
Fig. 2.14: Example of 3-points Circle command 37
Fig. 2.15: Example of tan-tan-radius Circle command 38
Fig. 2.16: Example of tan-tan-tan Circle command 38
Fig. 2.17: Arc command 39
Fig. 2.18: Different options for Arc command 39
Fig. 2.19: Example of 3-point Arc command 40
Fig. 2.20: Example of start-center-end Arc command 40
Fig. 2.21: Example of start-center-angle Arc command 40
Fig. 2.22: Example of start-center-length Arc command 41
Fig. 2.23: Example of start-end-angle Arc command 41
Fig. 2.24: Example of start-end-direction Arc command 41
Fig. 2.25: Example of start-end-radius Arc command 42
Fig. 2.26: Example of center-start-end Arc command 42
Fig. 2.27: Example of center-start-angle Arc command 42
Fig. 2.28: Example of center-start-length Arc command 43
Fig. 2.29: Example of continue Arc command 43
Fig. 2.30: Rectangle command 44
Fig. 2.31: Example of Rectangle command 44
Fig. 2.32: Polyline command 45
Fig. 2.33: Example of Polyline command 45
Fig. 2.34: Polygon command 46
Fig. 2.35: Example of polygon inscribed in circle 46
Fig. 2.36: Example of polygon circumscribed in circle 47
Fig. 2.37: Ellipse command 47
Fig. 2.38: Example of center Ellipse command 48
Fig. 2.39: Example of axis-end Ellipse command 48
Fig. 2.40: Example of elliptical-arc Ellipse command 49
Fig. 2.41: Spline command 49
Fig. 2.42: Example of Spline Fit method 50
Fig. 2.43: Example of Spline CV method 50
Fig. 2.44: Divide command 51

Fig. 2.45: Example of Divide command 51
Fig. 2.46: Measure command 52
Fig. 2.47: Example of Measure command 52
Fig. 2.48: Move command 53
Fig. 2.49: Copy command 54
Fig. 2.50: Stretch command 55
Fig. 2.51: Rotate command 56
Fig. 2.52: Mirror command 57
Fig. 2.53: Scale command 58
Fig. 2.54: Trim command 59
Fig. 2.55: Extend Command 60
Fig. 2.56: Fillet Command 61
Fig. 2.57: Chamfer Command 62
Fig. 2.58: Array Command 63
Fig. 2.59: Explode Command 64
Fig. 2.60: Offset Command 65
Fig. 2.61: Break Command 66
Fig. 2.62: Join Command 67
Fig. 2.63: Lengthen Command 68

Unit : 3 Basic Dimensioning and Annotations in AutoCAD

Fig. 3.1: Quick Dimensioning 77
Fig. 3.2: Example of Quick Dimensioning 78
Fig. 3.3: Menu Dimensioning 79
Fig. 3.4: Dimension Toolbar 80
Fig. 3.5: Different types of Dimension Tools 83
Fig. 3.6: Example of a part drawing with various dimensions 83
Fig. 3.7: Linear Dimension Tool 84
Fig. 3.8: Example of linear dimensioning 84
Fig. 3.9: Aligned Dimension Tool 85
Fig. 3.10: Example of aligned dimensioning 85
Fig. 3.11: Angular Dimension Tool 86
Fig. 3.12: Example of angular dimensioning 86
Fig. 3.13: Arc Length Dimension Tool 87
Fig. 3.14: Radius Dimension Tool 88
Fig. 3.15: Example of radius dimensioning 88
Fig. 3.16: Diameter Dimension Tool 89
Fig. 3.17: Ordinate Dimension Tool 89
Fig. 3.18: Example of ordinate dimensioning 90
Fig. 3.19: Jogged Dimension Tool 90
Fig. 3.20: Example of jogged dimensioning 91

Fig. 3.21: Example of addition of text to the Dimension 92
Fig. 3.22: Example of text rotated by a given angle 93
Fig. 3.23: Text Command 94
Fig. 3.24: Single line text command 94
Fig. 3.25: Multiline text command 95
Fig. 3.26: Font style 95
Fig. 3.27: Font Colour 96
Fig. 3.28: Space between characters of the text 96
Fig. 3.29: Text alignment 96

Unit : 4 Hatching, Blocks and Views in AutoCAD
Fig. 4.1: Hatch command 104
Fig. 4.2: Example of Hatching 104
Fig. 4.3: Hatch tool 105
Fig. 4.4: Properties of Hatch Lines 105
Fig. 4.5: Hatching with different colours 106
Fig. 4.6: Hatch pattern scale 107
Fig. 4.7: Hatch angle 107
Fig. 4.8: Island detection tool 108
Fig. 4.9: Island detection tool options 108
Fig. 4.10: Types of Island detection tool 109
Fig. 4.11: Example of Island detection tool 109
Fig. 4.12: Steps in hatching an object with Island detection tool 110
Fig. 4.13: Steps in hatching with text 110
Fig. 4.14: Remove boundary of text while hatching 111
Fig. 4.15: Block definition 112
Fig. 4.16: Inserting Blocks 113
Fig. 4.17: The Attribute Definition dialog box 115
Fig. 4.18: Exploding Blocks 115
Fig. 4.19: Purge tools in Cleanup panel 116
Fig. 4.20: Purgeable Items option in Purge tools 117
Fig. 4.21: Find Non-Purgeable Items option in Purge tools 117
Fig. 4.22: Layer command on the ribbon pane 118
Fig. 4.23: Various layer properties 119
Fig. 4.24: Additional layer properties 120
Fig. 4.25: Example of objects drawn on different layers 120

Unit : 5 Viewport and Isomeric Drawing in AutoCAD
Fig. 5.1: Creating a new layout 127
Fig. 5.2: Opening the created layout 128
Fig. 5.3: Components of a layout 128

Fig. 5.4: Output option 128
Fig. 5.5: Page setup manager 129
Fig. 5.6: Modifying the layout 129
Fig. 5.7: Select DWG to PDF.pc3 130
Fig. 5.8: Select plot style table to acad.std 130
Fig. 5.9: Select ISO expand A1(841.00 x 594.00 MM) 131
Fig. 5.10: Select plot scale to 1.1 131
Fig. 5.11: Select rectangular option 132
Fig. 5.12: Create first and second corner of the rectangle 132
Fig. 5.13: Activate the model space inside the viewport 133
Fig. 5.14: Set the viewport scale 133
Fig. 5.15: Lock/unlock the drawing inside the viewport 134
Fig. 5.16: Insert the created model into single rectangular viewport 134
Fig. 5.17: Modify the viewport using grip 135
Fig. 5.18: Multiple rectangular viewport 135
Fig. 5.19: Select polygonal option from the ribbon 136
Fig. 5.20: Specifying the starting point 136
Fig. 5.21: Specifying next point 136
Fig. 5.22: Creation of polygonal viewport 137
Fig. 5.23: Create the objects 137
Fig. 5.24: Select object option 138
Fig. 5.25: Select an object to clip viewport 138
Fig. 5.26: Objects converted into viewport 138
Fig. 5.27: Select the clip option from the ribbon 139
Fig. 5.28: Select the viewport for clipping 139
Fig. 5.29: Select clipping object 140
Fig. 5.30: Clipped viewport in the selected object 140
Fig. 5.31: Scaling viewports 140
Fig. 5.32: Examples of viewport scale 141
Fig. 5.33: Freeze layer 142
Fig. 5.34: Layer override 143
Fig. 5.35: Change the color of selected layers 143
Fig. 5.36: Isoplane 144
Fig. 5.37: Example of an isometric drawing 145
Fig. 5.38: Example of an isometric drawing (Step 1 and 2) 145
Fig. 5.39: Example of an isometric drawing (Step 3) 146
Fig. 5.40: Example of an isometric drawing (Step 4 and 5) 146
Fig. 5.41: Example of an isometric drawing (Step 6) 146
Fig. 5.42: Example of an isometric drawing (Step 7 and 8) 147
Fig. 5.43: Example of an isometric drawing (Step 9, 10 and 11) 147
Fig. 5.44: Example of an isometric drawing (Step 12, 13 and 14) 147

Fig. 5.45: Example of an isometric drawing (Step 15) 148
Fig. 5.46: Example of an isometric drawing (Step 16 and 17) 148
Fig. 5.47: Example of an isometric drawing (Step 18, 19 and 20) 148
Fig. 5.48: Example of an isometric drawing (Step 21, 22 and 23) 149
Fig. 5.49: Example of an isometric drawing (Step 24 and 25) 149
Fig. 5.50: Example of an isometric drawing (Step 26 and 27) 149
Fig. 5.51: Example of an isometric drawing (Step 28, 29 and 30) 150
Fig. 5.52: Example of an isometric drawing (Step 31, 32, 33 and 34) 150
Fig. 5.53: Example of an isometric drawing (Step 35, 36, 37 and 38) 150
Fig. 5.54: Example of an isometric drawing (Step 39) 151
Fig. 5.55: Example of an isometric drawing (Step 40 and 41) 151
Fig. 5.56: Example of an isometric drawing (Step 42, 43, 44 and 45) 151
Fig. 5.57: Example of an isometric drawing (Step 46 and 47) 152
Fig. 5.58: Example of an isometric drawing (Step 48, 49, 50 and 51) 152
Fig. 5.59: Example of an isometric drawing (Step 52 and 53) 152

Unit : 6 Printing and Plotting in AutoCAD
Fig. 6.1: Plot command 160
Fig. 6.2: Configuring plotters 160
Fig. 6.3: Page setup 161
Fig. 6.4: Options for configuring page setup 161
Fig. 6.5: Printer/plotter selection 161
Fig. 6.6: Page size selection 162
Fig. 6.7: Selection of area of plotting 162
Fig. 6.8: Plot offset 162
Fig. 6.9: Plot scale 163
Fig. 6.10: Creating plot style 163
Fig. 6.11: Plot style tables editor 164
Fig. 6.12: Plot style tables editor options 164
Fig. 6.13: Selecting an object color 165
Fig. 6.14: (a) Line type (b) Line weight 166
Fig. 6.15: (a) Line end style (b) Line join style (c) Line fill style 167
Fig. 6.16: Example of changing the plot style leads to change in graphics (colour) of the object 167
Fig. 6.17: Steps in direct printing 168
Fig. 6.18: Example of print by layout 169
Fig. 6.19: Selecting new sheet set 170
Fig. 6.20: Selecting predefined sets 170
Fig. 6.21: Assigning name and storage location 170
Fig. 6.22: Browsing drawing layouts for printing 171
Fig. 6.23: Previewing and finalizing 171

Fig. 6.24: Publish to PDF 172
Fig. 6.25: Example of drawing published to PDF for printing as per assigned layout 172

Unit : 7 Machine Drawing Practice Using AutoCAD
Fig. 7.1: Sleeve and cotter joint 179
Fig. 7.2: Detailed views of sleeve and cotter joint 180
Fig. 7.3: Assembly drawing of sleeve and cotter joint 181
Fig. 7.4: Cotter joint with socket and spigot 182
Fig. 7.5: Assembly drawing of cotter joint with socket and spigot 182
Fig. 7.6: Detailed views of cotter joint with socket and spigot 183
Fig. 7.7: Knuckle joint 184
Fig. 7.8: Detailed views of knuckle joint 185
Fig. 7.9: Assembly drawing of knuckle joint 186
Fig. 7.10: Protected type flanged coupling 187
Fig. 7.11: Assembly drawing of protected type flanged coupling 187
Fig. 7.12: Detailed views of protected type flanged coupling 188
Fig. 7.13: Universal coupling 189
Fig. 7.14: Detailed views of universal coupling 190
Fig. 7.15: Assembly drawing of universal coupling 191
Fig. 7.16: Plummer block 192
Fig. 7.17: Assembly drawing of plummer block 192
Fig. 7.18: Detailed views of plummer block 193
Fig. 7.19: Footstep bearing 194
Fig. 7.20: Detailed views of footstep bearing 195
Fig. 7.21: Assembly drawing of footstep bearing 196
Fig. 7.22: Simple eccentric 197
Fig. 7.23: Assembly drawing of simple eccentric 197
Fig. 7.24: Detailed views of simple eccentric 198
Fig. 7.25: Machine vice 199
Fig. 7.26: Detailed views of machine vice 200
Fig. 7.27: Assembly drawing of machine vice 201
Fig. 7.28: Connecting rod 202
Fig. 7.29: Representation of connecting rod 203
Fig. 7.30: Detailed views of connecting rod 203
Fig. 7.31: Stuffing box 204
Fig. 7.32: Detailed views of stuffing box 205
Fig. 7.33: Assembly drawing of stuffing box 206
Fig. 7.34: Three-dimensional model and sectional view of screw jack 207
Fig. 7.35: Assembly drawing of screw jack 208
Fig. 7.36: Detailed views of screw jack 209

CONTENTS

Foreword *iii*

Acknowledgement *v*

Preface *vii*

Outcome Based Education *ix*

Course Outcomes *x*

Guidelines for Teachers *xi*

Guidelines for Students *xii*

List of Figures *xiii*

1. Introduction to CAD Software 1-27

Unit Specifics 1

Rationale 1

Pre-Requisites 2

Unit Outcomes 2

1.1 Cad Software 2

1.1.1 Computer Aided Drafting 2

1.1.2 Concept 3

1.1.3 Various CAD Software3

1.1.4 Benefits of using AutoCAD 3

1.1.5 AutoCAD 2022 4

1.1.6 System Requirements for AutoCAD 5

1.2 Components Of Autocad Software Window 5

1.2.1 Title Bar 6

1.2.2 Ribbon Toolbar 6

1.2.3 Quick Access Toolbar 6

1.2.4 Viewport 7

1.2.5 Command Prompt 7

1.2.6 Draw Toolbar 7

1.2.7 Modify Toolbar 8

1.2.8 UCS Icon 8

1.2.9 Status Bar 9

1.3 File Features 9

1.3.1 Opening a New File 9

1.3.2 Saving a File 10

1.3.3 Opening an Existing Drawing File 10

1.3.4 Quitting a session 11

1.4 Setting Up New Drawing 11
1.4.1 Setting Units 11
1.4.2 Setting Limits 12
1.4.3 Creating Grid 13
1.4.4 Setting Snap 13
1.4.5 Dynmode 13
1.4.6 Orthomode 14
1.4.7 Polar Tracking 14
1.4.8 Snap Tracking 14
1.4.9 Object Snap 15
1.4.10 Isodraft 15
1.4.11 Lineweight Display 16
1.4.12 Transparency Display 17
1.4.13 Selection Cycling 18
1.4.14 Dynamic UCS 18
1.4.15 Graphic Configures 19
1.4.16 Undo and Redo Command 19
1.5 Coordinate System In Autocad 19
1.5.1 Absolute Coordinate 19
1.5.2 Relative Coordinate 20
1.5.3 Polar Coordinate 20
1.5.4 Cartesian Coordinate System 21
Keyboard shortcuts for various commands 22
Unit Summary 24
Exercise 24
Know more 25
Reference and Suggested Reading 27
A. Multiple Choice Questions 24
B. Subjective Questions 26
Dynamic QR Code for Further Reading 27

2. Drawing Aids and Editing.. 28-74
Unit Specifics 28
Rationale 28
Pre-Requisites 29
Unit Outcomes 29

2.1 Introduction 29
2.2 Draw Menu 29
2.2.1 Point Command 30
2.2.2 Line Command 31
2.2.3 Ray Command 32
2.2.4 Construction Line Command 33
2.2.5 Circle Command 35
2.2.6 Arc Command 39
2.2.7 Rectangle Command 44
2.2.8 Polyline Command 45
2.2.9 Polygon Command 46
2.2.10 Ellipse Command 47
2.2.11 Spline Command 49
2.2.12 Divide Command 51
2.2.13 Measure Command 52
2.3 Modify Menu 53
2.3.1 Move Command 53
2.3.2 Copy Command 54
2.3.3 Stretch Command 55
2.3.4 Rotate Command 56
2.3.5 Mirror Command 56
2.3.6 Scale Command 57
2.3.7 Trim Command 58
2.3.8 Extend Command 59
2.3.9 Fillet Command 60
2.3.10 Chamfer Command 61
2.3.11 Array Command 62
2.3.12 Explode Command 64
2.3.13 Offset Command 64
2.3.14 Break Command 65
2.3.15 Join Command 66
2.3.16 Lengthen Command 67
Unit Summary 68
Exercise 68
Multiple Choice Questions 69
Subjective Questions 70

Problems for Practice 70
Know more 73
Reference and Suggested Reading 74
Dynamic QR Code for Further Reading 74

3. Basic Dimensioning and Annotations in AutoCAD 75-101
Unit Specifics 75
Rationale 75
Pre-Requisites 76
Unit Outcomes 76
3.1 Introduction 76
3.2 Dimensioning In Autocad 77
3.2.1 Quick Dimensioning 77
3.2.2 Menu Dimensioning 78
3.2.3 Command Dimensioning 79
3.2.4 Toolbar Dimensioning 80
3.3. Dimension Types 82
3.3.1 Linear 84
3.3.2 Aligned 85
3.3.3 Angular 86
3.3.5 Radius 87
3.3.6 Diameter 88
3.3.7 Ordinate 89
3.3.8 Jogged 90
3.4. Dimension Editing 91
3.4.1. Adding Text to the dimension 92
3.4.2. Rotating the dimension text 92
3.5. Text Writing In Autocad 93
3.5.1. Single Line Text 93
3.5.2. Multiline Text 94
Unit Summary 97
Exercise 97
A. Multiple Choice Questions 97
B. Subjective Questions 98
C. Problems for Practice 99
Know More 100

Reference and Suggested Reading 101
Dynamic QR Code for Further Reading 101

4. Hatching, Blocks and Views in AutoCAD 102-124

Unit Specifics 102
Rationale 102
Pre-Requisites 103
Unit Outcomes 103
4.1 Introduction 103
4.2 Hatching Commands 104
4.2.1 Hatch Tool 105
4.2.2 Properties of Hatch Lines 105
4.2.3 Island Detection Tools 107
4.2.4 Hatching and Text 110
4.2.5 Hatch Line Edits 111
4.3 Block Making 112
4.3.1 Creating Block Definitions 112
4.3.2 Inserting Blocks 113
4.3.3 Attributes: Fill-in-the-Blank Blocks 114
4.3.4 Exploding Blocks 115
4.3.5 Purging Unused Block Definitions 116
4.4 Layers Command 117
Unit Summary 121
Exercise 121
A. Multiple Choice Questions 121
B. Subjective Questions 123
Know more 123
Reference and Suggested Reading 124
Dynamic QR Code for Further Reading 124

5. Viewport and Isomeric Drawing in AutoCAD 125-157

Unit Specifics 125
Rationale 125
Pre-Requisites 126
Unit Outcomes 126
5.1 Introduction 126
5.2 Viewports Creation 127

5.2.1 Layout creation 127
5.2.2 Single Rectangular Viewports 132
5.2.3 Multiple Rectangular Viewports 134
5.2.4 Polygonal Viewports 135
5.2.5 Converting Existing Objects 137
5.2.6 Clipping Existing Viewports 139
5.3 Scaling and Maximizing Viewports 140
5.4 Freezing Layers in Viewports 142
5.5 Layer Override in Viewports 142
5.6 Isometric Drawing 144
Unit Summary 153
Exercise 153
A. Multiple Choice Questions 153
B. Subjective Questions 155
Know more 156
Reference and Suggested Reading 156
Dynamic QR Code for Further Reading 157

6. Printing and Plotting in AutoCAD 158-176
Unit Specifics 158
Rationale 158
Pre-Requisites 159
Unit Outcomes 159
6.1 Introduction 159
6.2 Plotting 159
6.2.1 Configuring Plotters 160
6.2.2 Creating Plot Style Tables 163
6.3 Printing 167
6.3.1 File Formats for Printing 168
6.3.2 Direct Printing 168
6.3.3 Print by Layout 168
6.3.4 Sheet Set Manager 169
Unit Summary 173
Exercise 173
A. Multiple Choice Questions 173
B. Subjective Questions 173

Know more 175
Reference and Suggested Reading 176
Dynamic QR Code for Further Reading 176

7. Machine Drawing Practice Using AutoCAD 177-216

Unit Specifics 177
Rationale 177
Pre-Requisites 178
Unit Outcomes 178
7.1 Cotter Joint With Sleeve 178
7.2 Cotter Joint With Socket And Spigot 181
7.3 Knuckle Joint 184
7.4 Protected Type Flanged Coupling 186
7.5 Universal Coupling 189
7.6 Plummer Block 191
7.7 Footstep Bearing 194
7.8 Simple Eccentric 196
7.9 Machine Vice 199
7.10 Connecting Rod 201
7.11 Stuffing Box 204
7.12 Screw Jack 207
Unit Summary 210
Exercise 210
A. Multiple Choice Questions 210
B. Subjective Questions 211
Know more 212
Reference and Suggested Reading 212
Dynamic QR Code for Further Reading 212

Co And Po Attainment Table 215
Index 216

1 Introduction to CAD Software

UNIT SPECIFICS

This unit presents information related to the following topics.

- Essential system requirements for AutoCAD 2022;
- Components of the AutoCAD software window;
- Procedures for file management including opening, saving and setting up new drawings;
- Understanding and utilizing different coordinate systems in AutoCAD;

This unit is designed to familiarize readers with the foundational aspects of AutoCAD, starting with the system requirements necessary for optimal software performance. It provides a walkthrough of the various components within the AutoCAD software window, each serving specific functions to facilitate the user's design process. The unit also outlines the straightforward procedures for file management in AutoCAD, including the steps to open new files, save files, open existing drawing files and exit the application. Additionally, it introduces the different coordinate systems available in AutoCAD, which are fundamental for accurately defining and locating points in the drawing space.

Apart from this at the end of the unit, a succinct recapitulation of the overall broad concepts is provided in the form of a unit summary. Besides, a large number of multiple-choice questions as well as descriptive type questions with Bloom's taxonomy action verbs are included. A list of references and suggested readings are given in the unit so that one can go through them for practice. It is important to note that for getting more information on various topics of interest some QR codes have been provided in different sections which can be scanned for relevant supportive knowledge. Video resources along with QR codes are mentioned for getting more information on various topics of interest which can be surfed or scanned through mobile phones for viewing.

RATIONALE

With the advent of technology, computer-aided design (CAD) has become indispensable in various industries due to its precision and efficiency compared to traditional drafting methods. AutoCAD, a premier CAD software, is extensively used for creating detailed digital drawings that can be easily modified, shared and stored. Learning the basics of AutoCAD, including understanding its interface, file management procedures and coordinate systems, is crucial for students and professionals who aim to excel in fields that require computer-aided drafting and design. This unit is structured to provide learners with the fundamental skills and knowledge needed to navigate and utilize AutoCAD effectively, laying the groundwork for more advanced studies and applications of the software in the future.

PRE-REQUISITES

Before reading this unit, the students are advised to revisit the following:

- Mathematics: Coordinate and Plane geometry (Class XII)
- Engineering Graphics (ES101)
- Basic Computer Skills

UNIT OUTCOMES

After studying this unit students will be able to:

U1-O1: Understand and meet the system requirements for installing and running AutoCAD 2022 efficiently.

U1-O2: Identify and utilize the components of the AutoCAD software window effectively.

U1-O3: Execute file management procedures in AutoCAD, including opening, saving and setting up new drawings.

U1-O4: Apply different coordinate systems in AutoCAD for accurate point definition and location in the drawing space.

U1-O5: Develop a foundational understanding of AutoCAD.

Unit-1 Outcomes	EXPECTED MAPPING WITH COURSE OUTCOMES (1 – Weak Correlation; 2 – Medium correlation; 3 – Strong Correlation)				
	CO-1	CO-2	CO-3	CO-4	CO-5
U1-O1	1	1	1	1	1
U1-O2	1	1	1	1	1
U1-O3	1	1	1	1	1
U1-O4	1	1	1	1	1
U1-O5	1	1	1	1	1

1.1 CAD SOFTWARE

In the ever-evolving domains of design and engineering, Computer-Aided Design (CAD) software has established itself as an irreplaceable asset. They provide a foundation for countless design-centric tasks. The introduction of CAD software has ushered in a transformative era in the design field, facilitating levels of precision and efficiency that were previously inconceivable with traditional manual drafting methods. This unit aims to shed light on the core components of CAD, offering insights into its fundamental principles, diverse applications and the distinct advantages provided by AutoCAD 2022. AutoCAD, in particular, is renowned for its user-friendly interface and a wide array of features that simplify the design process, making it an invaluable tool for professionals in the industry.

1.1.1 Computer Aided Drafting

Computer Aided Drafting (CAD) epitomizes a paradigm shift in design and documentation methodologies. It is not merely a digital tool but a comprehensive system that facilitates the meticulous creation, modification, analysis and optimization of designs. CAD provides a digital canvas allowing for effortless replication, modification and storage of designs, thereby obviating the need for cumbersome paper drafts. The precision afforded by CAD is unparalleled, ensuring consistency and accuracy in every design while minimizing the risk of human errors that are often inherent in manual drafting.

1.1.2 Concept

The CAD software is a refined digital platform, meticulously crafted to replicate and enhance the functionalities of conventional drafting tools. Thus, they often serve as an indispensable resource for a diverse group of professionals. These include architects, engineers, drafters and artists, all of whom rely on CAD for various purposes. From drafting architectural blueprints and designing complex machinery to creating detailed technical drawings and bringing artistic concepts to life, CAD software's versatility is evident. It offers users the ability to visualize their designs in both 2D and 3D formats, providing a comprehensive perspective crucial for the in-depth analysis and review of projects. The software's ability to represent designs visually in multiple formats allows for a more nuanced understanding and interpretation of designs, making it an essential tool in various professional fields.

1.1.3 Various CAD Software

The spectrum of CAD software available in the market is extensive and diverse which are meticulously designed to cater to the specific needs and requirements of different sectors and projects. AutoCAD, a pioneering creation of Autodesk, stands prominently in the CAD arena, celebrated for its robust set of features and functionalities. It enjoys widespread acclaim and usage across a multitude of industries, serving as a reliable tool for various design and drafting needs.

However, the CAD software landscape is not limited to AutoCAD alone. There are other significant players in the market, each bringing something unique to the table. For instance, SolidWorks is another highly regarded software, particularly invaluable in the realm of mechanical design. It offers an intuitive interface and powerful tools that are crucial for creating complex mechanical parts and assemblies, making it a favourite among mechanical engineers and designers.

Similarly, CATIA is another potent CAD software that is predominantly utilized in the aerospace and automotive industries. It is renowned for its advanced surface modelling capabilities, allowing engineers to create complex, three-dimensional models with high precision. This makes it an indispensable tool for designing aerodynamic and automotive components, where accuracy and detail are paramount.

The selection of CAD software is often a meticulous process that depends heavily on the specific requirements of the project at hand and the demands of the industry it caters to. Each software boasts a unique set of features and specialities, designed to address different challenges and needs in the design process. Therefore, professionals must make an informed choice, selecting a tool that aligns best with their project needs and industry standards, ensuring efficiency and accuracy in their design and drafting endeavours.

1.1.4 Benefits of using AutoCAD

AutoCAD since its inception in 1982 has evolved continuously, adding new features and functionalities each year. Some of the specific advantages of using AutoCAD are:

Accurate and Error-Reducing: AutoCAD's dynamic engineering model minimizes errors, allowing for real-time design modifications and improvements.

- **Time and Cost Efficient:** With a user-friendly interface, AutoCAD expedites the design process, saving both time and money while enhancing productivity.

- **Easy Data Transfer:** AutoCAD facilitates the sharing of heavy files without data loss, simplifying the process of online data uploading and sharing among multiple users.
- **Controllable and Manageable:** It supports scanning features for data, aiding in material quantity measurement and cost calculation, which is crucial for production management.
- **Manufacturing Database Creation:** AutoCAD assists in generating essential manufacturing data, helping in the creation and management of extensive databases crucial for production.
- **Seamless Import/Export:** It supports easy and quick import and export of files, with compatibility with various tools and features that enhance productivity.
- **Point Cloud Application:** Users can utilize point clouds from 3D laser scanners as a starting point for their designs, creating a 3D representation of existing structures.
- **Layering Feature:** This allows users to hide or display specific details in complex drawings, facilitating a clearer understanding of the design.
- **Quick Calculations:** AutoCAD enables swift calculations of mass, area, volume and center of gravity, with auto-dimensioning features for easy and accurate sketching.
- **Command Features:** With commands like Fill, Hatch, Section lines, chamfer and fillet, AutoCAD serves as a comprehensive solution for designers.
- **Image Tracing:** This feature supports the digital tracing of traditional diagrams and drawings, aiding in the conversion process.
- **PDF Import Enhancements:** AutoCAD's SHF text recognition tool converts geometries from imported PDF text into editable text objects, simplifying the editing process.

1.1.5 AutoCAD 2022

AutoCAD 2022, one of the most recent versions, is distinguished by its unparalleled versatility and comprehensive suite of features, making it a standout choice in the crowded CAD software market. This iteration is a universal solution adeptly serving various sectors, including but not limited to architecture, electronics, civil engineering and manufacturing. Its widespread application is a testament to its adaptability and proficiency in addressing diverse design needs.

Designed with user efficiency in mind, AutoCAD 2022 simplifies and accelerates the design process from conception to completion. Users are greeted with an intuitive interface that is easy to navigate, making the software accessible to both new and experienced users. The interface is complemented by an extensive library of tools and resources, providing users with everything they need to bring their designs to life. These tools are not only abundant but also highly sophisticated, allowing for a high degree of customization and precision in every project.

One of the standout features of AutoCAD 2022 is its automation capabilities. Through intelligent, automated processes, the software minimizes the need for tedious, manual input, thereby speeding up the design timeline significantly. This automation does not compromise on quality; instead, it enhances the accuracy and precision of the final output, ensuring that users can produce top-quality designs with minimal effort.

Furthermore, AutoCAD 2022 is designed for seamless integration and compatibility with various other software and platforms. This feature is crucial in today's collaborative and interconnected work environments, where designs often need to be shared, reviewed and modified by different stakeholders using different systems. With advanced compatibility features, users can effortlessly export, share and integrate their AutoCAD designs, facilitating smoother collaboration and coordination among project teams. This ease of integration extends the software's utility and makes it a valuable asset for professionals in different fields.

1.1.6 System Requirements for AutoCAD

To harness the full potential of AutoCAD, users must adhere to the specific system requirements stipulated by Autodesk. These requirements are essential to ensure that the software functions smoothly and efficiently, providing a seamless user experience. The system requirements encompass various parameters including the operating system, processor speed, memory, display resolution etc. It is imperative for users to cross-verify their computer specifications with the system requirements outlined by Autodesk for AutoCAD to guarantee a smooth and hassle-free installation and operation.

To optimally run AutoCAD 2022, including its specialized toolsets, users must ensure their systems meet the requirements specified by Autodesk. For Windows users, the operating system should be a 64-bit Microsoft® Windows® 11 or Windows 10. The processor should be at least 2.5–2.9 GHz, with a recommended speed of 3+ GHz for optimal performance. The basic memory requirement is 8 GB, but 16 GB is recommended for smoother operation. The display resolution should be 1920 x 1080 with True Color for conventional displays, while high resolution and 4K displays should support resolutions up to 3840 x 2160 on Windows 10, with a capable display card. The display card should have a minimum of 1 GB GPU with 29 GB/s Bandwidth and be DirectX 11 compliant, with a recommended 4 GB GPU with 106 GB/s Bandwidth and DirectX 12 compliance for better performance. A minimum disk space of 10.0 GB is required, with additional requirements for large datasets, point clouds and 3D modelling.

For Mac users, the operating system should be Apple® macOS® Monterey v12 (requiring Update 2022.2), Big Sur v11, Catalina v10.15, or Mojave v10.14. The basic model requirements include Apple Mac Pro® 4.1, MacBook Pro 5.1, iMac® 8.1, Mac mini® 3.1, MacBook Air®, or MacBook® 5.1, with recommended models supporting Metal Graphics Engine. Mac models with M series chips are supported under Rosetta 2 mode. The CPU type should be a 64-bit Intel CPU, with an Intel Core i7 or higher recommended. The basic memory requirement is 4GB, with 8GB or higher recommended. The display resolution should be a basic 1280 x 800 display or a high resolution at 2880 × 1800 with Retina Display.

For specialized toolsets on Windows, additional disk space and memory are required, with specific requirements for each toolset, as outlined in the user manual. Understanding and adhering to these system requirements is imperative for users looking to harness the full potential of AutoCAD 2022 efficiently and effectively.

1.2 COMPONENTS OF AUTOCAD SOFTWARE WINDOW

The AutoCAD Software Window is composed of various components, each designed to facilitate different aspects of the drafting and design process. Below are the key components:

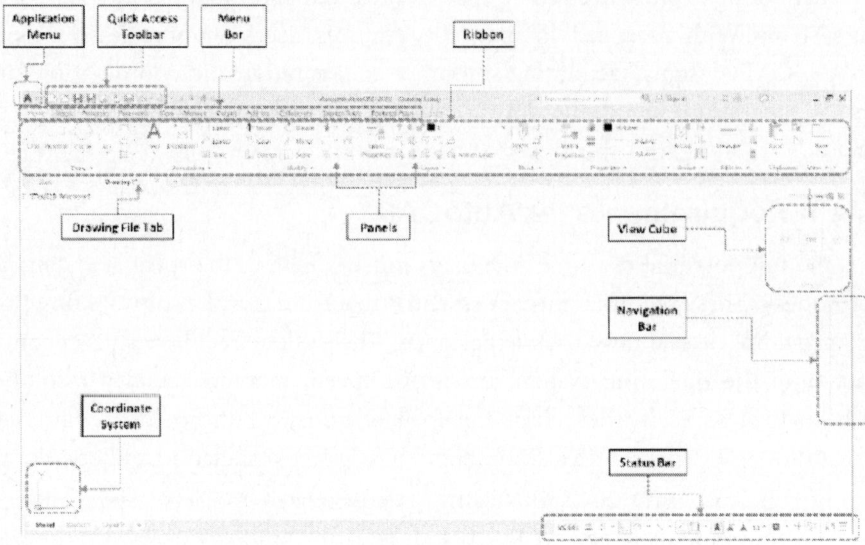

Fig. 1.1: AutoCAD Application window

1.2.1 Title Bar

The Title Bar is located at the top of the AutoCAD window and displays the name of the current drawing. It provides users with information about the active file and allows them to manage the AutoCAD window, including minimizing, maximizing and closing the application.

1.2.2 Ribbon Toolbar

The Ribbon Toolbar is a comprehensive set of tools organized into tabs and panels. Each tab corresponds to a type of activity, such as drafting or 3D modelling and the panels contain related tools and commands. This setup provides users with easy access to AutoCAD's extensive functionality in a structured and intuitive manner.

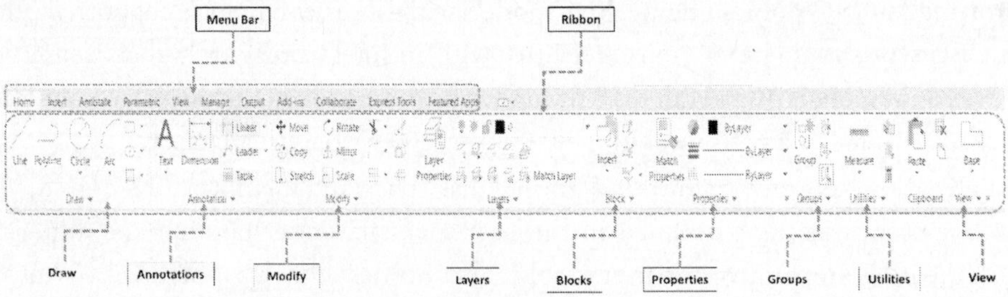

Fig. 1.2: Ribbon Toolbar

1.2.3 Quick Access Toolbar

The Quick Access Toolbar offers shortcuts to commonly used commands:

New: Initiates a new drawing file.

- **Open:** Allows users to open existing drawing files.
- **Save:** Saves the current drawing file.
- **Save As:** Saves the current drawing with a new name.
- **Open from Web and Mobile:** Opens existing files from AutoCAD web and mobile.
- **Save to Web and Mobile:** Saves files to AutoCAD web and mobile.
- **Plot:** Sends a drawing to a plotter or printer.
- **Undo:** Reverses the most recent action.
- **Redo:** Reverses the undo action.

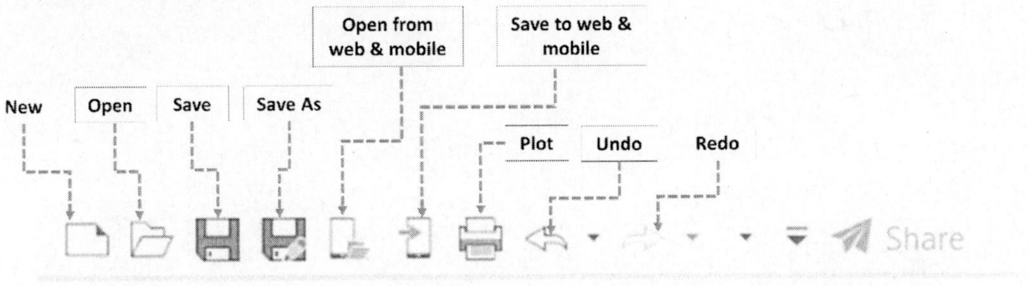

Fig. 1.3: Quick Access Toolbar

1.2.4 Viewport

The Viewport provides a window into the drawing area where users create and modify objects. Users can have multiple viewports, each displaying different views or perspectives of the drawing, facilitating a more efficient workflow.

1.2.5 Command Prompt

The Command Prompt is where users can enter commands and options. It also displays prompts, command history and error messages, serving as an interface for textual input and feedback.

1.2.6 Draw Toolbar

The Draw Toolbar contains tools and commands for creating objects, such as lines, circles and polygons. These fundamental tools are crucial for drafting and designing in AutoCAD.

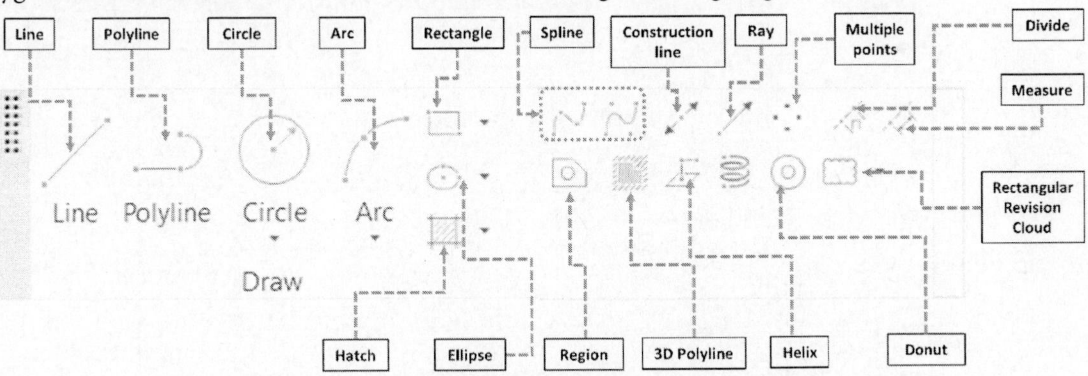

Fig. 1.4: Draw Toolbar

1.2.7 Modify Toolbar

The Modify Toolbar houses commands for editing and manipulating existing objects. This includes functions like move, copy, rotate and trim, which are essential for refining and adjusting designs.

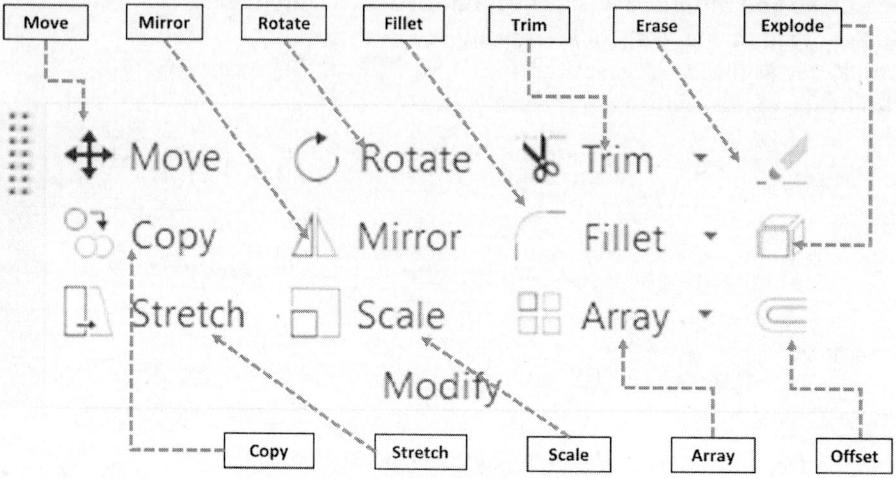

Fig. 1.5: Modify Toolbar

1.2.8 UCS Icon

The User Coordinate System (UCS) Icon establishes the location and orientation of a movable Cartesian coordinate system, essential for precision operations. The UCS defines:

- The XY plane (work or construction plane).
- Horizontal and vertical directions for various features.
- Alignment and angle of the grid, hatch patterns, text and dimensions.
- Origin and orientation for coordinate entry and absolute reference angles.
- For 3D operations, it defines work planes, projection planes and the Z-axis orientation.

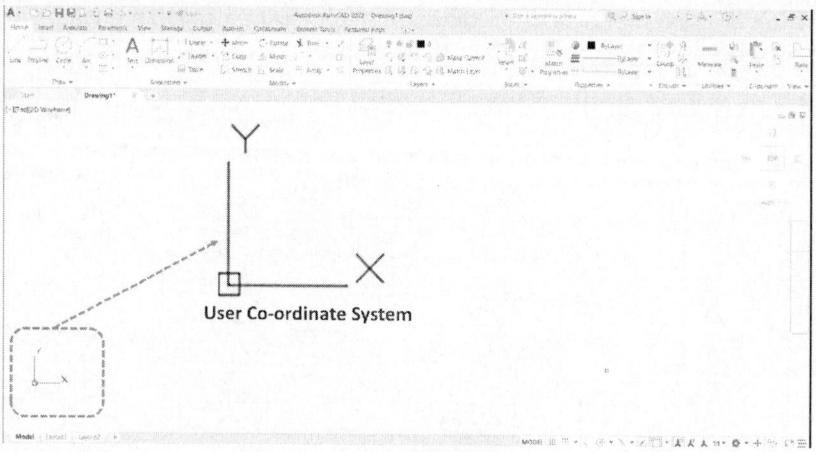

Fig. 1.6: UCS Icon

1.2.9 Status Bar

The Status Bar is located at the bottom of the AutoCAD window and provides information about the drawing and the current settings. It also offers access to drafting settings, object snap modes and workspace controls, helping users monitor and adjust their work environment efficiently.

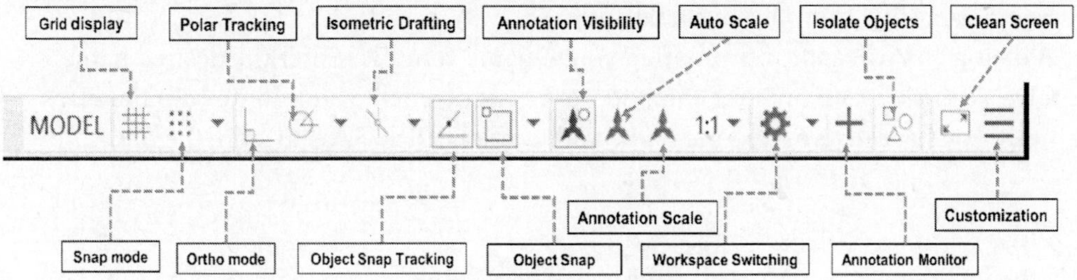

Fig. 1.7: Status Bar

1.3 FILE FEATURES

1.3.1 Opening a New File

Initiating a new project in AutoCAD begins with opening a new file. Follow these steps to create a fresh canvas for your design:

- Click on the 'New' option.
- Select 'Drawing' from the menu that appears.
- A list of templates will be displayed; choose the appropriate one (commonly, the 'acad' template is used for general purposes).
- After selecting the template, click 'Open' to access a new drawing area where you can start your drafting and designing process.

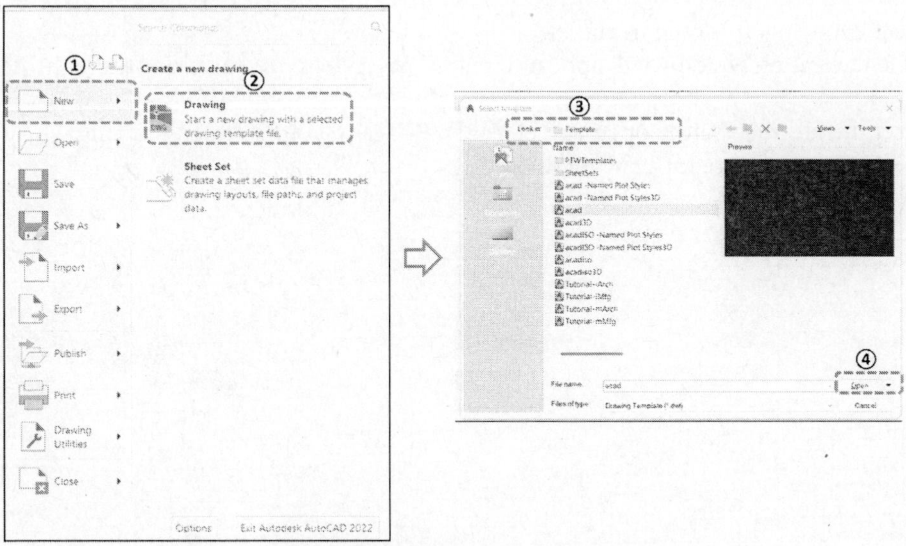

Fig. 1.8: Steps to open a new file

1.3.2 Saving a File

Preserving your work is crucial. To save your drawing file for the first time or to save a copy with a different name:

- Click on 'Save As' in the application menu.
- Select 'Drawing' from the available options.
- A dialog box will appear prompting you to name your file. Enter the desired name.
- Choose the location on your computer or network where you want to save the file.
- Once you have determined the name and location, click 'Save' to store the file.

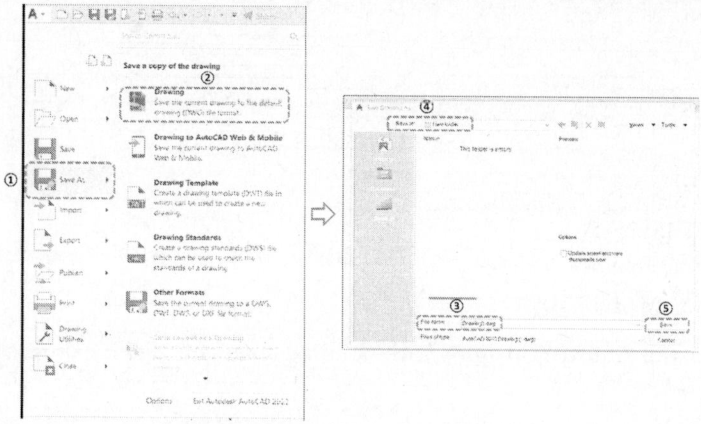

Fig. 1.9: Steps to save a file

1.3.3 Opening an Existing Drawing File

Accessing previously saved work is straightforward. To open an existing drawing file:

- Click on the 'Open' option in the application menu.
- Select 'Drawing' from the menu.
- A file navigation window will appear. Use it to navigate to the location where the desired file is stored.
- Select the file you wish to open and click 'Open' again. Your selected drawing will load into the AutoCAD workspace for viewing or editing.

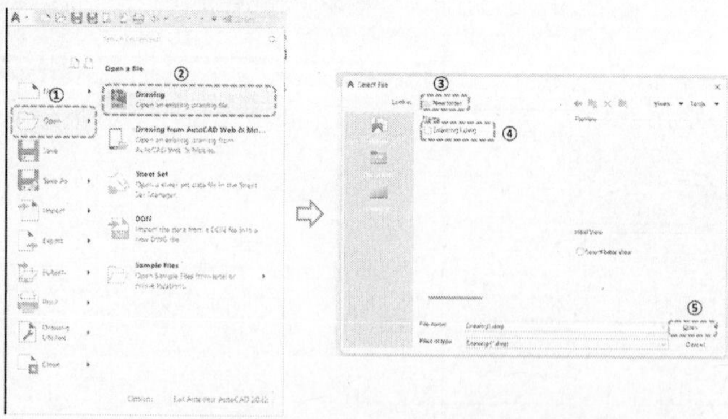

Fig. 1.10: Steps to open an existing drawing file

1.3.4 Quitting a session

When concluding your work session, you can easily exit AutoCAD:

- Click on the 'Close' option in the application menu.
- A dialog box will appear with options to close the 'Current Drawing' or 'All Drawings'.
- Select 'Current Drawing' if you wish to close only the active file. If you need to close the application entirely, along with all open files, select 'All Drawings'. Ensure you have saved all work before selecting this option.

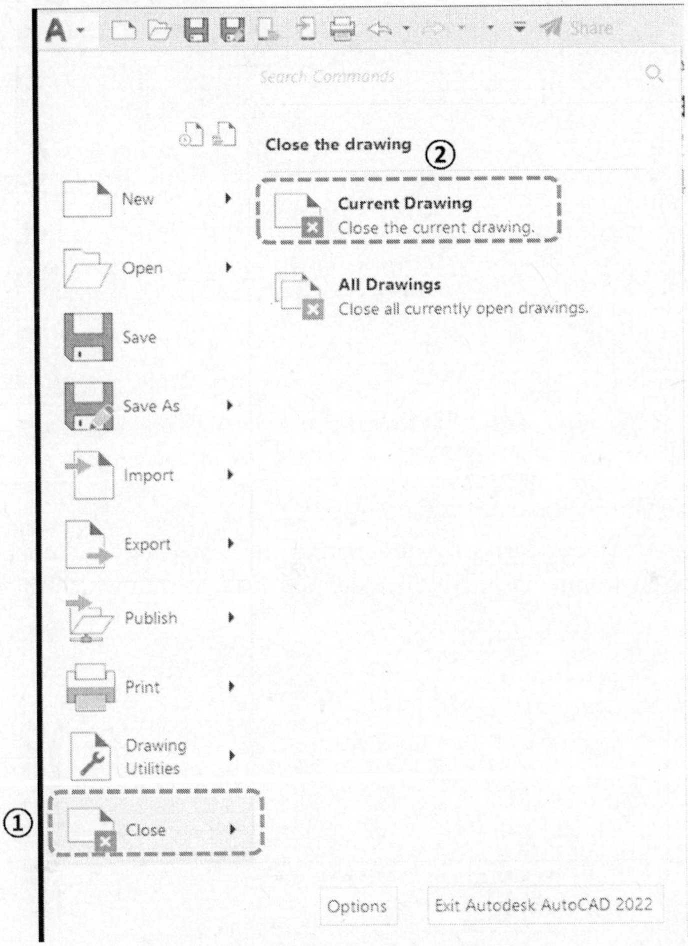

Fig. 1.11: Steps to quitting a session

1.4 SETTING UP NEW DRAWING

1.4.1 Setting Units

Before initiating a drawing, it is imperative to define the measurement units. This foundational step ensures that the drawing adheres to real-world scale and dimensions, facilitating accurate design and drafting.

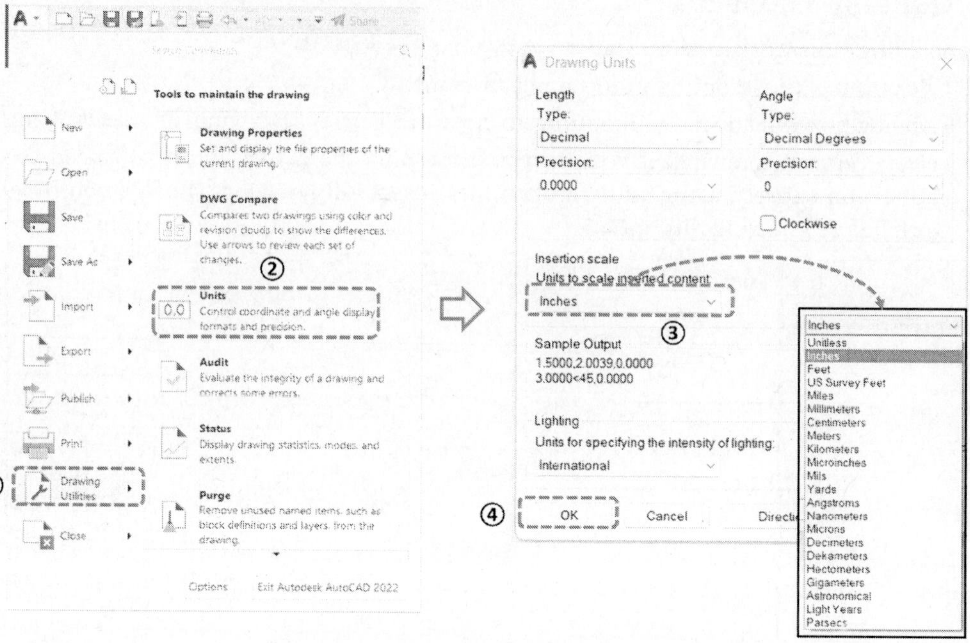

Fig. 1.12: Example of setting units

1.4.2 Setting Limits

Establish the drawing area boundaries by designating the lower-left and upper-right corners. This confines your drawing environment, providing a defined space within which to create and manipulate objects.

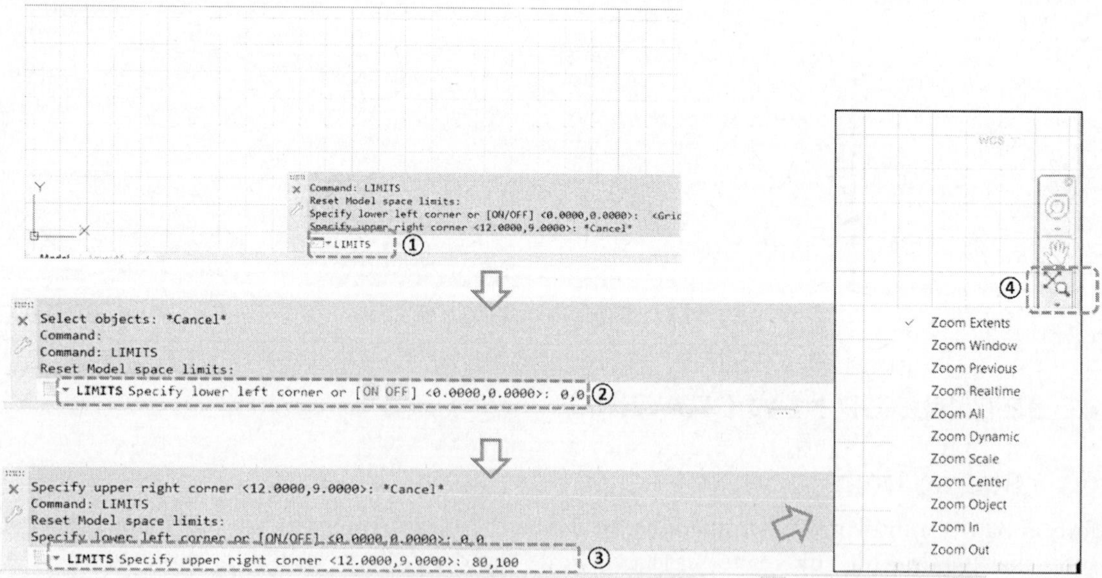

Fig. 1.13: Example of setting limits

1.4.3 Creating Grid

A grid serves as a crucial visual reference, simplifying the design process and aiding in the precise alignment of objects. It provides a structured backdrop against which drafting becomes more intuitive and accurate.

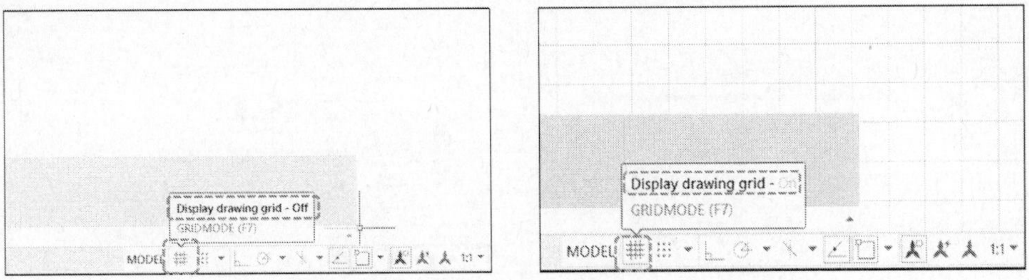

Fig. 1.14: Example of toggling grid on and off

1.4.4 Setting Snap

The Snap feature facilitates alignment by snapping objects to the nearest grid point. This ensures precision and consistency in placement, making the drafting process smoother and more controlled.

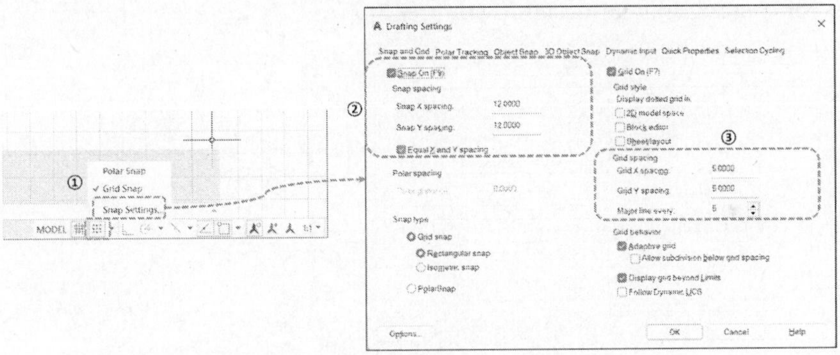

Fig. 1.15: Example of setting Snap and adjusting its properties

1.4.5 Dynmode

Dynamic Input (Dynmode) offers a convenient method for command and value input directly at the cursor's location. This feature enhances accuracy while allowing users to maintain focus on the drawing area.

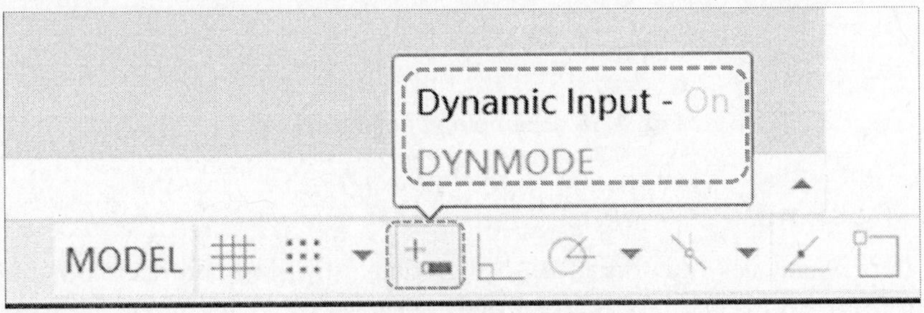

Fig. 1.16: Dynamic input

1.4.6 Orthomode

Orthomode restricts drawing movements to horizontal or vertical lines, guaranteeing straight lines and aiding in the creation of structured, orderly designs.

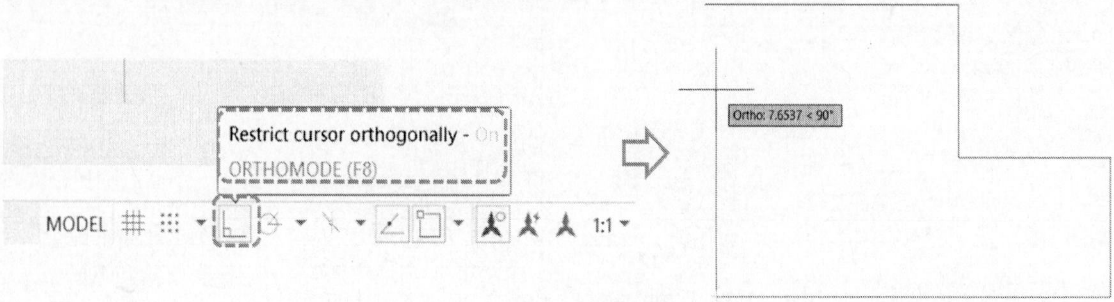

Fig. 1.17: Example of orthomode

1.4.7 Polar Tracking

Polar Tracking is invaluable for drawing or aligning objects at specified angles. It provides guidance and reference points for creating angular designs with precision.

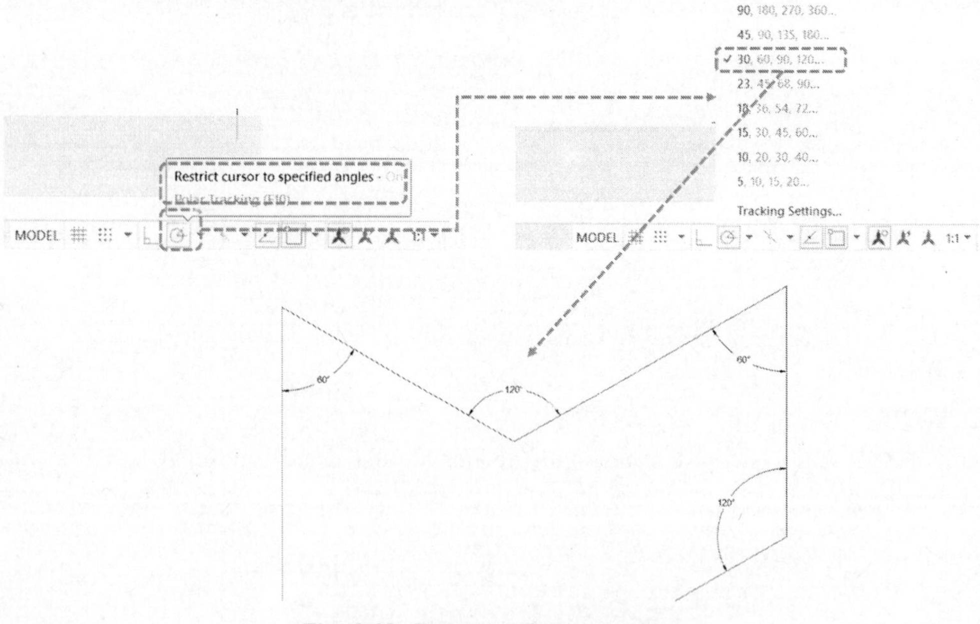

Fig. 1.18: Example of polar tracking

1.4.8 Snap Tracking

Snap Tracking offers visual cues and references to align objects with key points in other elements within the drawing. This feature is essential for maintaining consistency and alignment across various drawing elements.

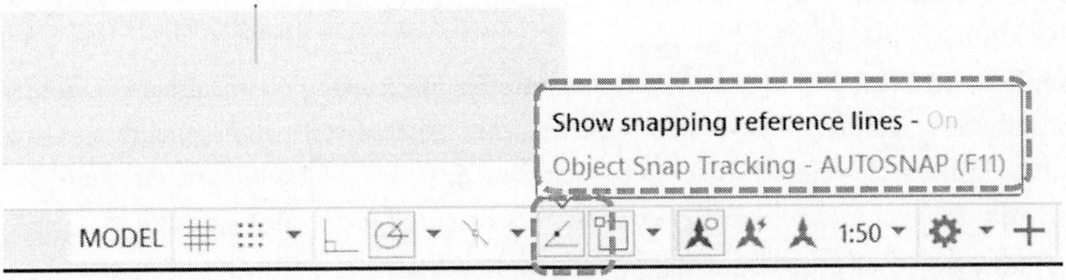

Fig. 1.19: Snap tracking

1.4.9 Object Snap

Object Snaps (OSNAP) are tools for specifying exact locations on objects, such as endpoints, midpoints and centers, providing precision in selecting and manipulating geometric points on objects.

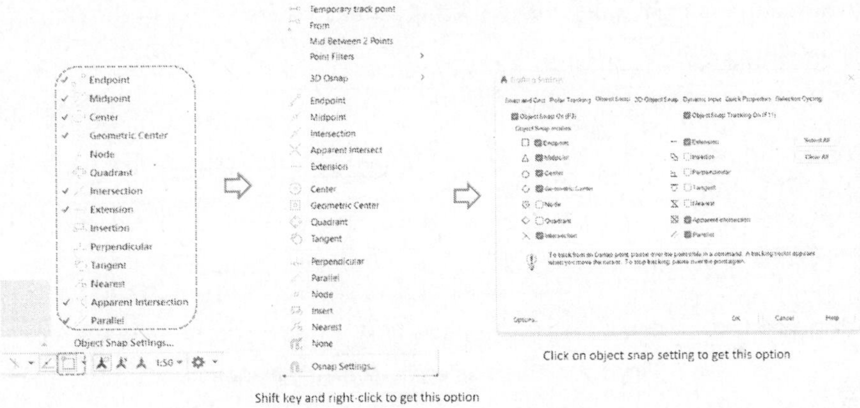

Fig. 1.20: Object snap options

1.4.10 Isodraft

The Isodraft button is a crucial tool in AutoCAD, primarily used for facilitating the drawing of isometric views by allowing users to seamlessly switch between different planes. When engaging in isometric drafting, this feature is invaluable as it provides a structured approach to creating accurate and proportional isometric drawings.

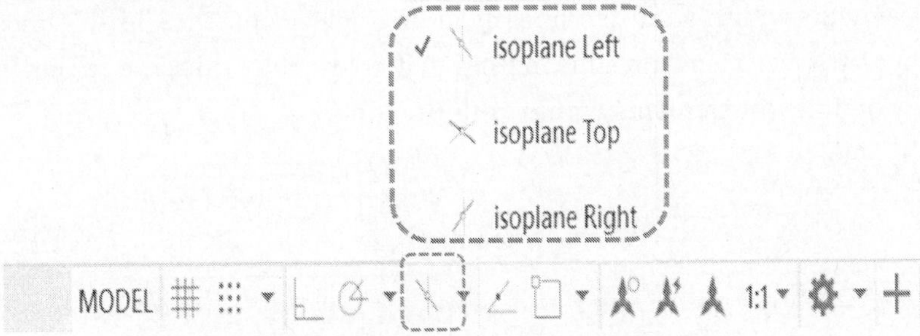

Fig. 1.21: Isodraft mode

When the Isodraft button is activated, it enables isometric drafting and allows users to choose from three distinct isometric planes:

- **Isoplane Left**: The Isoplane Left option is utilized for crafting planes that are oriented towards the left. These planes are defined by a pair of axes situated at 90 and 150 degrees. This setting is particularly useful when the user needs to create drawings that require a left-facing perspective, providing a clear and accurate view of the left-oriented facets of the design.
- **Isoplane Top**: Selecting the Isoplane Top option allows users to generate top-facing planes. These planes are characterized by the 30-degree and 150-degree axis pair. When a top-down perspective is necessary for the drawing, this setting offers a straightforward way to create top-facing isometric drawings, providing a bird's eye view of the design layout.
- **Isoplane Right**: The Isoplane Right option is designed for the creation of right-facing planes, which are defined by the axis pair set at 90 and 30 degrees. This option is essential when the drawing requires a right-sided view, offering a precise perspective of the right-sided elements of the design.

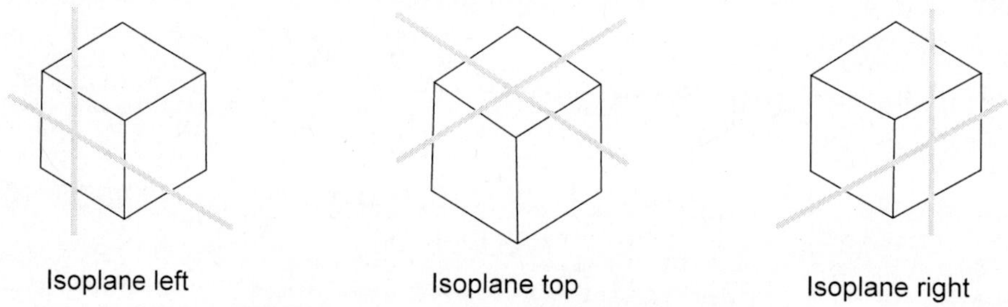

Fig. 1.22: Three different isometric planes

Orthographic Mode: The Orthographic mode within the Isodraft settings is designed to disable the isometric drafting function. When this mode is activated, the software reverts to the standard drafting mode, allowing for the creation of non-isometric, or orthographic drawings.

1.4.11 Lineweight Display

The LW (Lineweight) Display is a pivotal feature in AutoCAD, located within the status bar at the lower right-hand corner of the application window. Lineweight is a distinctive property that can be assigned to various graphical elements within your drawing, including objects, hatches, leader lines and dimension geometry. This property is crucial as it results in lines that are visually thicker and darker, enhancing the clarity and visibility of different elements within your drawing.

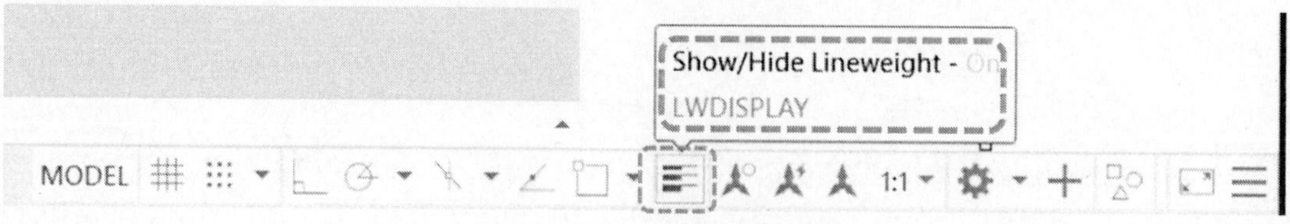

Fig. 1.23: Lineweight display

Lineweight is not merely a visual aid but a significant attribute that can be toggled on and off as per the user's requirements. The display of lineweight is subject to the space in which you are working, with differences noted between model space and paper space layout:

Model Space Display: Within the model space, a lineweight assigned a value of 0 is displayed with a width of one pixel. Lineweight other than 0 are displayed with a proportional pixel width, providing a visual differentiation between lines of varying weights. Importantly, the display of lineweight in model space remains constant regardless of the zoom level. For instance, a line assigned a lineweight that corresponds to four pixels will consistently be displayed with a width of four pixels, irrespective of the degree of zoom applied to the view.

Paper Space Layout Display: In contrast, within a paper space layout, lineweight are displayed using real-world units. This means that the display of lineweight dynamically adjusts with changes in the zoom factor, providing a more accurate representation of how the line will appear when plotted or printed. This dynamic adjustment aids in creating more accurate and visually coherent layouts for printing and presentation purposes.

Plotting Lineweight: By default, when a drawing is plotted, the lineweights are reproduced with the exact width corresponding to the assigned lineweight values. This ensures that the printed or plotted output accurately reflects the visual characteristics defined within the AutoCAD environment.

1.4.12 Transparency Display

The Transparency Display feature in AutoCAD is a valuable tool that allows users to adjust the visibility of individual objects or layers within a drawing. This feature is particularly useful when working with overlapping or densely populated drawings, as it helps in distinguishing between different elements without altering the actual structure or layout of the drawing.

The Transparency Display can be toggled ON or OFF using its dedicated button. When activated, it allows for the transparency settings applied to objects or layers to be visibly reflected within the drawing area. The process to adjust transparency is straightforward:

- To set or adjust the transparency of an object, you need to access the Layer Properties Manager. This can be done by clicking on the properties extension button located on the toolbar or by using the command line.
- Once the Layer Properties Manager is open, you will find options to adjust various properties of the selected layer or object. Locate the 'Transparency' setting. This setting allows you to define the degree of transparency applied to the objects within the selected layer. Transparency is set using a percentage scale, with 0% being completely opaque and 100% being completely transparent.
- In addition to setting transparency, you can also define the Line Weight and Line Type for the objects within the layer. Line Weight determines the thickness of the lines, while Line Type can be used to set different line styles (like dashed, dotted, etc.). Adjusting these settings in conjunction with transparency allows for more nuanced control over the visual appearance of your drawing.

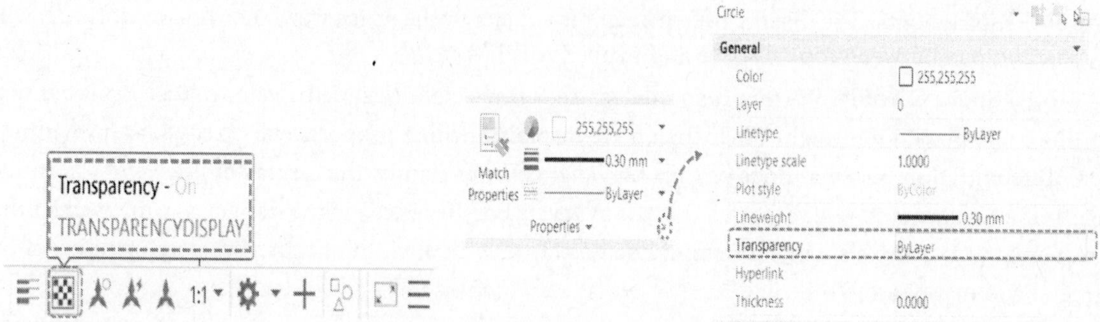

Fig. 1.24: Transparency display

1.4.13 Selection Cycling

Selection Cycling is a feature that facilitates the selection of overlapping objects. When multiple objects overlap, this feature provides a list of all objects at the selection point, allowing for precise selection.

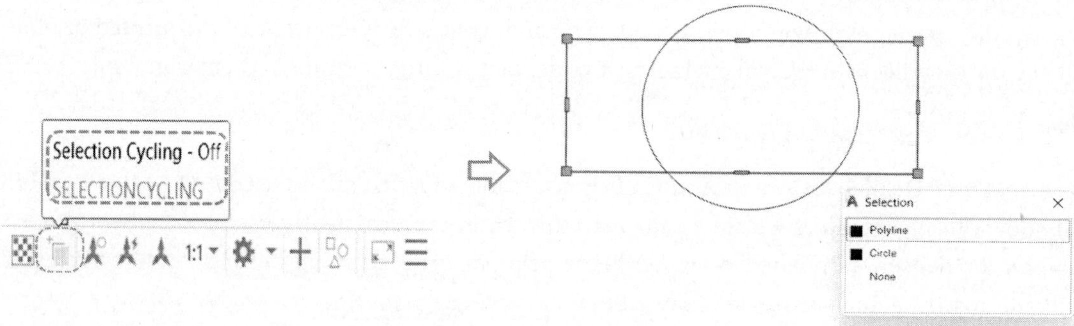

Fig. 1.25: Example of selection cycling

1.4.14 Dynamic UCS

Dynamic UCS temporarily aligns the drawing plane with 3D solids, planar meshes, or point cloud segments, providing a reference for creating objects with precision and ease.

Fig. 1.26: Dynamic UCS

1.4.15 Graphic Configures

Graphic Configures allow for the adjustment of graphical settings to optimize performance, often utilizing the capabilities of the computer's graphics card to enhance the drawing experience.

Fig. 1.27: Graphic Configures

1.4.16 Undo and Redo Command

The Undo command reverses the last action, while the Redo command reverses the last Undo. These commands are fundamental for correcting mistakes and experimenting with different design options without fear of losing work.

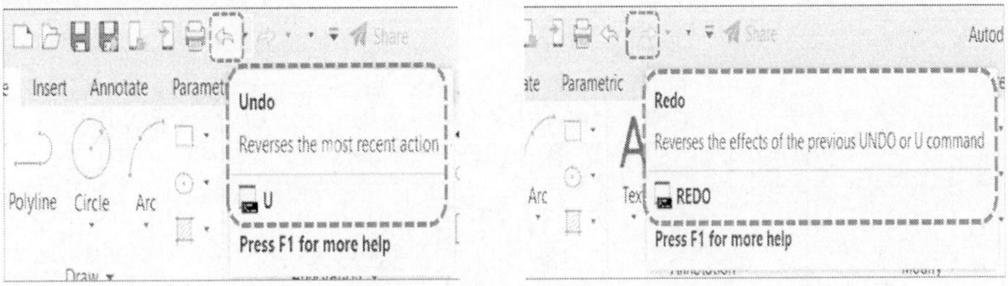

Fig. 1.28: Undo and Redo command

1.5 COORDINATE SYSTEM IN AUTOCAD

1.5.1 Absolute Coordinate

Absolute coordinates are essential reference tools in AutoCAD, directly relating to the User Coordinate System (UCS) origin. These coordinates provide a fixed, definitive point within the drawing space, serving as a reliable and constant point of reference for the creation and manipulation of objects within the drawing environment.

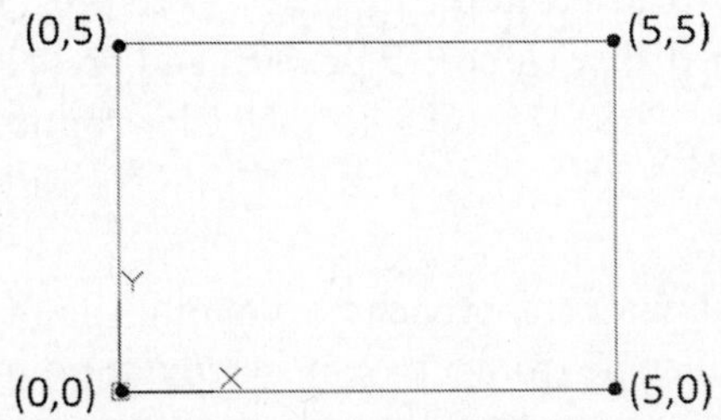

Fig. 1.29: Example of application of absolute coordinates

Example (using the Absolute Coordinate System):

Step 1: Click Home > Draw > Line on the ribbon or enter LINE or L in the command line.

Step 2: Specify first point: Type 0,0 and press ENTER.

Step 3: Specify next point or [Undo]: Type 5,0 and press ENTER.

Step 4: Specify next point or [Undo]: Type 5,5 and press ENTER.

Step 5: Specify next point or [Close Undo]: Type 0,5 and press ENTER.

Step 6: Specify next point or [Close Undo]: Select the Close option.

1.5.2 Relative Coordinate

Relative coordinates offer dynamic input options, as they are inherently based on the last point entered during the drawing process. This allows for incremental and intuitive input, facilitating the creation of objects in relation to previously defined points, thereby providing flexibility and ease of use during the drafting process.

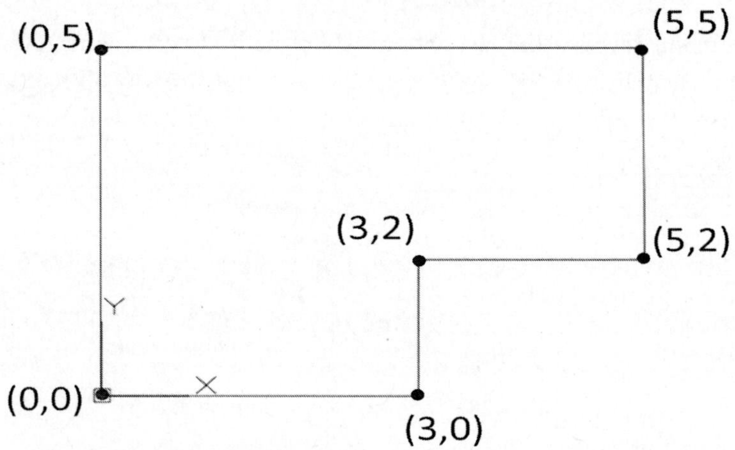

Fig. 1.30: Example of application of relative coordinates

Example (using the Relative Coordinate System):

Step 1: Click Home > Draw > Line on the ribbon or enter LINE or L in the command line.

Step 2: Specify first point: Type 0,0 and press ENTER.

Step 3: Specify next point or [Undo]: Type @3,0 and press ENTER.

Step 4: Specify next point or [Undo]: Type @3,2 and press ENTER.

Step 5: Specify next point or [Close Undo]: Type @5,2 and press ENTER.

Step 6: Specify next point or [Close Undo]: Type @5, 5 and press ENTER.

Step 7: Specify next point or [Close Undo]: Type @0,5 and press ENTER.

Step 8: Specify next point or [Close Undo]: Select the Close option.

1.5.3 Polar Coordinate

Polar coordinates introduce a different approach to defining points within the drawing space. By specifying both a distance and an angle from a base or reference point, users can accurately locate and define points in a way that is often more intuitive and appropriate for certain types of drawings and design scenarios.

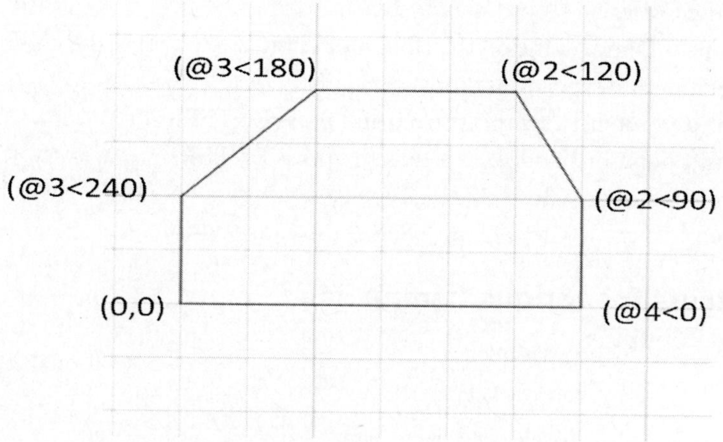

Fig. 1.31: Example of application of polar coordinates

Example (using the Polar Coordinate System):

Step 1: Click Home > Draw > Line on the ribbon or enter LINE or L in the command line.

Step 2: Specify first point: Type 0,0 and press ENTER.

Step 3: Specify next point or [Undo]: Type @4<0 and press ENTER.

Step 4: Specify next point or [Undo]: Type @2<90 and press ENTER.

Step 5: Specify next point or [Close Undo]: Type @2<120 and press ENTER.

Step 6: Specify next point or [Close Undo]: Type @3<180 and press ENTER.

Step 7: Specify next point or [Close Undo]: Type @3<240 and press ENTER.

Step 8: Specify next point or [Close Undo]: Select the Close option.

1.5.4 Cartesian Coordinate System

The Cartesian coordinate system is foundational to AutoCAD, utilizing the X, Y and Z axes to precisely define points within a three-dimensional space. This system is prevalent and widely used for both 2D and 3D drawings in AutoCAD due to its straightforward and universally understood methodology for locating points and creating objects within the drawing environment. The Cartesian system provides a structured and systematic approach to drafting, making it an indispensable tool for professionals working within AutoCAD.

Fig. 1.32: Example of application of cartesian coordinates

Example (using the Cartesian Coordinate System):

Step 1: Click Home > Draw > Line on the ribbon or enter LINE or L in the command line.

Step 2: Specify first point: Type 0,0 and press ENTER.

Step 3: Specify next point or [Undo]: Type 6,0 and press ENTER.

Step 4: Specify next point or [Undo]: Type 6,4 and press ENTER.

Step 5: Specify next point or [Close Undo]: Type 0,4 and press ENTER.

Step 6: Specify next point or [Close Undo]: Select the Close option.

Keyboard shortcuts for various commands

Shortcut	Description
F1	Display help
F2	Toggle text screen
F3 or OSNAP	Toggle object snap mode
F4	Toggle 3DOsnap
F5 or ISOPLANE	Toggles between isometric and orthographic projection
F6	Toggles dynamic UCS
F7 or GRID	Toggles grid display
F8	Toggles orthogonal mode
F9	Toggles snap mode
F10	Toggles polar tracking
F11	Toggles object snap tracking
F12	Toggles dynamic input
A or ARC	Creates an arc
B or BLOCK	Creates a block definition from selected objects
C or CIRCLE	Creates a circle
D or DIMSTYLE	Creates and modifies dimension style
E or ERASE	Removes objects from a drawing
F or FILLET	Rounds and fillets the edges of objects
G or GROUP	Creates and manages saved sets of objects called groups
H or HATCH	Fills an enclosed area or selected object with hatch pattern, solid fill or gradient fill
I or INSERT	Inserts a block or drawing into the current drawing
J or JOIN	Joins similar objects to form a single, unbroken object
L or LINE	Creates straight line segments
M or MOVE	Moves objects a specified distance in a specified direction
O or OFFSET	Creates concentric circles, parallel lines and parallel curves
P or PAN	Adds a parameter wit grip to a dynamic block definition

Q or QSAVE or CTRL + S	Save the current drawing
R or REDRAW	Refreshes the display in the current viewport
S or STRETCH	Stretches objects crossed by a selection window or polygon
T or MTEXT	Creates a multiline text object
U or CTRL + Z	Reverses the last action
V or VIEW	Saves and restores named views, camera views, layout views and preset views
W or WBLOCK	Writes objects or a block to a new file
X or EXPLODE	Breaks a compound object into its component objects
Z or ZOOM	Increase or decrease magnification
CTRL + E	Toggle coordinate display
CTRL + G	Toggle Grid
CTRL + E	Cycle isometric planes
CTRL + F	Toggle running object snaps
CTRL + H	Toggle Pick Style
CTRL + Shift+ H	Toggle Hide pallets
CTRL + I	Toggle Coords
CTRL + Shift + I	Toggle Infer Constraints
CTRL + 0 (zero)	Clean Screen
CTRL + 1	Property Palette
CTRL + 2	Design Centre Palette
CTRL + 3	Tool Palette
CTRL + 4	Sheet Set Palette
CTRL + 6	DBConnect Manager
CTRL + 7	Markup Set Manager Palette
CTRL + 8	Quick Calc
CTRL + 9	Command Line
CTRL + N	New Drawing
CTRL + O	Open drawing
CTRL + P	Plot dialog box
CTRL + Tab	Switch to next
CTRL + Shift + Tab	Switch to previous drawing
CTRL + Page Up	Switch to previous tab in current drawing
CTRL + Page Down	Switch to next tab in current drawing
CTRL + Q	Exit
CTRL + A	Select all objects
CTRL + C	Copy object
CTRL + X	Cut object
CTRL + V	Paste object

CTRL + Shift+ C	Copy to clipboard with base point
CTRL + Shift+ V	Paste data as block
CTRL + Y	Redo last action
CTRL + [or CTRL + \ or ESC	Cancel current command

UNIT SUMMARY

Based on the contents covered in this unit you should remember that:

- AutoCAD is a key tool in design and engineering, helping you draw with more accuracy and speed than drawing by hand.
- CAD helps you create, change, analyze and optimize designs on your computer, making your drawings more accurate and consistent while reducing mistakes.
- There are many different CAD software options available, each with its special features. The right one for you depends on your project and industry.
- AutoCAD is very useful because it can be used in many different industries. It is easy to use, has many tools, can automate some parts of the drawing process and works well with other software.
- To use AutoCAD smoothly and efficiently, your computer must meet the system requirements set by Autodesk.
- The AutoCAD window has different parts like the Title Bar, Ribbon Tool Bar, Quick Access Tool Bar, View Port, Command Prompt, Draw Tool Bar, Modify Tool Bar, UCS Icon and Status Bar. Each part has a specific function to help you draw.
- Managing files in AutoCAD, like opening new files, saving files, opening existing files and closing the application, is simple but important.
- When starting a new drawing, you need to set up things like units, limits and grids. You will also use features like Dynmode, Ortho, Polar Tracking, Snap Tracking, Object Snap, Isodraft, LW Display, Transparency Display, Selection Cycling, Dynamic UCS, Graphic Configures and the Undo and Redo commands.
- It is important to understand and use different coordinate systems in AutoCAD, like Absolute Coordinates, Relative Coordinates, Polar Coordinates and the Cartesian Coordinate System, to accurately define and locate points in your drawing.

EXERCISES

A. Multiple Choice Questions

2.1 What is the primary function of AutoCAD?

(a) Text Editing (b) Web Browsing

(c) Design and Drafting (d) Data Analysis

1.2 Which component of AutoCAD window allows quick access to common tools?

(a) Title Bar (b) Ribbon Tool Bar

(c) Quick Access Tool Bar (d) Command Prompt

1.3 What does the UCS Icon represent in AutoCAD?
(a) Universal Coordinate System (c) Unified Command System
(b) User Coordinate System (d) Ultimate Creation Space

1.4 Which coordinate system uses X, Y and Z axes in AutoCAD?
(a) Polar Coordinate (b) Absolute Coordinate
(c) Cartesian Coordinate System (d) Relative Coordinate

1.5 What is the purpose of setting limits in a new drawing?
(a) To limit the number of tools (c) To limit the file size
(b) To set the drawing boundaries (d) None of the above

1.6 What does the Snap feature ensure in AutoCAD?
(a) Snaps pictures of the drawing (b) Quick saving of files
(c) Objects align with the nearest grid point (d) Smooth transition between files

1.7 Which feature in AutoCAD helps in drawing objects at specific angles?
(a) Dynmode (b) Ortho (c) Polar Tracking (d) Snap Tracking

1.8 What does the Object Snap (OSNAP) feature help with?
(a) Snapping pictures of objects (b) Quickly saving objects
(c) Specifying precise locations on objects (d) Object rotation

1.9 What is the function of the Isodraft mode?
(a) Drafting isotopes (b) Creating isometric drawings
(c) Drafting isolated objects (d) None of the above

1.10 What does the Lineweight (LW) display control in AutoCAD?
(a) Line length (b) Line colour
(c) Line visibility (d) Line weight visibility

1.11 What is the purpose of the Transparency Display feature?
(a) Making objects invisible (b) Adjusting object transparency
(c) Changing object colour (d) None of the above

1.12 Which feature allows the selection of overlapping objects?
(a) Object Snap (b) Selection Cycling
(c) Snap Tracking (d) Polar Tracking

1.13 What does the Dynamic UCS feature align with temporarily?
(a) 2D plane (b) 3D solid face (c) Object Snap (d) Polar Tracking

1.14 What does the Undo command do in AutoCAD?
(a) Redo the last action (b) Save the file
(c) Reverse the last action (d) Delete the file

1.15 Which of the following is NOT a coordinate system in AutoCAD?
(a) Absolute Coordinate (b) Relative Coordinate
(c) Polar Coordinate (d) Global Coordinate

1.16 What is the primary benefit of using AutoCAD for design and drafting?
- (a) Internet Browsing
- (b) Precision and Efficiency
- (c) File Compression
- (d) Data Analysis

1.17 What does the Quick Access Tool Bar in AutoCAD primarily provide?
- (a) Fast internet access
- (b) Quick access to common tools
- (c) Access to external files
- (d) Quick saving options

1.18 What is the function of the Draw Tool Bar in AutoCAD?
- (a) Drawing management
- (b) Access to drawing tools
- (c) File saving and opening
- (d) Access to modification tools

1.19 What is the purpose of setting units in a new drawing in AutoCAD?
- (a) To define measurement units
- (b) To set drawing speed
- (c) To define drawing color
- (d) None of the above

1.20 What does the Ribbon Tool Bar provide in AutoCAD?
- (a) Access to Internet tools
- (b) Access to drawing tools
- (c) File saving options
- (d) System settings access

Answers to Multiple-Choice Questions
1.1 (c), 1.2 (c), 1.3 (b), 1.4 (c), 1.5 (b), 1.6 (c), 1.7 (c), 1.8 (c), 1.9 (b), 1.10 (d), 1.11 (b), 1.12 (b), 1.13 (b), 1.14 (c), 1.15 (d), 1.16 (b), 1.17 (b), 1.18 (b), 1.19 (a), 1.20 (b)

B. Subjective Questions

1.1 Explain the significance and applications of the UCS in AutoCAD.

1.2 Describe the process of setting up a new drawing in AutoCAD.

1.3 Elaborate on the different coordinate systems available in AutoCAD.

1.4 Explain the importance of the Quick Access Tool Bar in AutoCAD.

1.5 List and describe the functions of the tools available in the Quick Access Tool Bar.

1.6 Describe the Snap feature in AutoCAD.

1.7 Elaborate on the Isodraft mode in AutoCAD. How does it assist in creating isometric drawings and what are its key features?

1.8 Explain the Object Snap (OSNAP) feature in AutoCAD. How does it enhance precision in designing and what are the different precise locations it can specify on objects?

1.9 Discuss the Dynamic UCS feature in AutoCAD. How does it aid in creating objects?

1.10 Describe the Transparency Display feature in AutoCAD. How does it improve visibility and emphasis on certain elements in a drawing

KNOW MORE

- *Customization of Interface*: AutoCAD allows users to fully customize the interface to suit their specific needs and preferences. This includes the ability to rearrange toolbars, create custom commands and even change the colour scheme of the interface to reduce eye strain during long working sessions.

- *AutoCAD Mobile App*: For professionals on the go, AutoCAD offers a mobile app that provides core drawing and drafting tools. With this app, users can view, create, edit and share AutoCAD drawings on mobile devices, allowing for productive work even when away from the workstation.
- *Cloud Storage Integration*: AutoCAD integrates seamlessly with Autodesk's cloud platform, allowing users to save and access their drawings from anywhere with an internet connection. This facilitates easy collaboration between team members located in different geographical areas.
- *3D Printing Support*: While this chapter primarily focused on 2D drafting, AutoCAD also supports 3D modelling and printing. Users can create 3D models and prepare them for 3D printing directly within the software, bridging the gap between digital design and physical prototypes.
- *Dynamic Blocks*: Dynamic Blocks offer more flexibility than regular blocks. They can be adjusted and manipulated without having to create new block definitions. This feature significantly speeds up the drafting process as users can create configurations on the fly.
- *Quick Measure Tool*: The Quick Measure tool automatically displays measurements, like distances and angles, when you hover over certain geometry. This feature is a time-saver for users who need to quickly reference dimensions without manually measuring each time.
- *AutoCAD Shortcuts*: To increase efficiency, AutoCAD has a wide range of keyboard shortcuts. Users can also create shortcuts for commands they frequently use, which can significantly speed up the drafting and design process.
- *AutoLISP*: For users who wish to automate and customize AutoCAD further, the software supports AutoLISP, a small, dynamically typed, LISP-like programming language. With AutoLISP, users can create custom functions, commands and even automate repetitive tasks.

REFERENCES AND SUGGESTED READINGS

1. Cadfolks. AutoCAD 2020 For Beginners, 7th ed.; Kishore, 2019.
2. Palm, B.S. Introduction to AutoCAD 2020: 2D and 3D Design; Routledge: London, England, 2020.
3. Hawkes, B. Succeeding with AutoCAD; McGraw-Hill Publishing: London, England, 1995.

Dynamic QR Code for Further Reading

System Requirements for AutoCAD

Navigating AutoCAD

File Management in AutoCAD

Setting up a New Drawing

Coordinate Systems in AutoCAD

Draw Toolbar in AutoCAD

Modify Toolbar in AutoCAD

Object Snaps in AutoCAD

Isodraft in AutoCAD

2 Drawing Aid and Editing Commands in AutoCAD

UNIT SPECIFICS

This unit presents information related to the following topics:

- basic sketching entities like lines, circles, arcs, polygons, ellipses etc.;
- edit commands like trim, extend, delete, copy, offset, array etc.;
- drawing various shapes using both GUI and command line;

This unit introduces the readers to the basic sketching tools available in AutoCAD. These tools are quite diverse ranging from lines to polylines to circles to polygons. The readers are presented with details on invoking the tools from the GUI as well as the command line. Modify commands are also covered in this unit. The readers are introduced to various edit and modify tools ranging from simple copy command to powerful array command.

Apart from this at the end of the unit, a succinct recapitulation of the overall broad concepts is provided in form of a unit summary. Besides, a large number of multiple-choice questions as well as descriptive type questions with Bloom's taxonomy action verbs are included. A list of references and suggested readings are given in the unit so that one can go through them for practice. It is important to note that for getting more information on various topics of interest some QR codes have been provided in different sections which can be scanned for relevant supportive knowledge. Video resources along with QR codes are mentioned for getting more information on various topics of interest which can be surfed or scanned through mobile phones for viewing. Practicals with detailed dimensions are provided to hone the drawing and drafting skills.

RATIONALE

Computer-generated drawings are now a standard practice in every industry. There are many benefits to using computer-generated drawings, including convenience in storing, retrieving, modifying, transmitting, etc. And the most important consideration is the precision with which the whole design is made, especially its curves, a quality that can't be achieved through hand drawing. Many engineering schools throughout the world have started using this method to increase their students' employability quotient. Digital drawings are typically made by merging and altering many elementary shapes, such as lines, circles and arcs to make more complicated shapes. Digital drawings are typically made with the help of the popular design and drafting program AutoCAD. The objectives of this unit are performance-based, so once you have finished the unit, you will be able to carry out all of the tasks it specifies. In this lesson, students will learn the fundamentals of drawing by constructing and changing basic shapes like lines, arcs, circles, etc. This section includes detailed explanations of how to execute each command, as well as exercises to hone your skills.

PRE-REQUISITES

Before reading this unit, the students are advised to revisit the following:

- Mathematics: Coordinate and Plane geometry (Class XII)
- Engineering Graphics (ES101)
- AutoCAD commands mentioned in Unit-1

UNIT OUTCOMES

After studying this unit students will be able to:

U2-O1: Apply Draw commands of AutoCAD to draw simple 2D entities

U2-O2: Understand and appreciate the differences and similarities of Circle and Arc command

U2-O3: Apply Modify commands of AutoCAD to modify and edit 2D entities

U2-O4: Create multiple copies of object(s) using commands like array, copy, mirror etc.

U2-O5: Manipulate the size of object(s) using commands like scale, stretch, lengthen etc.

Unit-2 Outcomes	EXPECTED MAPPING WITH COURSE OUTCOMES (1 – Weak Correlation; 2 – Medium correlation; 3 – Strong Correlation)				
	CO-1	CO-2	CO-3	CO-4	CO-5
U2-O1	1	2	3	1	-
U2-O2	-	2	3	1	-
U2-O3	-	2	3	1	-
U2-O4	-	2	3	1	-
U2-O5	-	2	2	1	-

2.1 INTRODUCTION

This unit demonstrates how to work with AutoCAD commands. You can execute AutoCAD commands by using command tools and a series of prompts. The prompts appear in the command line window and ask the user for a selection or numeric input so that a command sequence can be completed. This unit uses the draw menu and modify menu commands to show the various input and prompt sequences that are typical with AutoCAD commands.

2.2 DRAW MENU

The Draw menu commands can be used to create new objects such as lines, circles etc. Most AutoCAD drawings are composed purely and simply from these basic components. A good understanding of the Draw menu commands is mandatory for the efficient use of AutoCAD.

In common with most AutoCAD commands, the Draw menu commands can be started in several ways. Command names or shortcuts can be entered using the keyboard. Commands can also be started from the Draw pull-down menu. The method you use is dependent upon the type of work you are doing and how experienced a user you are. Do not worry too much about this, just use whatever method feels easiest or most convenient at the time. Your drawing technique will improve over time and with experience. So, do not expect to be working very quickly at first.

2.2.1 Point Command

In AutoCAD, the Point command is used to create a single point object at a specific location in the drawing. Points are commonly used to mark the location of an object or to reference a specific location in the drawing.

To draw a point, click **Home > Draw > Point** on the ribbon.

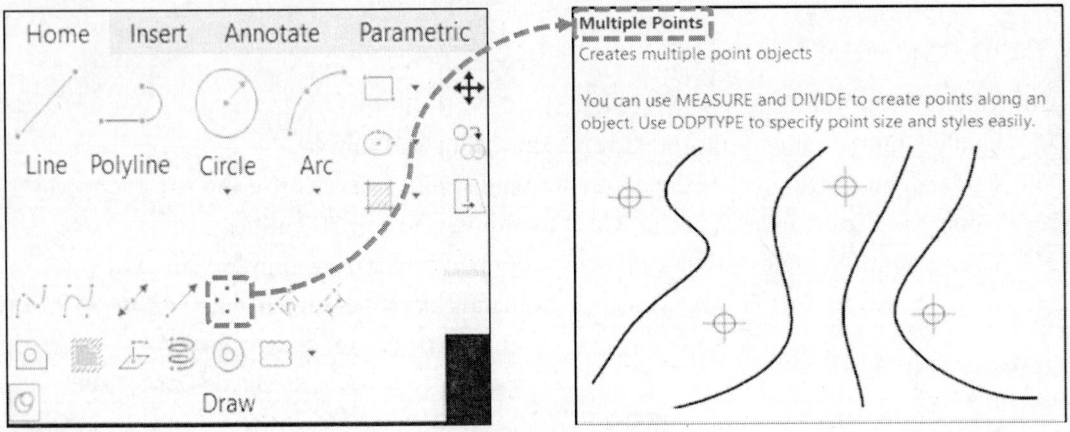

Fig. 2.1: Point command

The following steps can be followed to use the **Point** command —

Step 1: Click on Drawing

Step 2: Click on Multiple Point command

Step 3: Click on Drawing Grid and create points as per the requirement or dimension

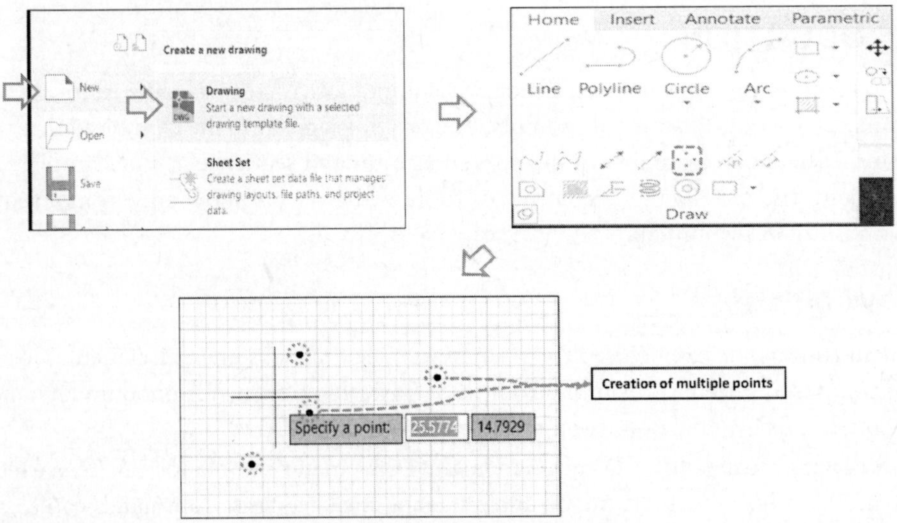

Fig. 2.2: Example of Point command

You can also use the following method to use the **Point** command —

Step 1: Type **POINT** or **PO** on the command line and press Enter.

Step 2: Select the location for the point in the drawing area.

Step 3: Press Enter to place the point at the selected location.

You can also specify the coordinates of the point using the command line. For example, to create a point at the coordinates (10,20), you would type "POINT 10,20" on the command line and press Enter.

You can customize the appearance of the point using the Properties palette or the Layer Properties Manager. You can also use the Point Style command to change the point style for all points in the drawing.

2.2.2 Line Command

In AutoCAD, the Line command is used to draw a straight line between two points in the drawing. You can use the Line command to create a variety of different types of lines, including horizontal, vertical and diagonal lines.

To draw a line, click **Home > Draw > Line** on the ribbon.

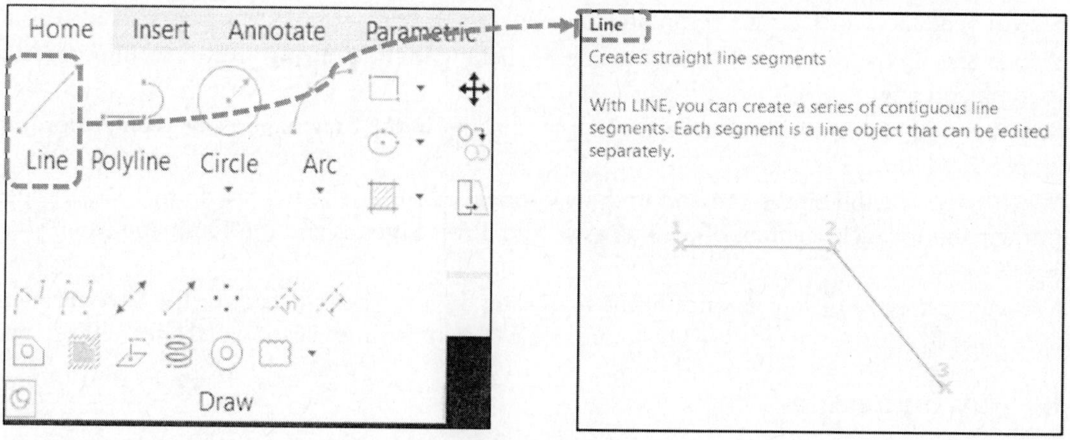

Fig. 2.3: Line command

The following steps can be followed to use the **Line** command —

Step 1: Click on Drawing

Step 2: Click on Line

Step 3: Specify the first point (press the left mouse button)

Step 4: Specify the second point (press the left mouse button)

To end the **Line** command, press **Enter** key or press the right mouse button. When the right mouse button (i.e., right-click) is pressed, a *pop-up menu* appears. Select **Enter** on the pop-up menu using the left mouse button.

The **Line** command can be restarted by pressing the right mouse button immediately after selecting **Enter**. A *pop-up menu* appears and by selecting **Repeat LINE** on the pop-up menu the **Line** command can be restarted.

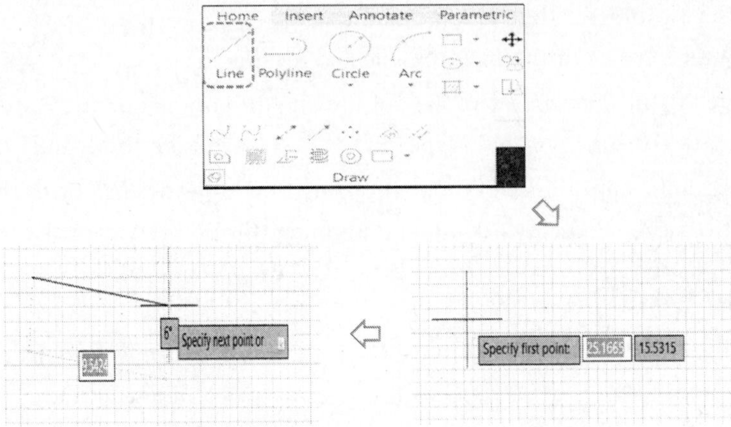

Fig. 2.4: Example of Line command

You can also use the following method to use the **Line** command —

Step 1: Type **LINE** or **L** on the command line and press Enter.

Step 2: Specify the starting point for the line by clicking in the drawing area or typing coordinates on the command line.

Step 3: Specify the ending point for the line by clicking in the drawing area or typing coordinates on the command line.

You can also use the Line command to draw a series of connected lines. To do this, simply specify the ending point for each segment of the line as you go. Press Enter to end the command when you are finished drawing the line.

You can customize the appearance of the line using the Properties palette or the Layer Properties Manager. You can also use the Line Weight command to change the line weight for all lines in the drawing.

2.2.3 Ray Command

In AutoCAD, the Ray command is used to draw a half-line that extends indefinitely from a starting point in a specific direction. A ray is similar to a line, but it has only one endpoint and extends indefinitely in one direction.

To draw a ray, click **Home** > **Draw** > **Ray** on the ribbon.

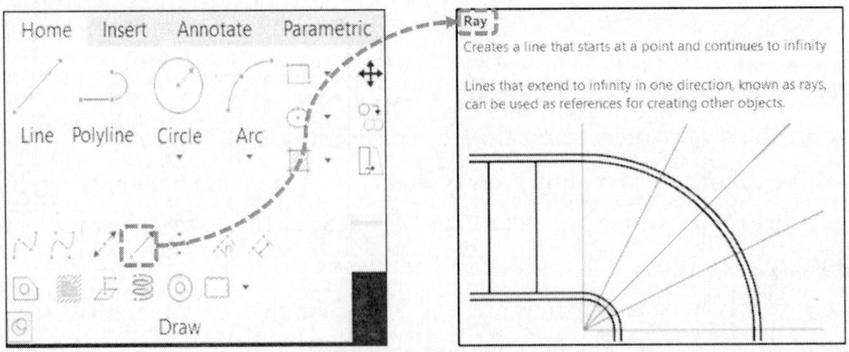

Fig. 2.5: Ray command

The following steps can be followed to use the **Ray** command —

Step 1: Click on Draw

Step 2: Click on Ray

Step 3: Specify the first point (press the left mouse button)

Step 4: Specify the second point (press the left mouse button)

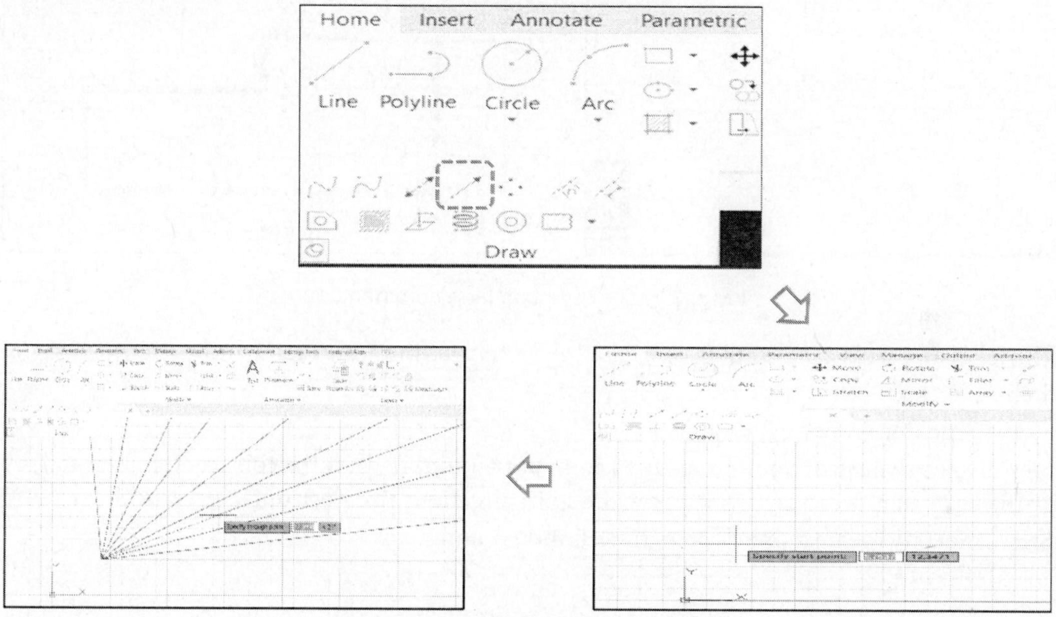

Fig. 2.6: Example of Ray command

You can also use the following method to use the **Ray** command —

Step 1: Type **RAY** on the command line and press Enter.

Step 2: Specify the starting point for the ray by clicking in the drawing area or typing coordinates on the command line.

Step 3: Specify the direction for the ray by clicking in the drawing area or typing coordinates on the command line.

You can also specify the length of the ray using the command line. For example, to create a ray that is 100 units long, you would type "RAY 100" on the command line and press Enter.

You can customize the appearance of the ray using the Properties palette or the Layer Properties Manager. You can also use the Line Weight command to change the line weight for all rays in the drawing.

The ray is extended to the display's edge in the direction indicated by the starting point and the through point. More than one ray can be made by redisplaying the prompt for a through point.

2.2.4 Construction Line Command

In AutoCAD, the Construction Line command is used to draw a temporary line that is used for construction or reference purposes. Construction lines are often used as a guide to help draw other objects, such as circles or arcs, but they are not typically used as part of the finished drawing.

To draw a construction line, click **Home** > **Draw** > **Construction Line** on the ribbon.

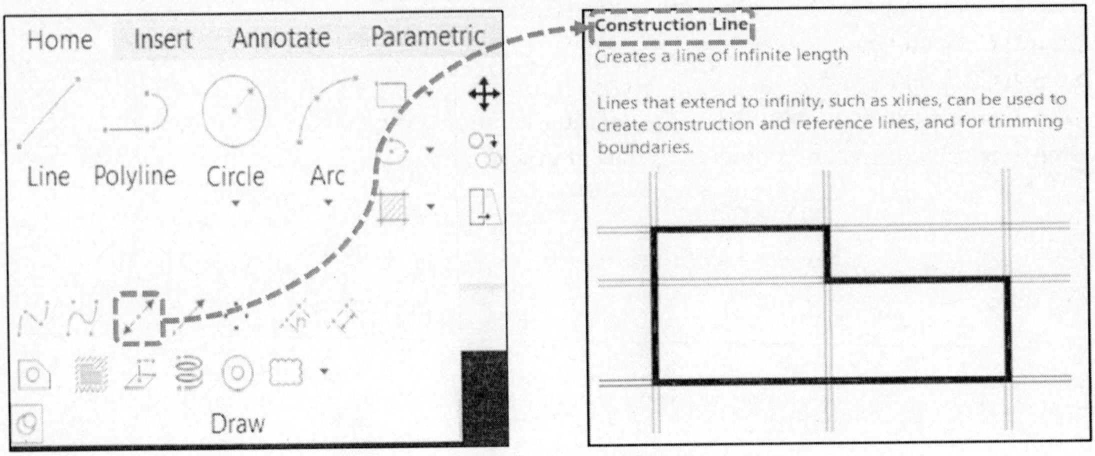

Fig. 2.7: Construction Line command

The following steps can be followed to use the **Construction Line** command —

Step 1: Click on Draw

Step 2: Click on Construction Line

Step 3: Specify the first point (press the left mouse button) to draw the construction line and further specify the next point which symbolises the different directions like horizontal or vertical.

Step 4: Specify the next point (press the left mouse button)

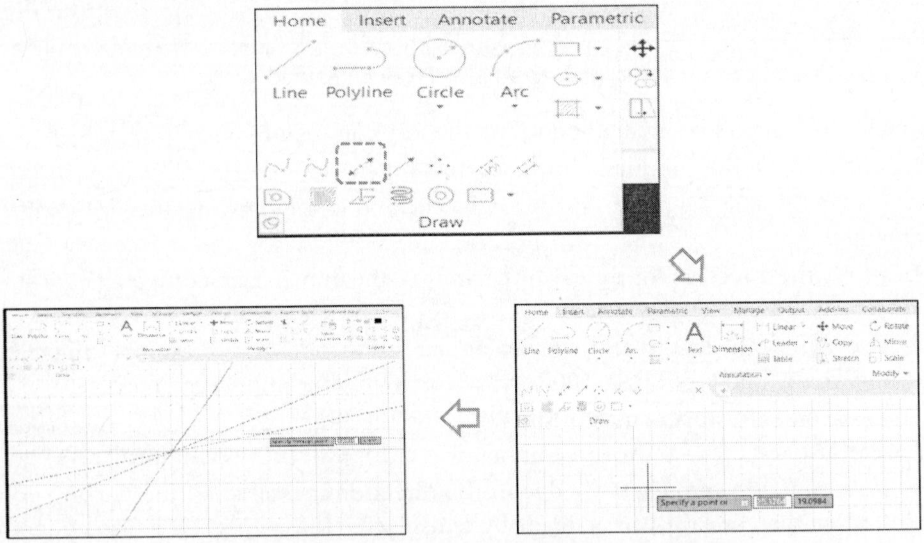

Fig. 2.8: Example of Construction Line command

You can also use the following method to use the **Construction Line** command —

Step 1: Type **XLINE** or **XL** on the command line and press Enter.

Step 2: Specify the starting point for the construction line by clicking in the drawing area or typing coordinates on the command line.

Step 3: Specify the ending point for the construction line by clicking in the drawing area or typing coordinates on the command line.

You can also use the Construction Line command to draw a series of connected lines. To do this, simply specify the ending point for each segment of the construction line as you go. Press Enter to end the command when you are finished drawing the construction line.

Construction lines are typically drawn using a thin, dashed line style to distinguish them from other objects in the drawing. You can customize the appearance of construction lines using the Properties palette or the Layer Properties Manager.

2.2.5 Circle Command

In AutoCAD, the Circle command is used to draw a circle with a specified radius and center point. You can use the Circle command to create circles of any size and at any location in the drawing.

To draw a circle, click **Home** > **Draw** > **Circle** on the ribbon.

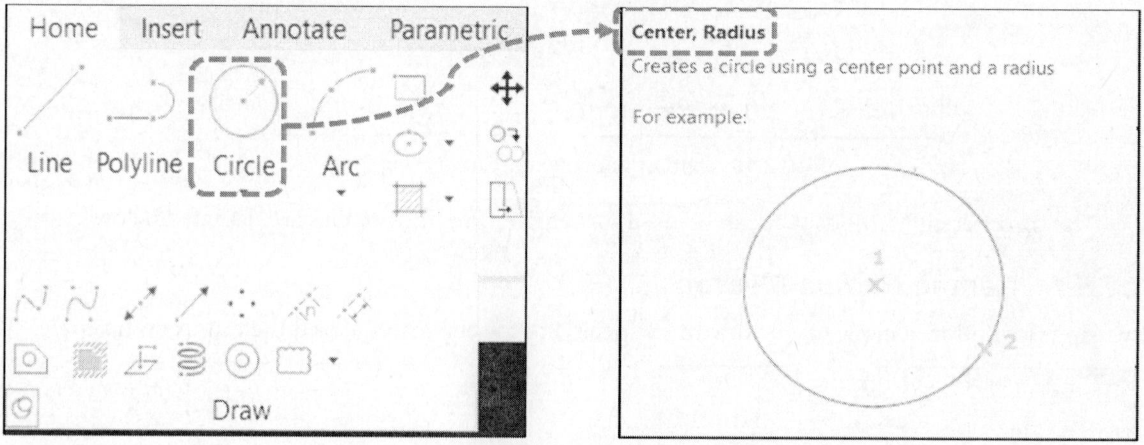

Fig. 2.9: Circle command

The following steps can be followed to use the **Circle** command —

Step 1: Click on Draw

Step 2: Click on Circle

There are several ways a circle can be drawn in AutoCAD. The default method is to pick the centre point and then either pick a second point on the circumference of the circle or enter the circle radius using the keyboard. The following six different methods are available in AutoCAD to draw a circle—

1. Center, Radius
2. Center, Diameter
3. 2 points
4. 3 points
5. Tan, Tan, Radius
6. Tan, Tan, Tan

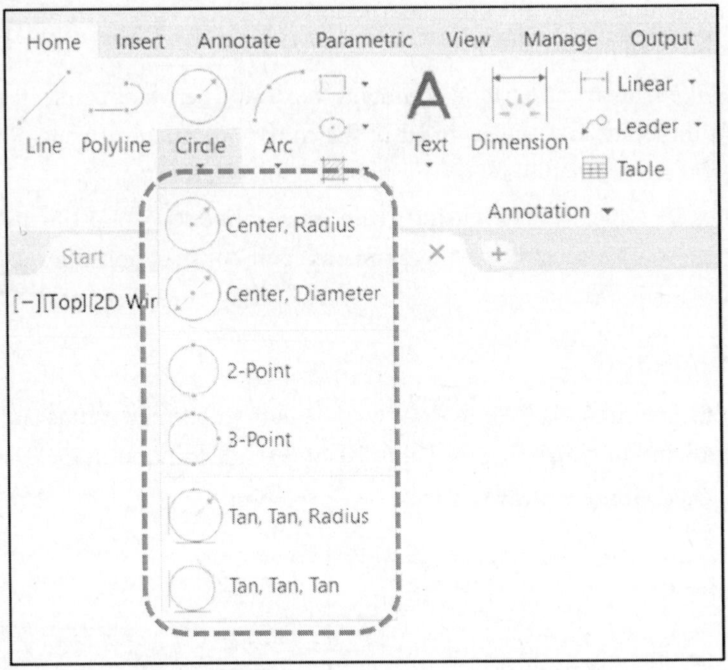

Fig. 2.10: Different options for Circle command

The six different methods that can be used in AutoCAD to draw circles are depicted below.

2.2.5.1 Center, Radius Method

By using this option, a circle can be drawn by specifying the center point and the radius of the circle.

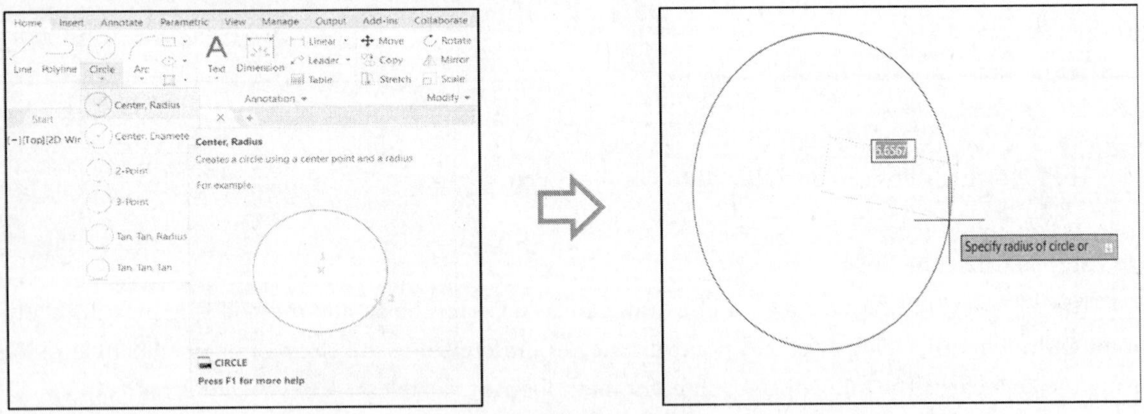

Fig. 2.11: Example of center-radius Circle command

2.2.5.2 Center, Diameter Method

By using this option, a circle can be drawn by specifying the center point and the diameter of the circle.

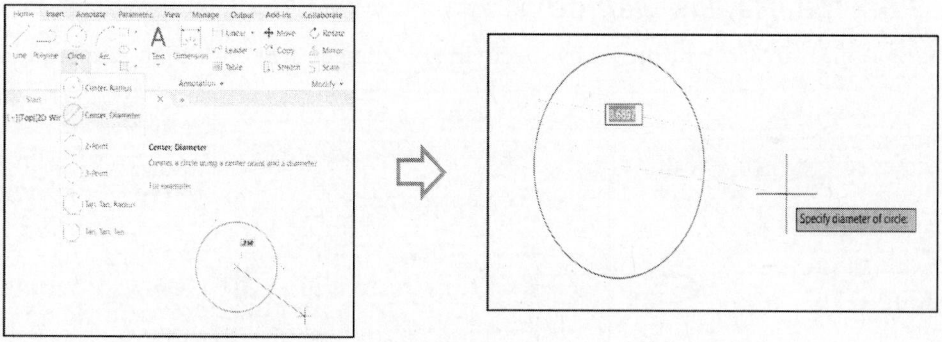

Fig. 2.12: Example of center-diameter Circle command

2.2.5.3 2-Points Method

By using this option, a circle can be drawn by specifying any two endpoints of the circle.

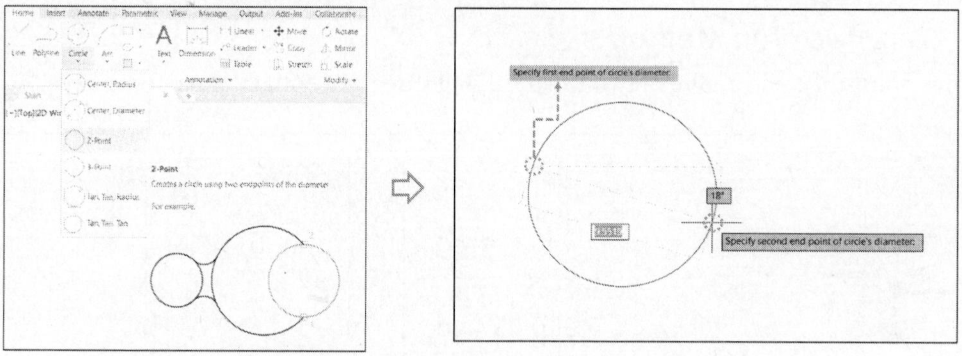

Fig. 2.13: Example of 2-points Circle command

2.2.5.4 3-Points Method

By using this option, a circle can be drawn by specifying any three points on the circumference of the circle.

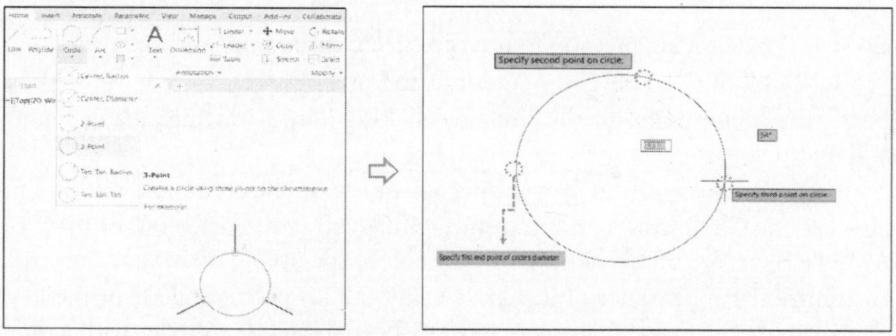

Fig. 2.14: Example of 3-points Circle command

2.2.5.5 Tan, Tan, Radius Method

By using this option, a circle can be drawn by specifying two tangential points to the circle and its radius.

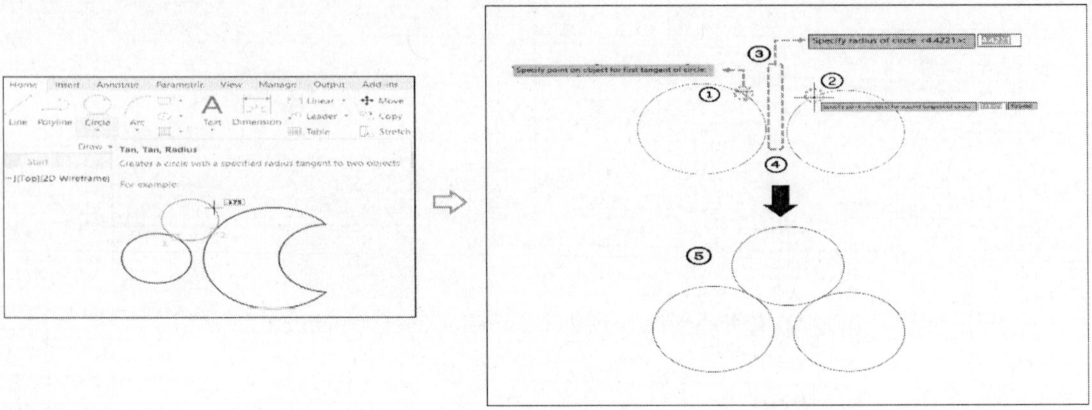

Fig. 2.15: Example of tan-tan-radius Circle command

2.2.5.6 Tan, Tan, Tan Method

By using this option, a circle can be drawn by specifying three tangential points to the circle.

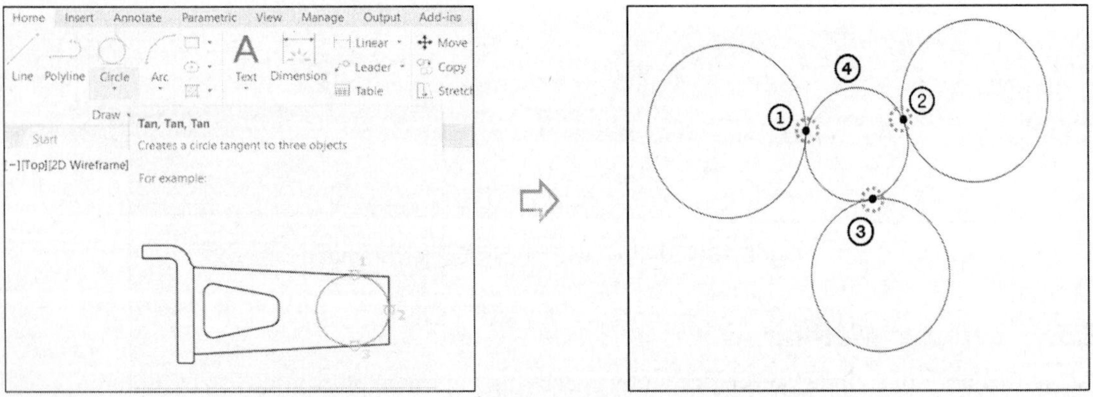

Fig. 2.16: Example of tan-tan-tan Circle command

You can also use the following method to use the **Circle** command —

Step 1: Type **CIRCLE** or **C** on the command line and press Enter.

Step 2: Specify the center point for the circle by clicking in the drawing area or typing coordinates on the command line.

Step 3: Specify the radius of the circle by typing a distance on the command line and pressing Enter.

You can also use the Circle command to draw a circle with a diameter rather than a radius. To do this, type "DIAMETER" or "D" on the command line before specifying the size of the circle.

You can customize the appearance of the circle using the Properties palette or the Layer Properties Manager. You can also use the Circle command to draw partial circles or circles with a different number of segments.

2.2.6 Arc Command

In AutoCAD, the Arc command is used to draw a curved line that is defined by a center point, a starting point and an ending point. Arcs are commonly used to create curves or rounded corners in drawings.

To draw an arc, click **Home > Draw > Arc** on the ribbon.

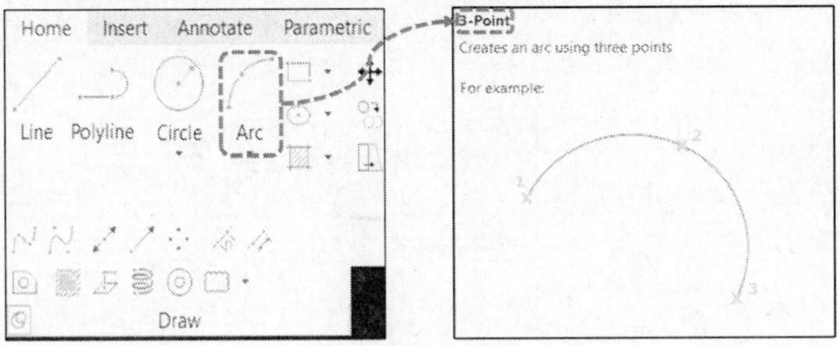

Fig. 2.17: Arc command

There are in total 11 ways in which an arc can be drawn in AutoCAD. The default method is to specify three points. The following 11 different methods are available in AutoCAD to draw an arc—

1. 3-Point
2. Start, Center, End
3. Start, Center, Angle
4. Start, Center, Length
5. Start, End, Angle
6. Start, End, Direction
7. Start, End, Radius
8. Center, Start, End
9. Center, Start, Angle
10. Center, Start, Length
11. Continue

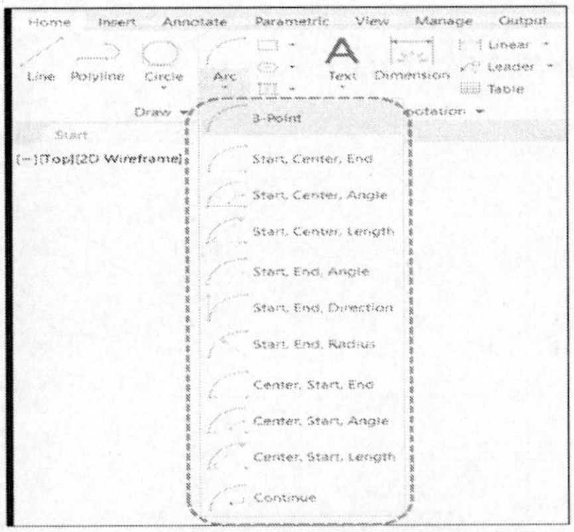

Fig. 2.18: Different options for Arc command

The 11 different methods that can be used in AutoCAD to draw arcs are depicted below.

2.2.6.1 3-Point Method

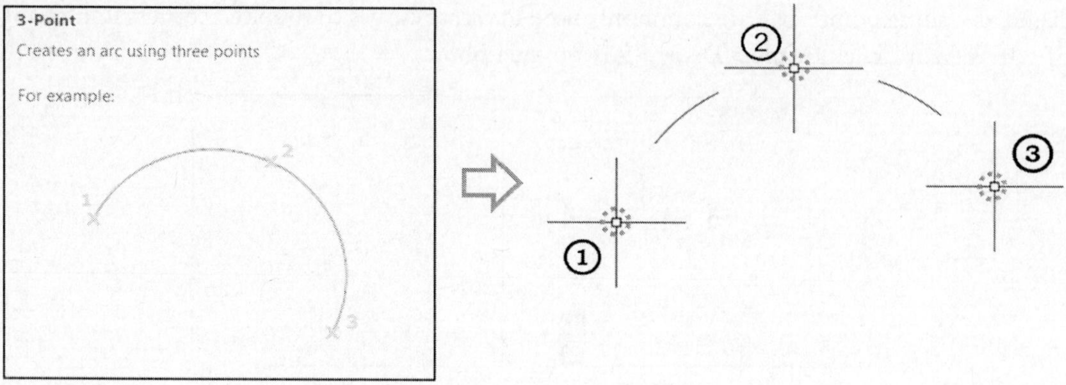

Fig. 2.19: Example of 3-point Arc command

2.2.6.2 Start, Center, End Method

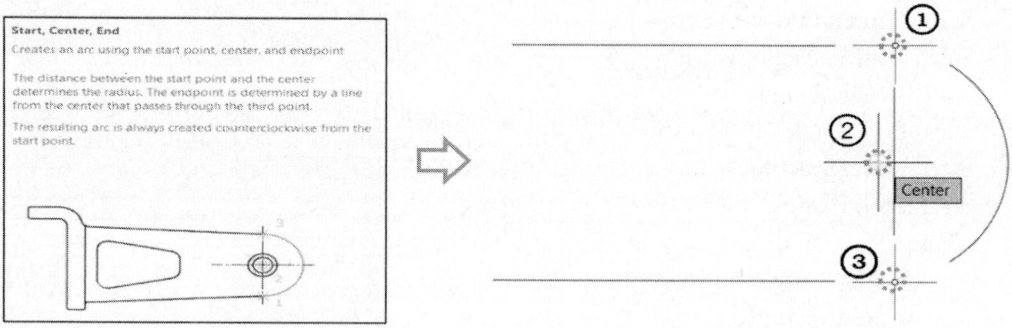

Fig. 2.20: Example of start-center-end Arc command

2.2.6.3 Start, Center, Angle Method

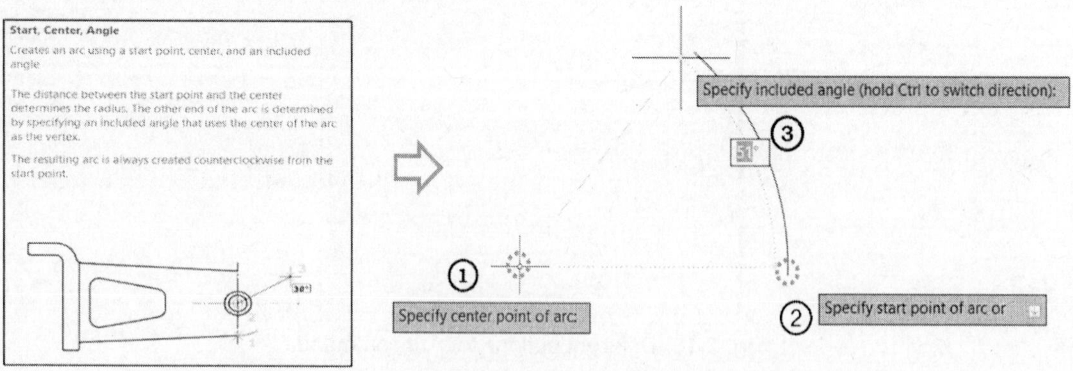

Fig. 2.21: Example of start-center-angle Arc command

2.2.6.4 Start, Center, Length Method

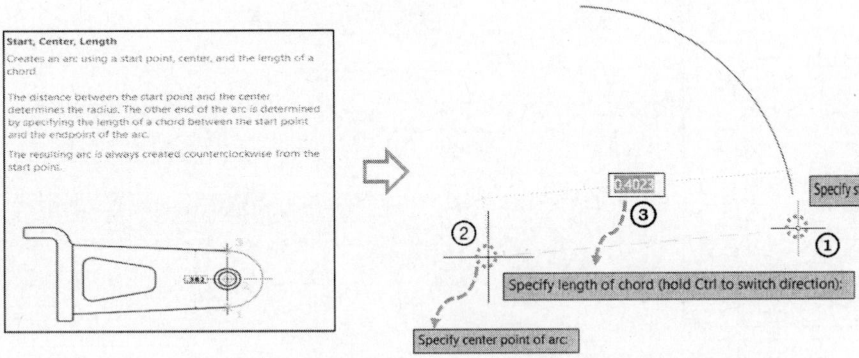

Fig. 2.22: Example of start-center-length Arc command

2.2.6.5 Start, End, Angle Method

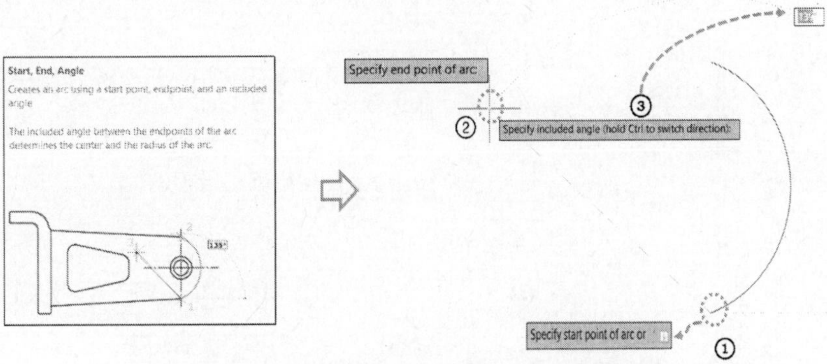

Fig. 2.23: Example of start-end-angle Arc command

2.2.6.6 Start, End, Direction Method

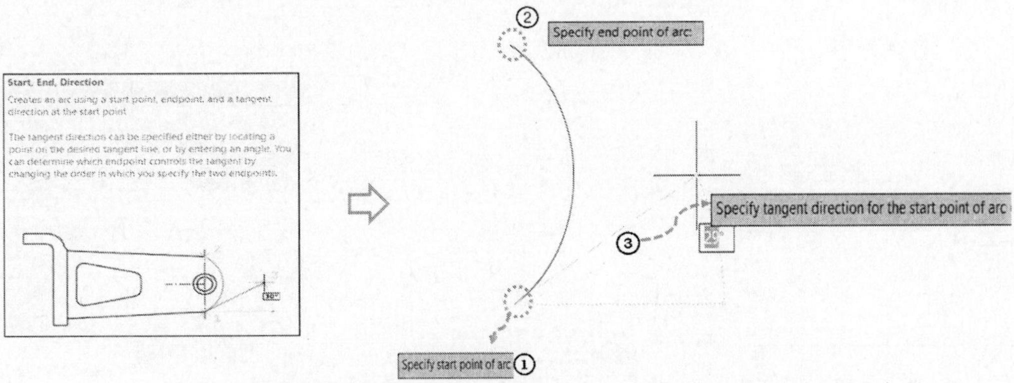

Fig. 2.24: Example of start-end-direction Arc command

2.2.6.7 *Start, End, Radius Method*

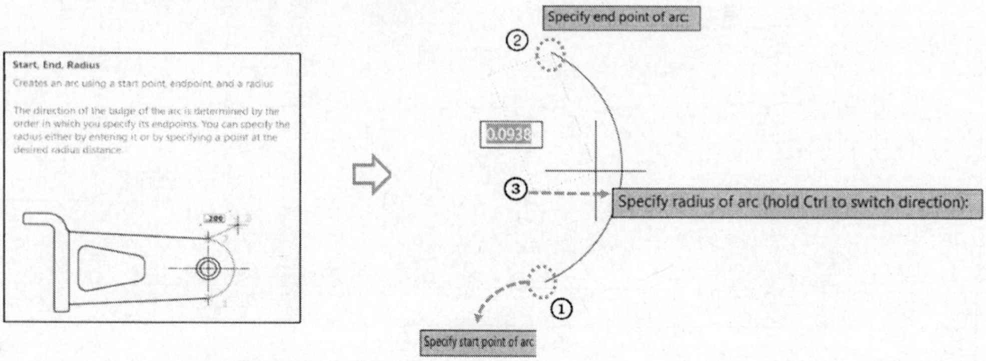

Fig. 2.25: Example of start-end-radius Arc command

2.2.6.8 *Center, Start, End Method*

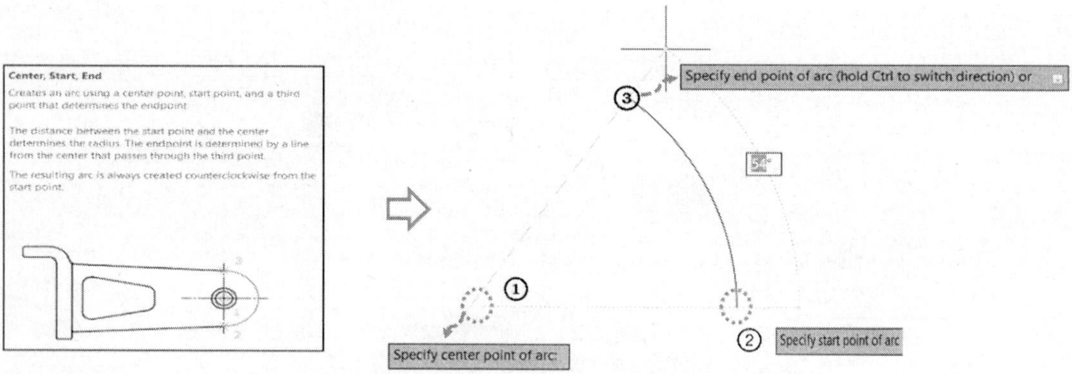

Fig. 2.26: Example of center-start-end Arc command

2.2.6.9 *Center, Start, Angle Method*

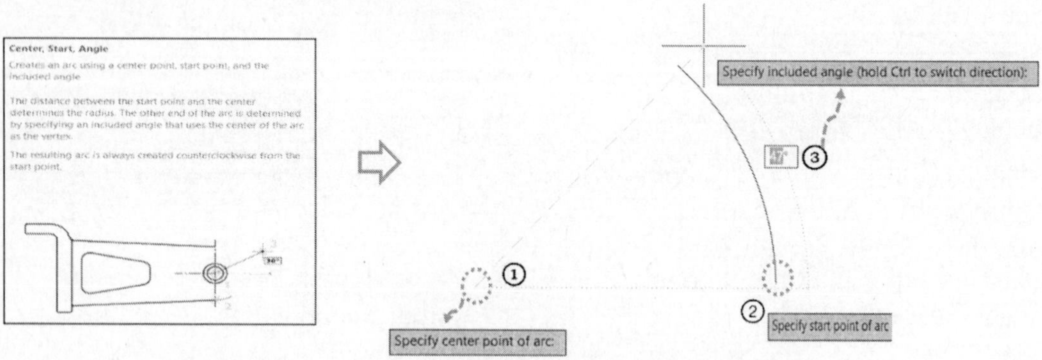

Fig. 2.27: Example of center-start-angle Arc command

2.2.6.10 Center, Start, Length Method

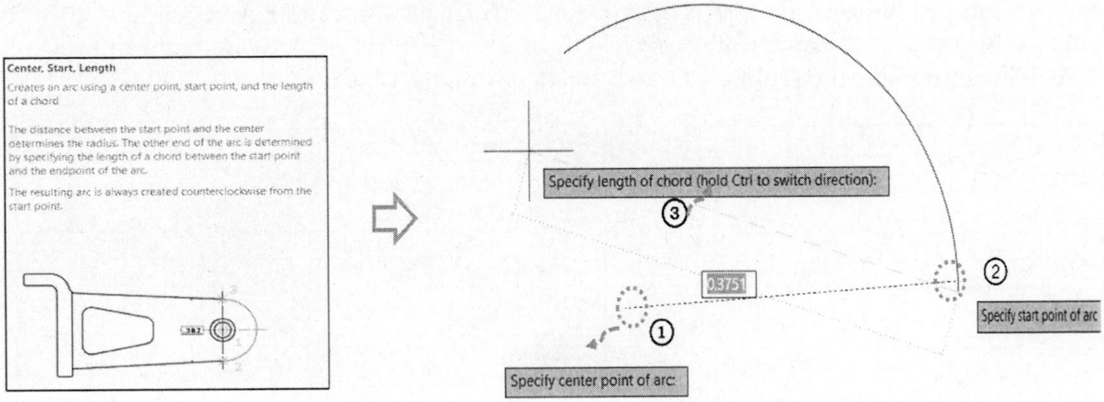

Fig. 2.28: Example of center-start-length Arc command

2.2.6.11 Continuous Arc Method

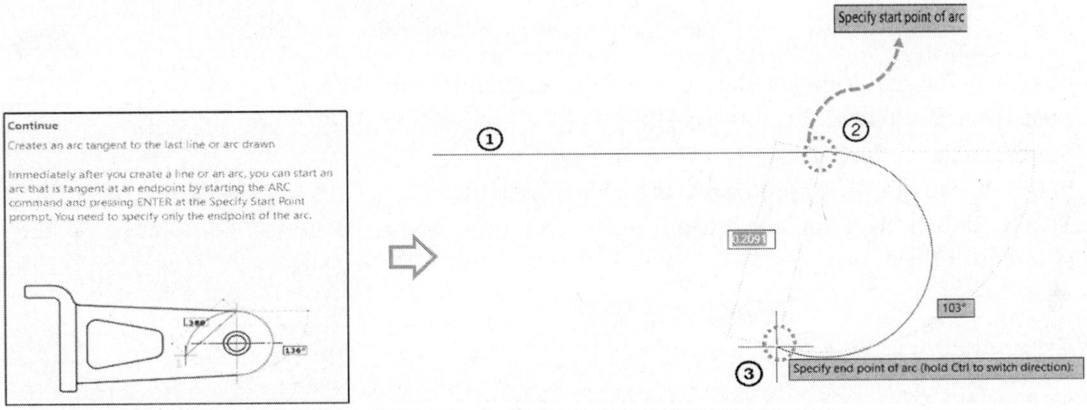

Fig. 2.29: Example of continue Arc command

You can also use the following method to use the **Arc** command —

Step 1: Type **ARC** or **A** on the command line and press Enter.

Step 2: Specify the center point for the arc by clicking in the drawing area or typing coordinates on the command line.

Step 3: Specify the starting point for the arc by clicking in the drawing area or typing coordinates on the command line.

Step 4: Specify the ending point for the arc by clicking in the drawing area or typing coordinates on the command line.

You can customize the appearance of the arc using the Properties palette or the Layer Properties Manager. You can also use the Arc command to draw partial arcs or arcs with a different number of segments.

2.2.7 Rectangle Command

In AutoCAD, the Rectangle command is used to draw a rectangle with a specified width and height. You can use the Rectangle command to create rectangles of any size and at any location in the drawing.

To draw a rectangle, click **Home** > **Draw** > **Rectangle** on the ribbon.

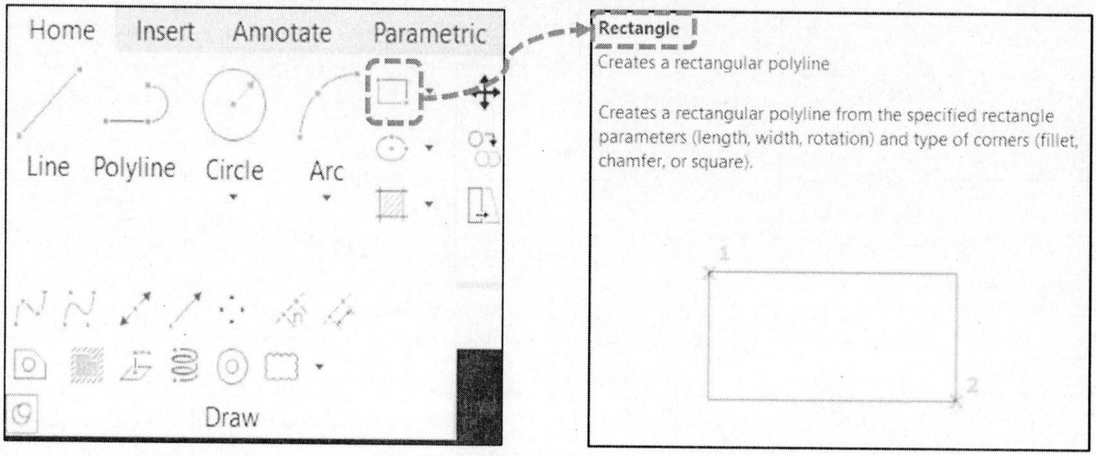

Fig. 2.30: Rectangle command

The following steps can be followed to use the **Rectangle** command—

Step 1: Click on Draw

Step 2: Click on Rectangle

Step 3: Specify the first point (press the left mouse button)

Step 4: Specify the length and width of the rectangle or specify the second point (press the left mouse button at the desired point)

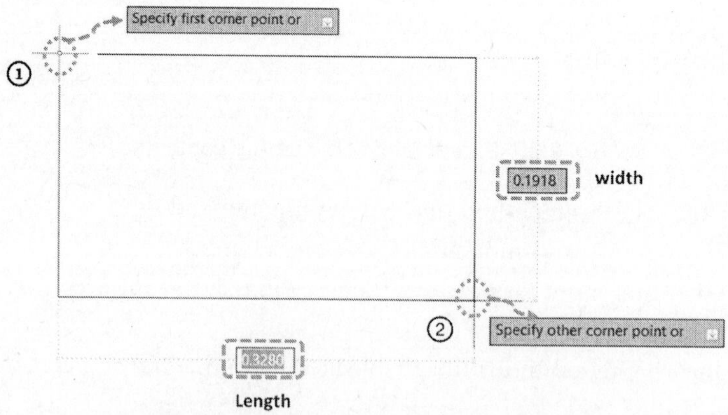

Fig. 2.31: Example of Rectangle command

You can also use the following method to use the **Rectangle** command —

Step 1: Type **RECTANGLE** or **REC** on the command line and press Enter.

Step 2: Specify the first corner of the rectangle by clicking in the drawing area or typing coordinates on the command line.

Step 3: Specify the opposite corner of the rectangle by clicking in the drawing area or typing coordinates on the command line.

You can also use the Rectangle command to draw a square by specifying the same width and height for the rectangle.

2.2.8 Polyline Command

In AutoCAD, the Polyline command is used to draw a series of connected lines or curves that form a single, continuous object. Polylines are often used to create complex shapes or to connect multiple lines and arcs together.

To draw a polyline, click **Home** > **Draw** > **Polyline** on the ribbon.

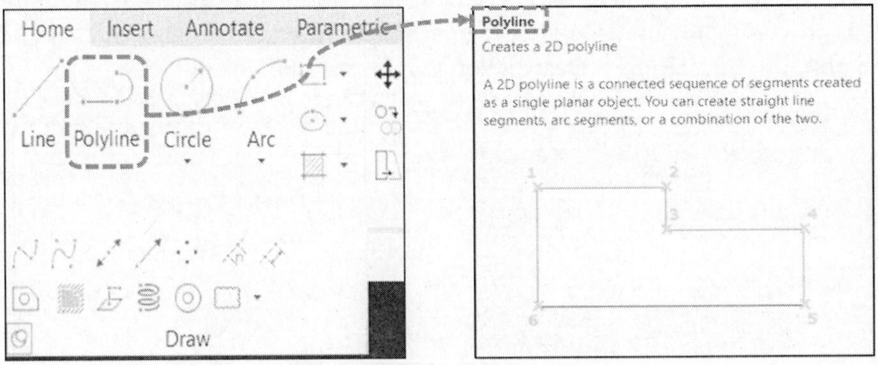

Fig. 2.32: Polyline command

The following steps can be followed to use the **Polyline** command —

Step 1: Click on Draw

Step 2: Click on Polyline

Step 3: Specify the first point (press the left mouse button)

Step 4: Specify the next point (press the left mouse button)

Step 5: Specify the arc using the right click of mouse

Step 6: Specify the end point of arc

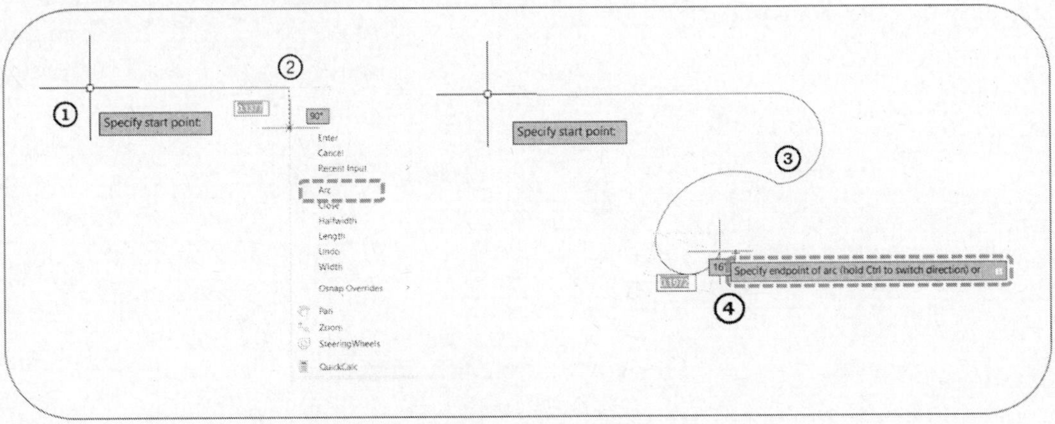

Fig. 2.33: Example of Polyline command

You can also use the following method to use the **Polyline** command –

Step 1: Type **PLINE** or **PL** on the command line and press Enter.

Step 2: Specify the starting point for the polyline by clicking in the drawing area or typing coordinates on the command line.

Step 3: Specify the ending point for the first segment of the polyline by clicking in the drawing area or typing coordinates on the command line.

Step 4: Specify the ending point for each additional segment of the polyline as needed.

Step 5: Press Enter to end the command when you are finished drawing the polyline.

2.2.9 Polygon Command

In AutoCAD, the Polygon command is used to draw a closed shape with a specified number of sides and radius. Polygons are commonly used to create regular shapes, such as triangles, squares and hexagons.

To draw a polygon, click **Home** > **Draw** > **Polygon** on the ribbon.

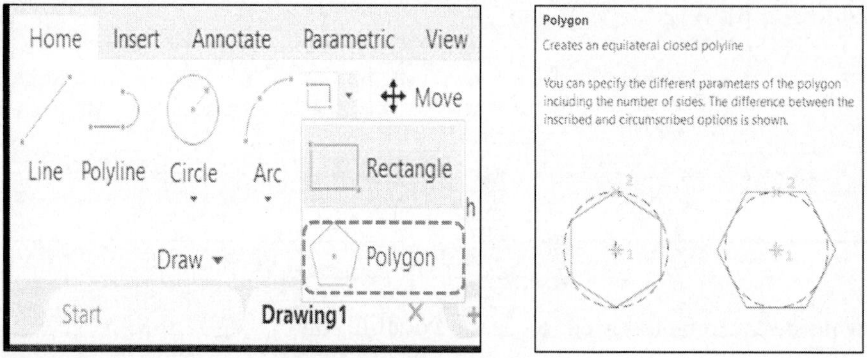

Fig. 2.34: Polygon command

Method 1: Draw a Polygon inscribed in circle

Step 1: Enter the number of sides

Step 2: Specify the end point of arc using the left mouse button.

Step 3: Choose the first option i.e., Inscribed in circle

Step 4: Specify the radius of circle

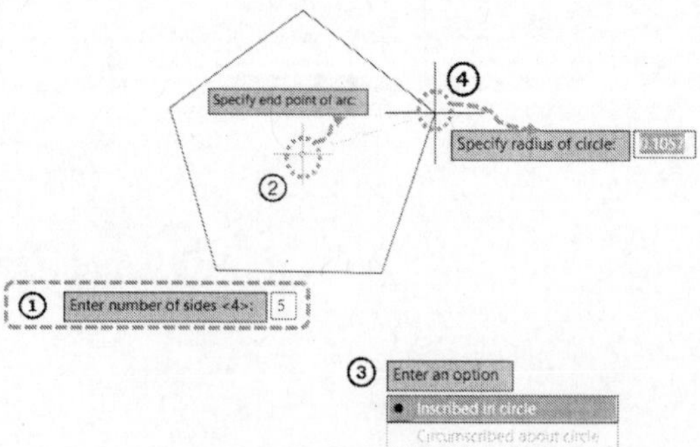

Fig. 2.35: Example of polygon inscribed in circle

Method 2: Draw a Polygon circumscribed in circle

Step 1: Enter the number of sides

Step 2: Specify the end point of arc using the left mouse button

Step 3: Choose the second option i.e., Circumscribed in circle

Step 4: Specify the radius of circle

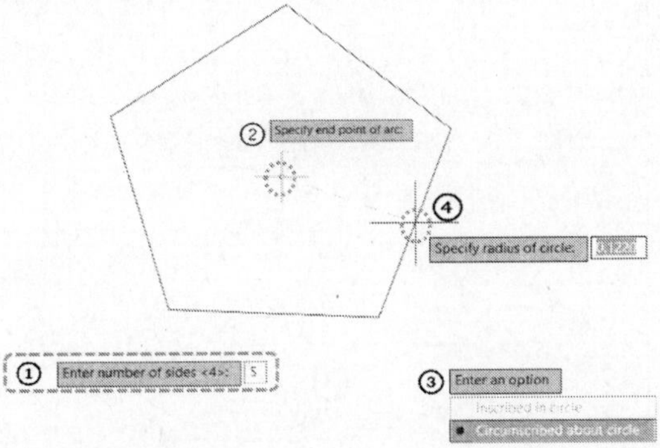

Fig. 2.36: Example of polygon circumscribed in circle

You can also use the following method to use the **Polygon** command —

Step 1: Type **POLYGON** or **POL** on the command line and press Enter.

Step 2: Specify the center point for the polygon by clicking in the drawing area or typing coordinates on the command line.

Step 3: Specify the number of sides for the polygon by typing a number on the command line and pressing Enter.

Step 4: Specify the radius of the polygon by typing a distance on the command line and pressing Enter.

2.2.10 Ellipse Command

In AutoCAD, the Ellipse command is used to draw an oval or ellipse shape with a specified major and minor axis. Ellipses are often used to create circular or oval shapes in drawings.

To draw an Ellipse, click **Home > Draw > Ellipse** on the ribbon.

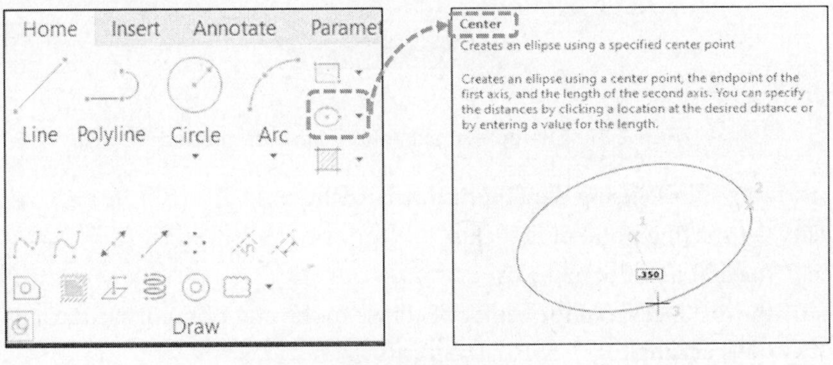

Fig. 2.37: Ellipse command

Method 1: Draw an Ellipse using the Center method.

Step 1: Specify the center of ellipse

Step 2: Specify the endpoint to first axis

Step 3: Specify the distance to the second axis

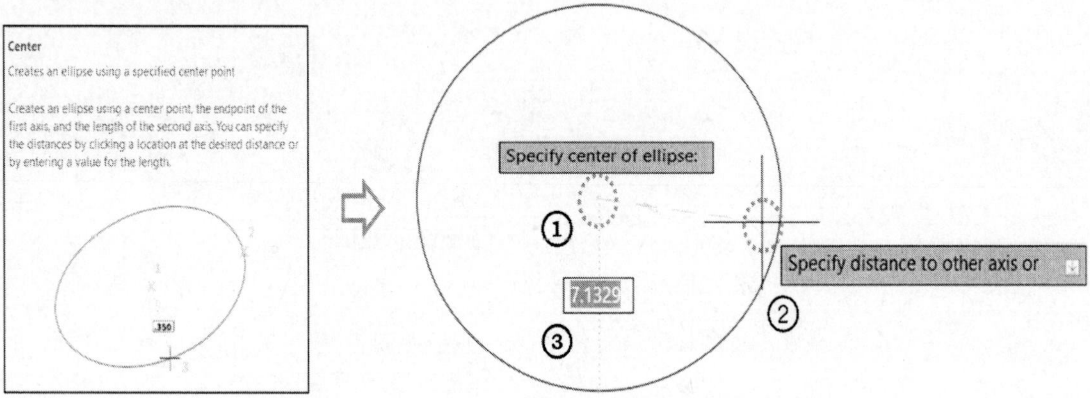

Fig. 2.38: Example of center Ellipse command

Method 2: Draw an Ellipse using the Axis, End method.

Step 1: Specify the starting point of first axis

Step 2: Specify the end point of first axis

Step 3: Specify the distance from the center of ellipse to the end point of second axis

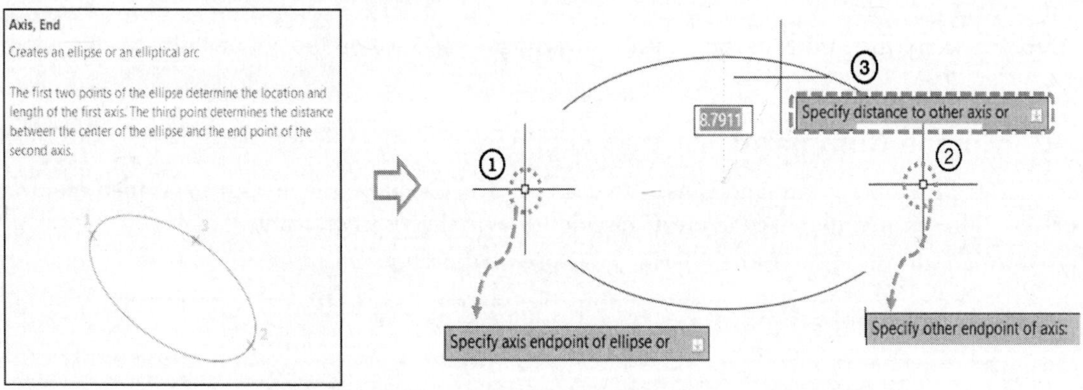

Fig. 2.39: Example of axis-end Ellipse command

Method 3: Draw an Ellipse using the Elliptical Arc method.

Step 1: Specify the starting point of first axis

Step 2: Specify the end point of first axis

Step 3: Specify the distance from the center of ellipse to the end point of second axis

Step 4: Specify the start angle

Step 5: Specify the end angle

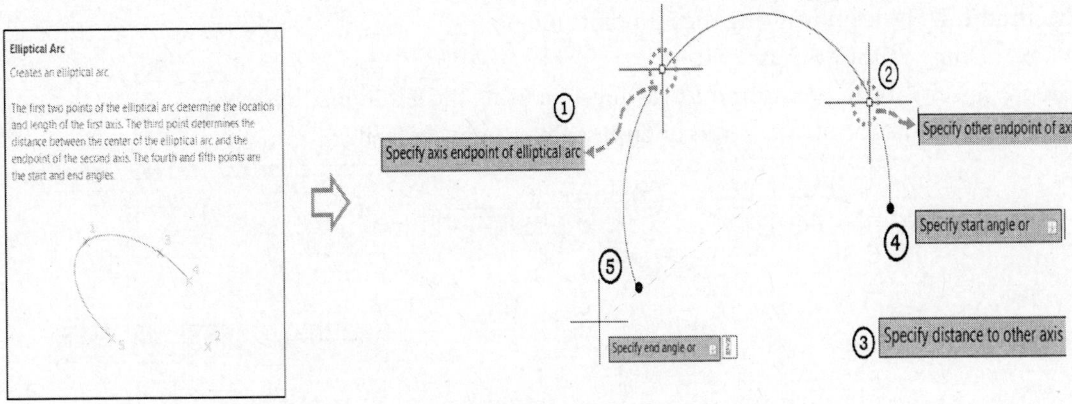

Fig. 2.40: Example of elliptical-arc Ellipse command

You can also use the following method to use the **Ellipse** command —

Step 1: Type **ELLIPSE** or **EL** on the command line and press Enter.

Step 2: Specify the center point for the ellipse by clicking in the drawing area or typing coordinates on the command line.

Step 3: Specify the major axis of the ellipse by clicking in the drawing area or typing coordinates on the command line.

Step 4: Specify the minor axis of the ellipse by clicking in the drawing area or typing coordinates on the command line.

2.2.11 Spline Command

In AutoCAD, the Spline command is used to draw a smooth, curved line that passes through a series of control points. Splines are often used to create complex, organic shapes or to smooth out the corners of a polyline.

To draw a Spline, click **Home > Draw > Spline** on the ribbon.

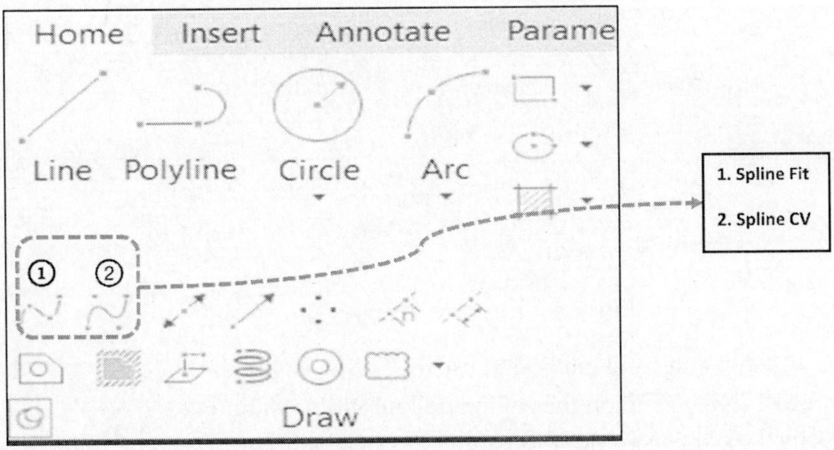

Fig. 2.41: Spline command

Method 1: Draw a Spline using the Fit points method.
Step 1: Home > Draw > Spline Fit
Step 2: Specify the fit points in the drawing area using the left mouse button
Step 3: Continue specifying points and press Enter or Esc to exit

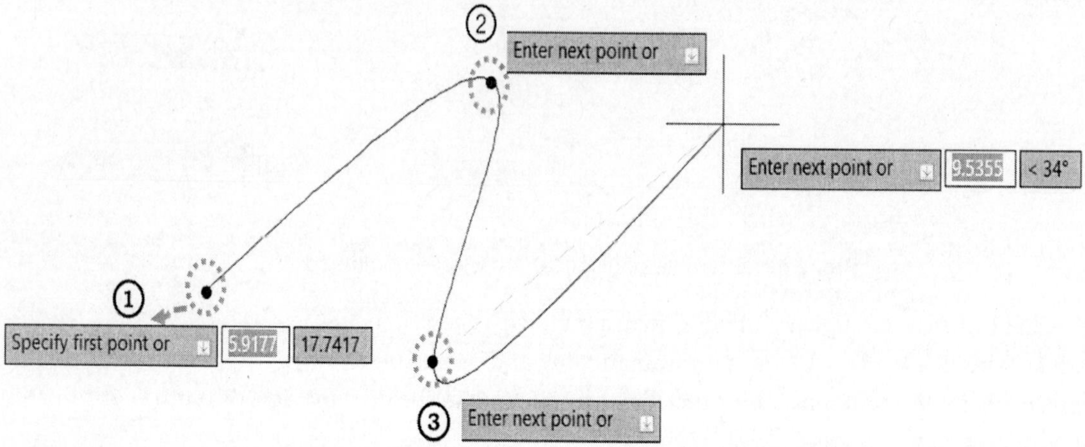

Fig. 2.42: Example of Spline Fit method

Method 2: Draw a Spline using the Control Vertices method.
Step 1: Home > Draw > Spline CV
Step 3: Specify the control vertices in the drawing area using the left mouse button
Step 3: Continue specifying points and press Enter or Esc to exit

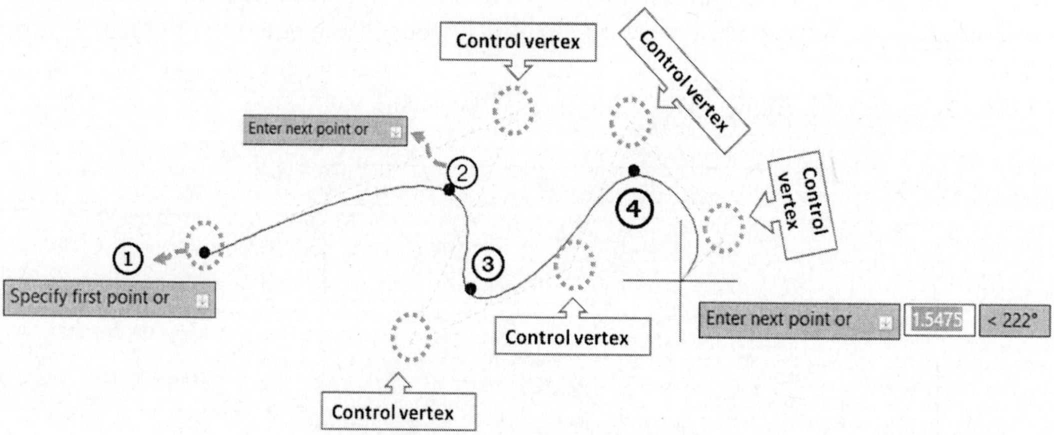

Fig. 2.43: Example of Spline CV method

You can also use the following method to use the **Spline** command —
Step 1: Type **SPLINE** or **SPL** on the command line and press Enter.
Step 2: Specify the starting point for the spline by clicking in the drawing area or typing coordinates on the command line.

Step 3: Specify the ending point for the first segment of the spline by clicking in the drawing area or typing coordinates on the command line.

Step 4: Specify the control points for the spline as needed by clicking in the drawing area or typing coordinates on the command line.

Step 5: Press Enter to end the command when you are finished drawing the spline.

2.2.12 Divide Command

In AutoCAD, the Divide command is used to divide an object into a specified number of equally spaced segments. The Divide command can be used on a variety of different object types, including lines, polylines, circles and arcs.

To create objects by using the Divide command, click **Home** > **Draw** > **Divide** on the ribbon.

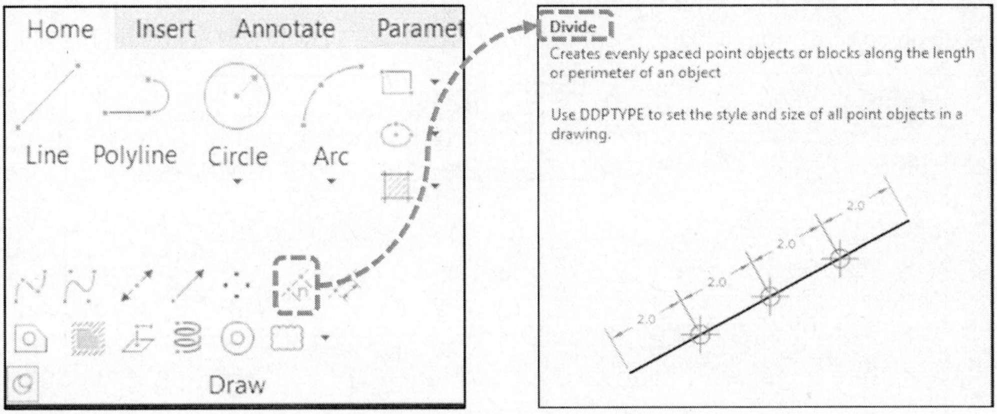

Fig. 2.44: Divide command

The following steps can be followed to use the **Divide** command —

Step 1: Click on Draw

Step 2: Click on Divide

Step 3: Select the object to divide (press the left mouse button)

Step 4: Enter the number of segments to divide

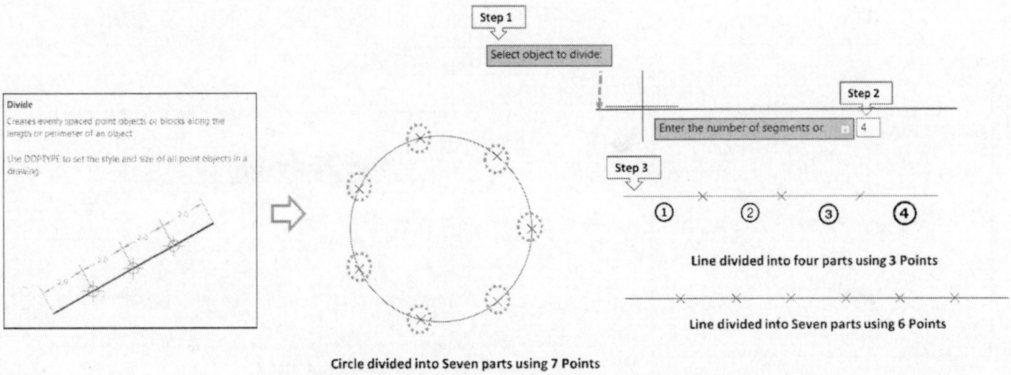

Fig. 2.45: Example of Divide command

You can also use the following method to use the **Divide** command —

Step 1: Type **DIVIDE** or **DIV** on the command line and press Enter.

Step 2: Select the object to divide.

Step 3: Specify the number of segments to create by typing a number on the command line and pressing Enter.

The Divide command will create a series of evenly spaced points along the selected object. You can use these points as reference points or as the starting or ending points for other objects.

2.2.13 Measure Command

In AutoCAD, the Measure command is used to create point objects or blocks at measured intervals along the length or perimeter of an object. The Measure command can be used on a variety of different object types, including lines, polylines, circles and arcs.

To create objects by using the Divide command, click **Home** > **Draw** > **Measure** on the ribbon.

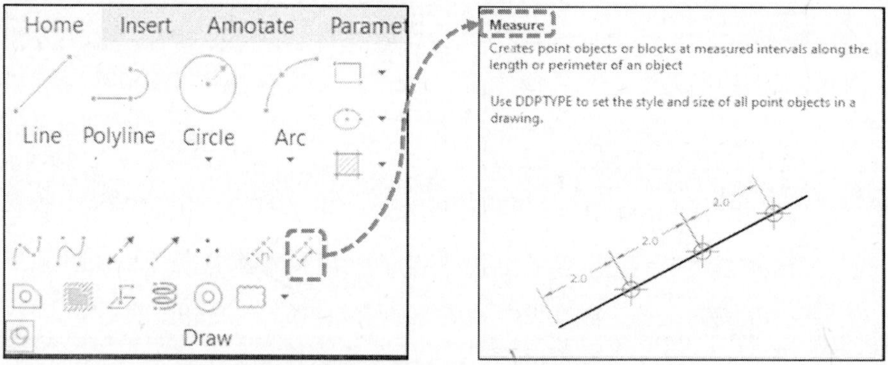

Fig. 2.46: Measure command

The following steps can be followed to use the **Measure** command —

Step 1: Click on Draw

Step 2: Click on Measure

Step 3: Select the object to divide (press the left mouse button)

Step 4: Specify the length of the segment for dividing

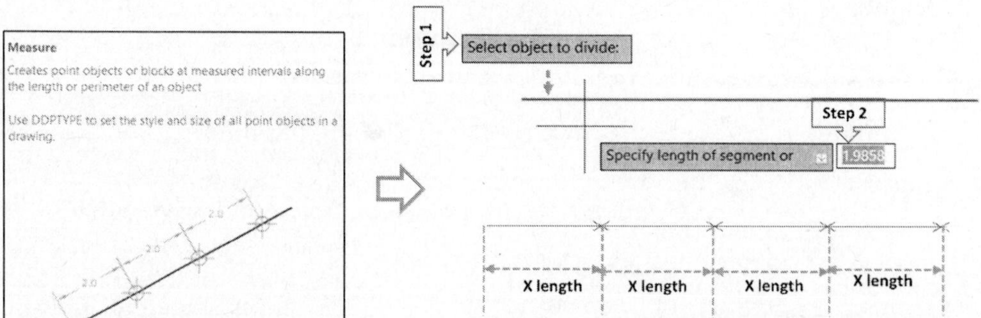

Fig. 2.47: Example of Measure command

You can also use the following method to use the **Measure** command –

Step 1: Type **MEASURE** on the command line and press Enter.

Step 2: Select the object to measure.

Step 3: Specify the length of the segment by typing a number on the command line and pressing Enter.

2.3 MODIFY MENU

AutoCAD drawings are rarely completed simply by using draw menu commands viz lines, circles etc. Most likely you will need to Modify these basic drawing objects in some way to create the image you need. AutoCAD provides a large range of modify tools such as Move, Copy, Rotate, Mirror etc. The Modify command tools can be accessed in one of three ways, from the keyboard, from the pull-down menu and from the toolbar. All of the Modify tools are available from the Modify pull-down and the Modify toolbar.

2.3.1 Move Command

The Move command is used to move a selected object from one location to another without changing its orientation. To move objects, select this tool and select the objects from the drawing area. After selecting objects, specify the 'base point' and the 'destination point'.

To use this command, click **Home** > **Modify** > **Move** on the ribbon or enter **MOVE** or **M** in the command line.

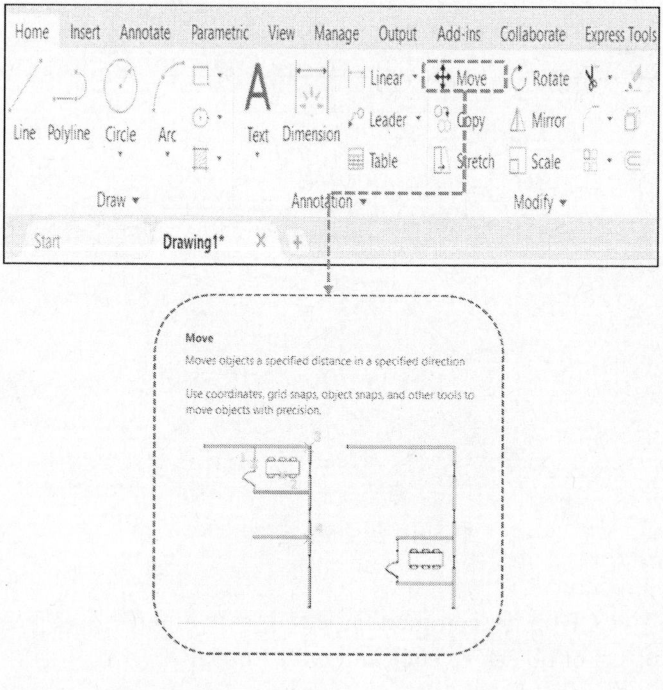

Fig. 2.48: Move command

To use the **Move** command —

Step 1: Type **MOVE** or **M** on the command line and press Enter.

Step 2: Select the object or objects to move and press Enter.

Step 3: Specify the base point for the move operation by clicking in the drawing area or typing coordinates on the command line.

Step 4: Specify the destination point for the move operation by clicking in the drawing area or typing coordinates on the command line.

2.3.2 Copy Command

The Copy command can be used to produce one or more copies of any previously created objects. Copy is a particularly helpful and time-saving command since it enables the creation of extremely complicated drawing elements that may then be copied as many times as necessary. This tool is similar to the Move tool, except that the original object will remain in its original location and a copy will be placed at the new location.

To use this command, click **Home > Modify > Copy** on the ribbon or enter **COPY** or **CP** in the command line.

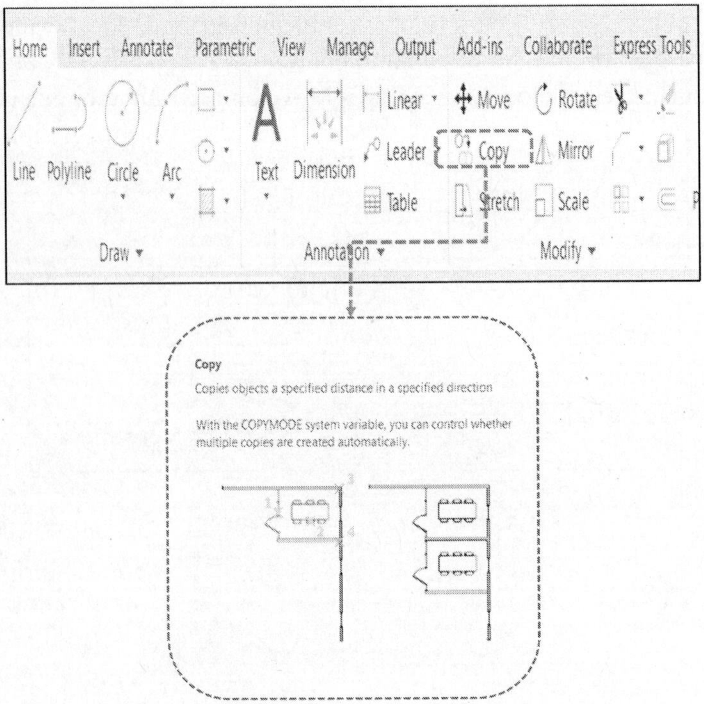

Fig. 2.49: Copy command

To use the **Copy** command —

Step 1: Type **COPY** or **CP** on the command line and press Enter.

Step 2: Select the object or objects to copy and press Enter.

Step 3: Specify the base point for the copy operation by clicking in the drawing area or typing coordinates on the command line.

Step 4: Specify the destination point for the copy operation by clicking in the drawing area or typing coordinates on the command line.

2.3.3 Stretch Command

Stretch command is used to alter the shape of a basic object like rectangle or polygon. Stretch can be used to relocate one or more of an object's vertices while leaving the remainder of the object unaltered. However, circles cannot be stretched using this tool.

To use this command, click **Home > Modify > Stretch** on the ribbon or enter **STRETCH** or **S** in the command line.

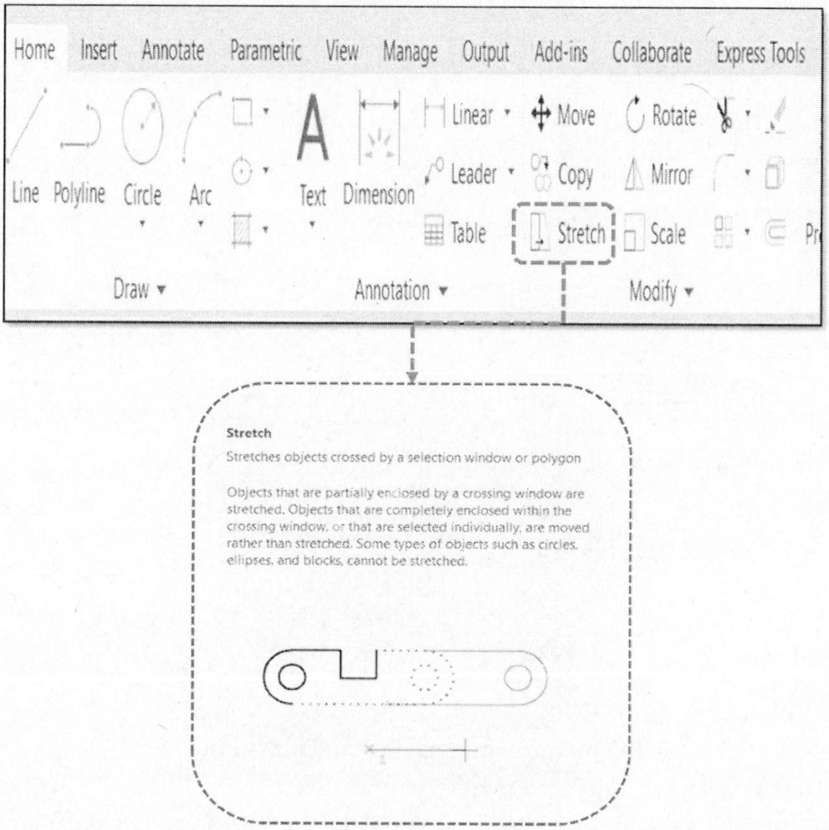

Fig. 2.50: Stretch command

To use the **Stretch** command –

Step 1: Type **STRETCH** or **S** on the command line and press Enter.

Step 2: Select the object or objects to stretch and press Enter.

Step 3: Specify the base point for the stretch operation by clicking in the drawing area or typing coordinates on the command line.

Step 4: Specify the destination point for the stretch operation by clicking in the drawing area or typing coordinates on the command line.

2.3.4 Rotate Command

Using the Rotate tool, an object or set of objects can be rotated about a given point. To rotate objects, activate this tool and pick the desired objects in the drawing window. After selecting items, the base point and rotation angle must be specified. The objects will be rotated relative to the specified base point.

To use this command, click **Home** > **Modify** > **Rotate** on the ribbon or enter **ROTATE** or **RO** in the command line.

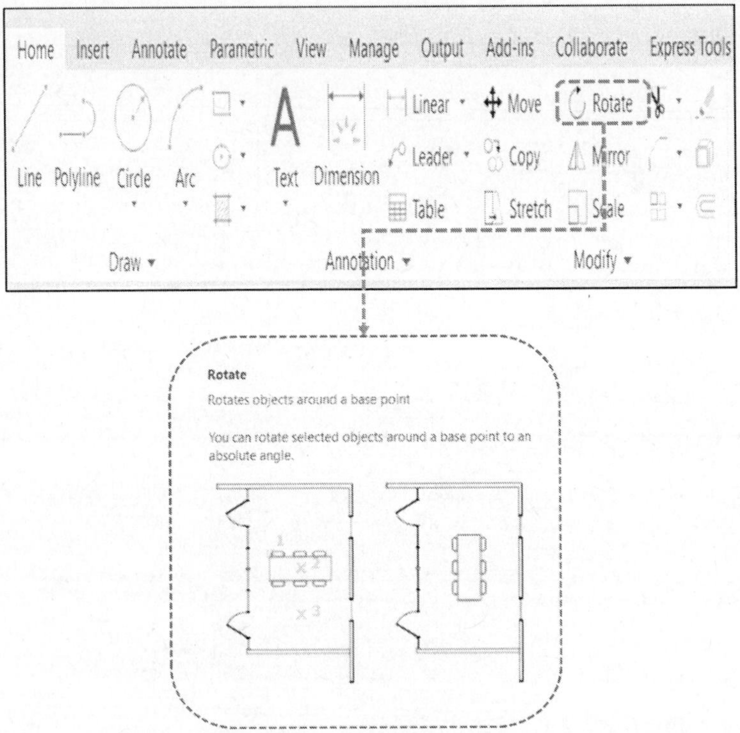

Fig. 2.51: Rotate command

To use the **Rotate** command —

Step 1: Type **ROTATE** or **RO** on the command line and press Enter.

Step 2: Select the object or objects to rotate and press Enter.

Step 3: Specify the base point for the rotation by clicking in the drawing area or typing coordinates on the command line.

Step 4: Specify the angle of rotation by typing a value on the command line and pressing Enter.

2.3.5 Mirror Command

Using the Mirror command, a mirror image of an object can be created. With this tool, symmetrical graphics can be effortlessly generated. To mirror objects, a "mirror line" along which they will be reflected must be specified. Either a new line can be created or an existing line can be specified as the mirror line.

To use this command, click **Home** > **Modify** > **Mirror** on the ribbon or enter **MIRROR** or **MI** in the command line.

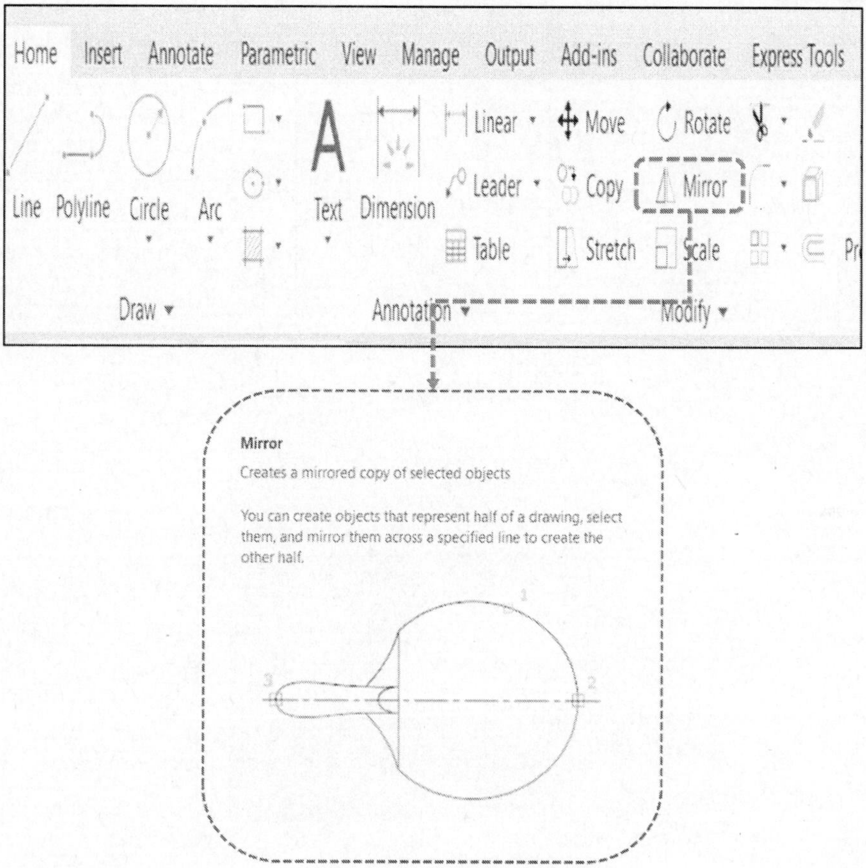

Fig. 2.52: Mirror command

To use the **Mirror** command —

Step 1: Type **MIRROR** or **MI** on the command line and press Enter.

Step 2: Select the object or objects to mirror and press Enter.

Step 3: Specify the first point of the mirror axis by clicking in the drawing area or typing coordinates on the command line.

Step 4: Specify the second point of the mirror axis by clicking in the drawing area or typing coordinates on the command line.

Step 5: Specify the option to keep or erase the source objects.

2.3.6 Scale Command

The Scale command allows for the resizing of objects. An object's shape is unaffected by whether it is scaled up or down. It is necessary to choose a 'base point' and a scale factor after selecting the objects to be scaled. The object's scale factor is the ratio of its final size to its initial size. Thus, if an object is to be doubled in size, a scale factor of 2 should be used.

To use this command, click **Home > Modify > Scale** on the ribbon or enter **SCALE** or **SC** in the command line.

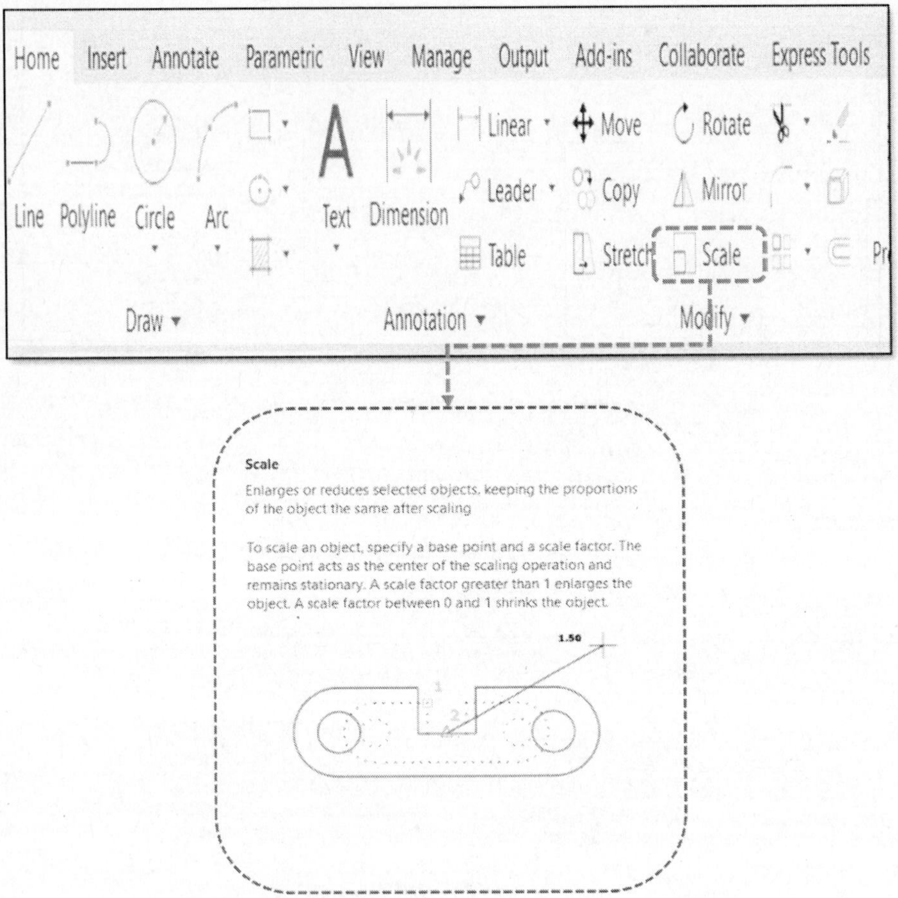

Fig. 2.53: Scale command

To use the **Scale** command –

Step 1: Type **SCALE** or **SC** on the command line and press Enter.

Step 2: Select the object or objects to scale and press Enter.

Step 3: Specify the base point for the scale operation by clicking in the drawing area or typing coordinates on the command line.

Step 4: Specify the scale factor by typing a value on the command line and pressing Enter.

2.3.7 Trim Command

Using the Trim tool, any undesired segments of an item that are overlapping with other objects can be removed. Invoking the Trim tool is the initial step in trimming an object; after that, the cutting edge (the object it intersects) and the area to be trimmed should be selected. A simple command line selection of "select all" will use all drawing objects as "cutting edges" provided that multiple intersection points exist.

To use this command, click **Home** > **Modify** > **Trim** on the ribbon or enter **TRIM** or **TR** in the command line.

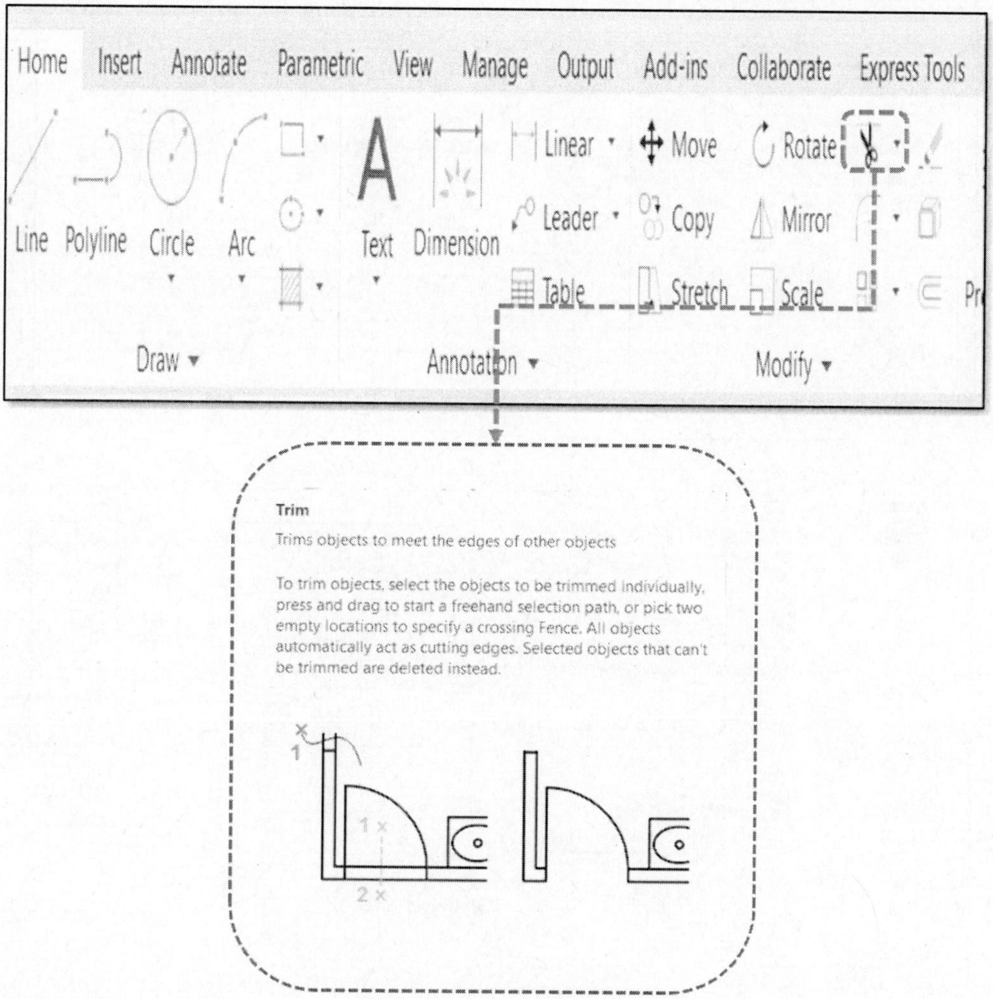

Fig. 2.54: Trim command

To use the **Trim** command —

Step 1: Type **TRIM** or **TR** on the command line and press Enter.

Step 2: Select the objects to trim.

Step 3: Select the cutting edge or edges by clicking on them or by selecting them from a list.

Step 4: Press Enter to end the command.

2.3.8 Extend Command

With this command, a line, polyline or arc can be made longer so that it can join with another shape in the drawing (known as the boundary edge). While the Extend command looks and functions similarly to the Trim tool, its intended usage is the complete opposite.

To use this command, click **Home > Modify > Trim > Extend** on the ribbon or enter **EXTEND** or **EX** in the command line.

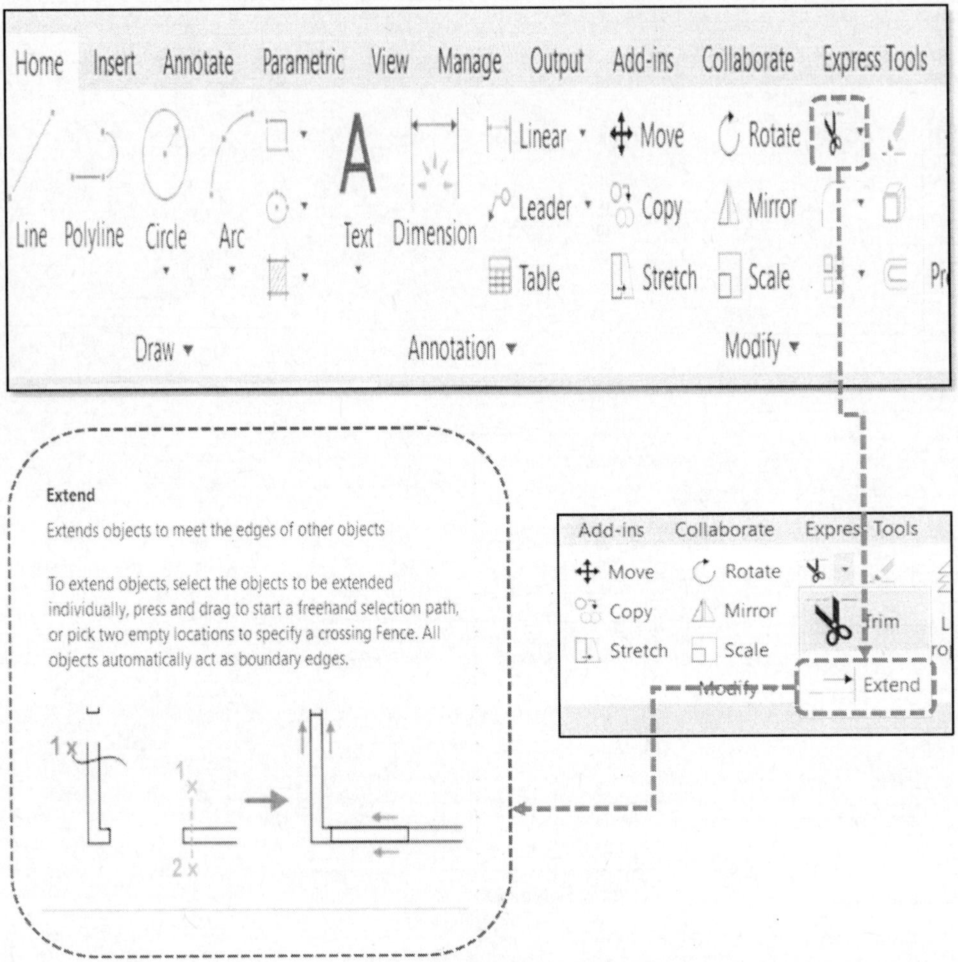

Fig. 2.55: Extend Command

To use the **Extend** command —

Step 1: Type **EXTEND** or **EX** on the command line and press Enter.

Step 2: Select the objects to extend and press Enter.

Step 3: Select the boundary or cutting edge by clicking on it or by selecting it from a list.

Step 4: Press Enter to end the command.

2.3.9 Fillet Command

An arc between two intersecting lines or neighbouring polyline segments can be drawn with the Fillet command. For doing so, first, the fillet radius is specified and then the lines to be filleted are selected.

To use this command, click **Home** > **Modify** > **Fillet** on the ribbon or enter **FILLET** or **F** in the command line.

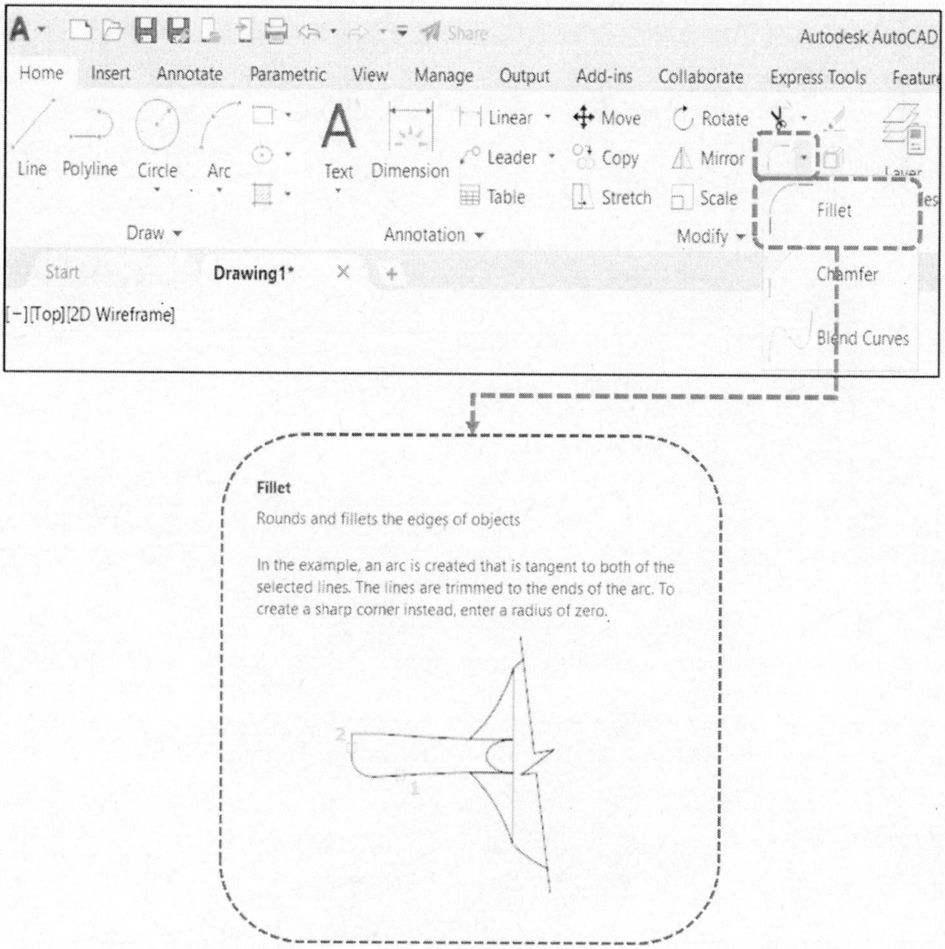

Fig. 2.56: Fillet Command

To use the **Fillet** command —

Step 1: Type **FILLET** or **F** on the command line and press Enter.

Step 2: Specify the radius of the fillet by typing a value on the command line and pressing Enter.

Step 3: Select the first object to fillet.

Step 4: Select the second object to fillet.

2.3.10 Chamfer Command

The Chamfer command is utilized to remove a sharp corner with an angled line, typically at a 45-degree angle to the neighbouring face. The Chamfer command creates a chamfer between any two non-parallel lines or adjacent polyline segments.

To use this command, click **Home > Modify > Chamfer** on the ribbon or enter **CHAMFER** or **CHA** in the command line.

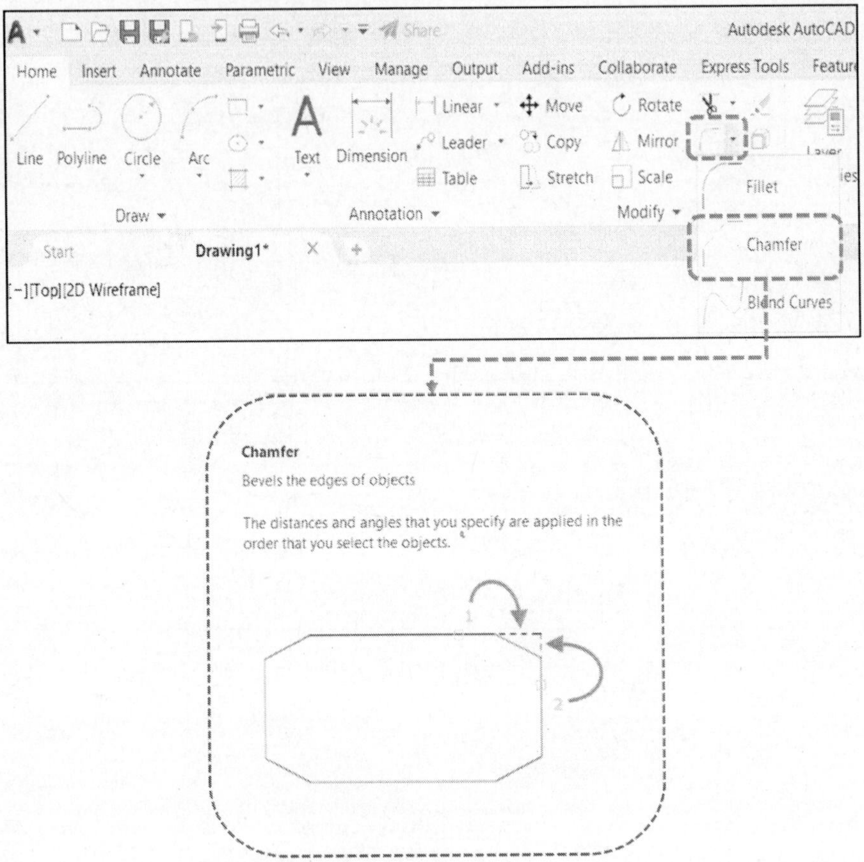

Fig. 2.57: Chamfer Command

To use the **Chamfer** command —

Step 1: Type **CHAMFER** or **CHA** on the command line and press Enter.

Step 2: Specify the first chamfer distance or angle by typing a value on the command line and pressing Enter.

Step 3: Specify the second chamfer distance or angle by typing a value on the command line and pressing Enter.

Step 4: Select the first line to chamfer.

Step 5: Select the second line to chamfer.

2.3.11 Array Command

The Array command duplicates selected objects and arrange them in a rectangular matrix, circular pattern or route array. It can be considered a completely automated upgrade of the Copy command. The command has three options— Rectangular array, Path array and Polar array.

To use this command, click **Home > Modify > Array** on the ribbon or enter **ARRAY** or **AR** in the command line.

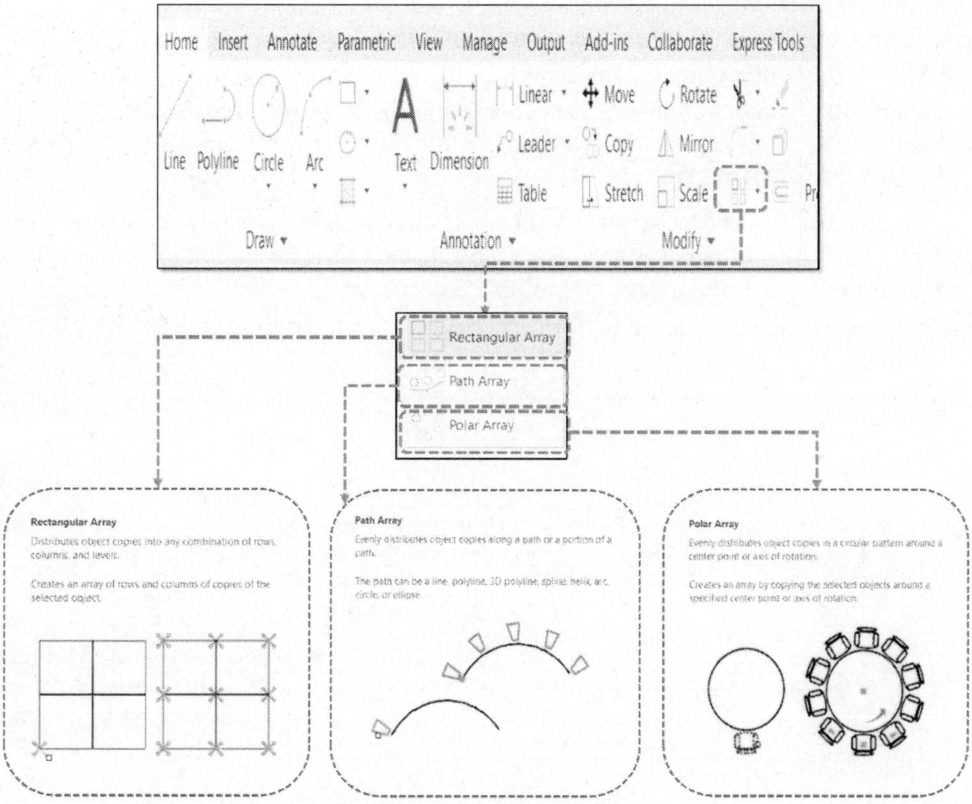

Fig. 2.58: Array Command

Rectangular array: A rectangular array in AutoCAD is a series of copies of an object arranged in a grid pattern with a specified number of rows and columns.

To create a rectangular array —

Step 1: Type **ARRAY** or **AR** on the command line and press Enter.

Step 2: Select the object to array.

Step 3: Choose the "Rectangular" option on the command line.

Step 4: Specify the number of rows and columns in the array.

Step 5: Specify the spacing and orientation of the objects in the array.

Path array: A path array in AutoCAD is a series of copies of an object arranged along a specified path.

To create a path array —

Step 3: Choose the "Path" option on the command line.

Step 4: Select the path for the array (such as a line, arc, or spline).

Step 5: Specify the number of objects in the array and the spacing between them.

Polar array: A polar array in AutoCAD is a series of copies of an object arranged in a circular pattern around a center point.

To create a polar array —

Step 3: Choose the "Polar" option on the command line.

Step 4: Specify the center point for the array.

Step 5: Specify the number of objects in the array and the angle between them.

2.3.12 Explode Command

The Explode tool is used to split a set of items into several separate objects. For instance, a drawing created with the Polyline tool behaves as a single object. Using the Explode tool, users can explode a polyline, rectangle, or any set of objects into multiple objects.

To use this command, click **Home** > **Modify** > **Explode** on the ribbon or enter **EXPLODE** or **X** in the command line.

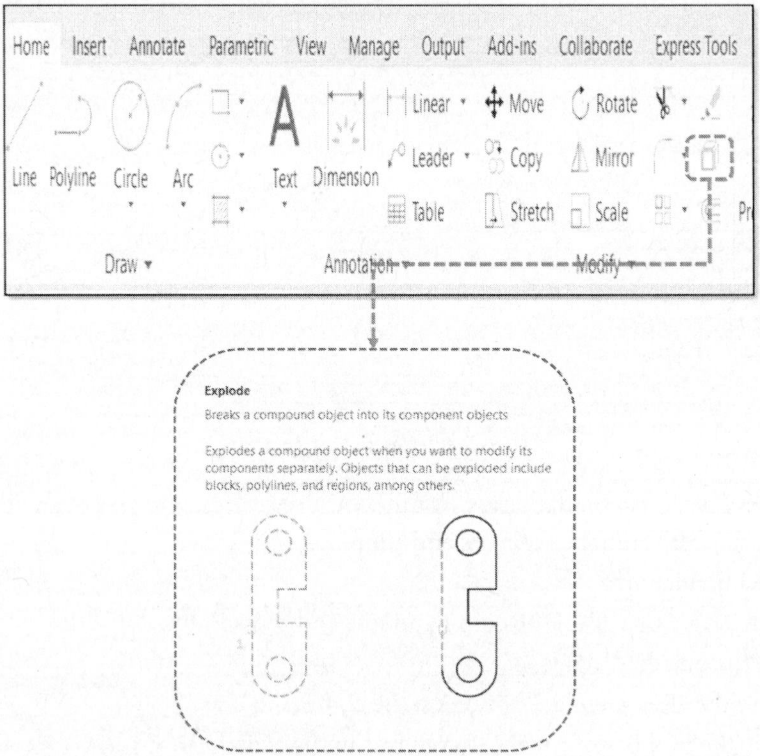

Fig. 2.59: Explode Command

To use the **Explode** command —

Step 1: Type **EXPLODE** or **X** on the command line and press Enter.

Step 2: Select the compound object to explode and press Enter.

2.3.13 Offset Command

The Offset command is used to make copies of parallel lines, polylines, circles, arcs etc. To make a parallel clone of an object, first, the offset distance is specified and then the object is chosen. Next, the user must choose the side on which the parallel duplicate will be placed.

To use this command, click **Home** > **Modify** > **Offset** on the ribbon or enter **OFFSET** or **O** in the command line.

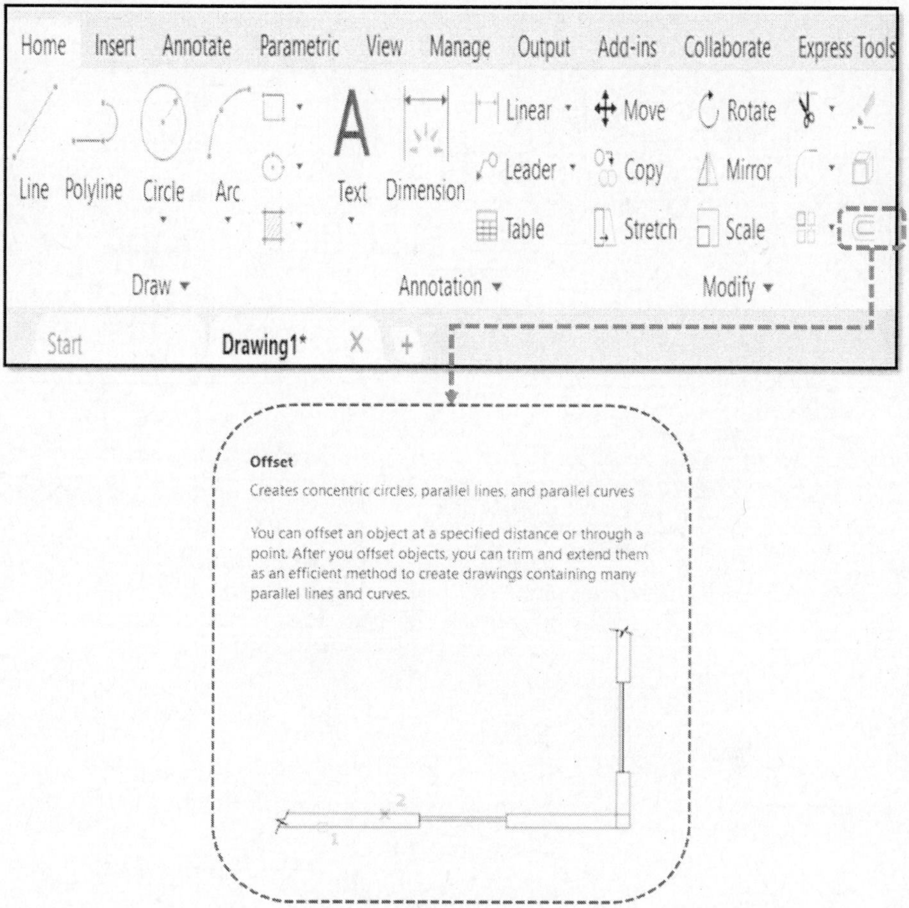

Fig. 2.60: Offset Command

To use the **Offset** command —

Step 1: Type **OFFSET** or **O** on the command line and press Enter.

Step 2: Specify the distance of the offset by typing a value on the command line and pressing Enter.

Step 3: Select the object or objects to offset and press Enter.

Step 4: Specify point on side to offset.

Step 5: Press Enter to end the command.

2.3.14 Break Command

By specifying two break points, the Break command allows the user to break (or remove a portion of) an object. Lines, polylines, circles, arcs, ellipses, splines, xlines and rays can all be broken with the Break command.

To use this command, click **Home** > **Modify** > **Break** on the ribbon or enter **BREAK** or **BR** in the command line.

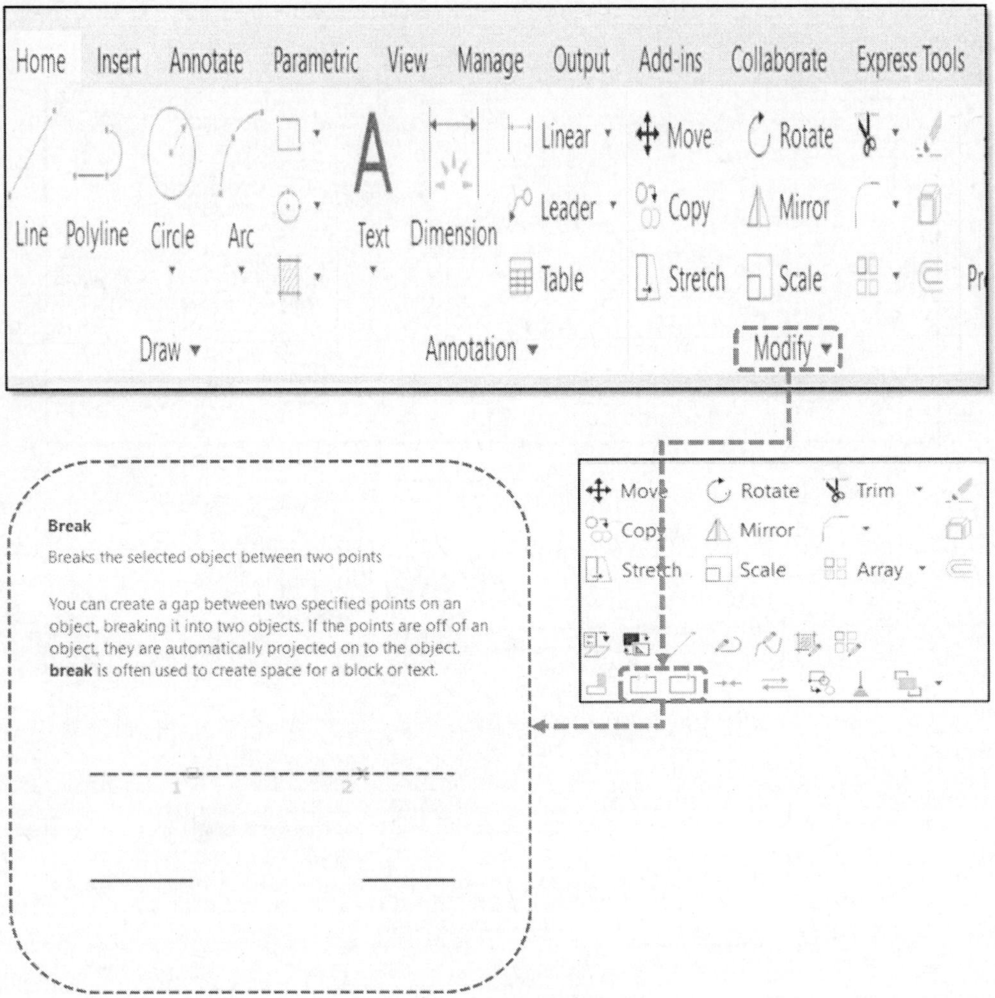

Fig. 2.61: Break Command

To use the **Break** command —

Step 1: Type **BREAK** or **BR** on the command line and press Enter.

Step 2: Select the object to break.

Step 3: Specify the point where the object will be broken by clicking in the drawing area or typing coordinates on the command line.

2.3.15 Join Command

The join command can be used to join the endpoints of two linear or curved objects. It can be considered the opposite of the break command.

To use this command, click **Home** > **Modify** > **Join** on the ribbon or enter **JOIN** or **J** in the command line.

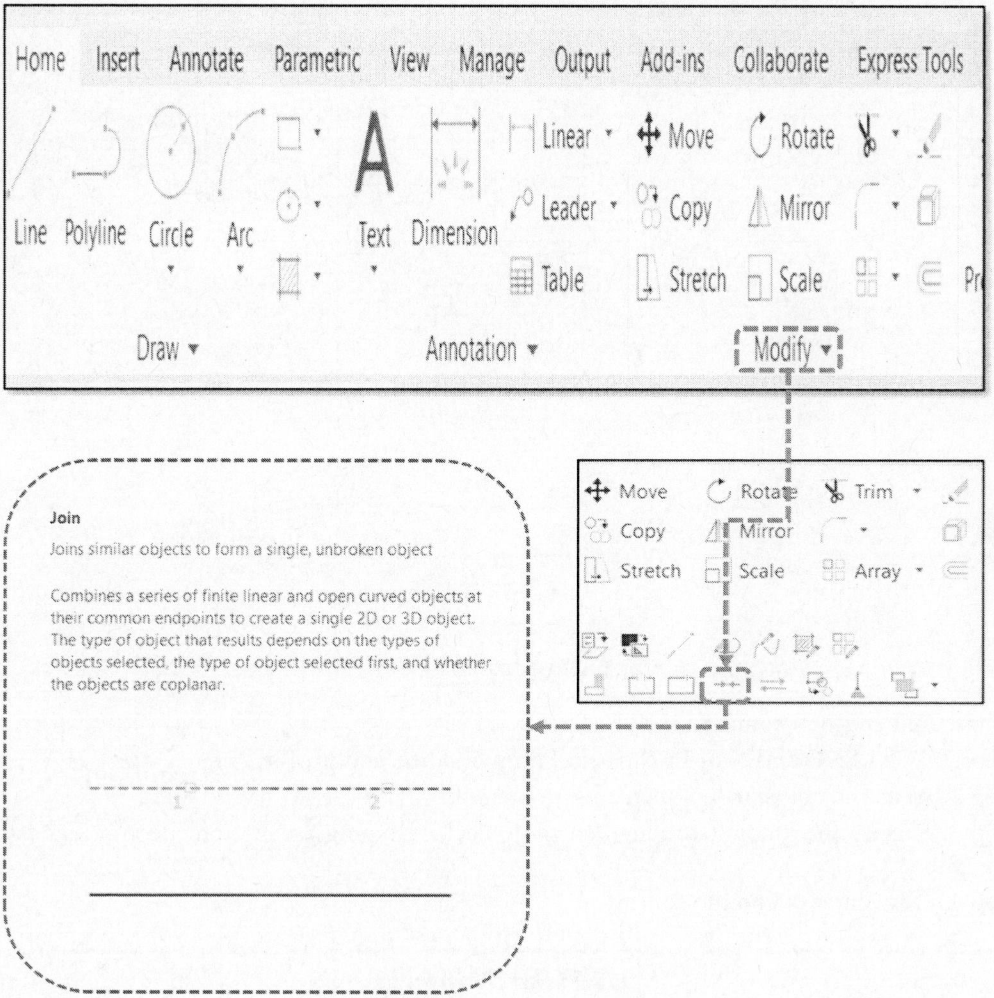

Fig. 2.62: Join Command

To use the **Join** command —

Step 1: Type **JOIN** or **J** on the command line and press Enter.

Step 2: Select the objects to join and press Enter.

2.3.16 Lengthen Command

Lengthen command can be used to increase or decrease the length of objects. It can often be used instead of Trim or Extend commands. Lines, Arcs, open Polylines, elliptical Arcs and open Splines can all be modified in length with the Lengthen command without using cutting or boundary edges.

To use this command, click **Home > Modify > Lengthen** on the ribbon or enter **LENGTHEN** or **LEN** in the command line.

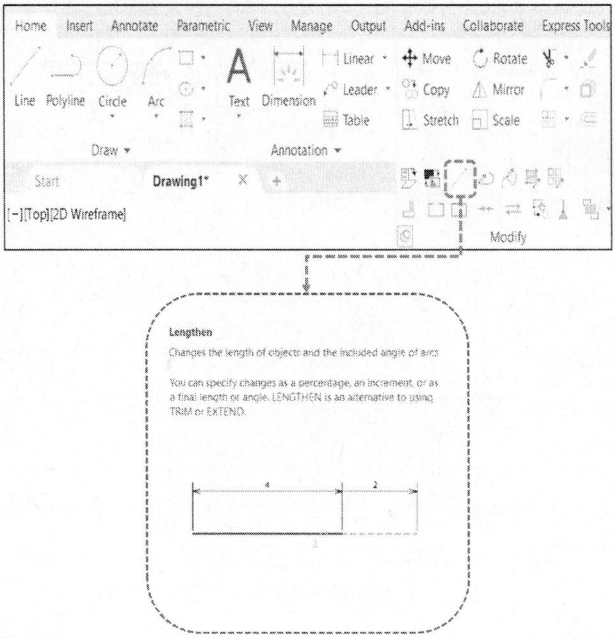

Fig. 2.63: Lengthen Command

To use the **Lengthen** command —

Step 1: Type **LENGTHEN** or **LEN** on the command line and press Enter.

Step 2: Select the object to lengthen and press Enter.

Step 3: Specify the amount to lengthen the object by typing a value on the command line and pressing Enter.

Step 4: Press Enter to end the command.

UNIT SUMMARY

In this unit you have learned about:

- Basic draw commands like lines, polylines, circles etc.
- The difference between similar commands like line, polyline, ray, construction line etc.
- Circle drawing methods (Center-Radius – Diameter; 2-points; 3-points; Tan-Tan-Radius; Tan-Tan-Tan).
- Ellipse drawing methods (3-Point; Start-Center-End; Start-Center-Angle; Start-Center-Length; Start-End-Angle; Start-End-Direction; Start-End-Radius; Center-Start-End; Center-Start-Angle; Center-Start-Length; Continue).
- Polygon (Inscribed-in-circle; circumscribed-in-circle)
- Modify commands like offset, mirror, copy, array etc. for creating multiple copies of object(s).
- Modify commands like scale, stretch, lengthen etc. for changing relative size of object(s).
- Array commands (Rectangular array, Path array and Polar array).
- Exploding and joining of object(s).

EXERCISES

A. Multiple Choice Questions

2.1 Which command can be used to break a polyline into individual lines and arcs?
(a) EXPLODE (b) BREAK (c) TRIM (d) ARRAY

2.2 The AutoCAD drawing space has its origins in
(a) 0,1 (b) 0,0 (c) 1,0 (d) 1,1

2.3 Which of the following actions can be performed by scrolling the mouse?
(a) scale (b) extents/all
(c) pan (d) Zoom in/zoom out

2.4 The number of workspaces in AutoCAD is
(a) 1 (b) 2 (c) 3 (d) 4

2.5 What is the Zoom command's shortcut?
(a) Z (b) ZOOM (c) Rotate mouse wheel (d) All

2.6 What is the shortcut for Trim command?
(a) X (b) T (c) TR (d) TL

2.7 What is the shortcut for Fillet command?
(a) FILL (b) FI (c) F (d) None

2.8 The command for converting a block into a drawing object.
(a) REDEFINE (b) BREAK (c) TRIM (d) ARRAY

2.9 Which command is used to convert separate entities into a polyline?
(a) Union (b) Polyline (c) Join (d) Subtract

2.10 Which command is used to save a drawing in AutoCAD?
(a) SHIFT + S (b) ALT + S (c) CTRL + S (d) CTRL + ALT + S

2.11 Which command is used for rectangle?
(a) REC (b) R (c) RET (d) RE

2.12 Which command is to access the AutoCAD stretch function?
(a) ST (b) S (c) SR (d) CTRL + S

2.13 Which command is used for erase?
(a) E (b) ER (c) CTRL + E (d) EL

2.14 On the command line, type _____ and click Enter to create a polygon.
(a) P (b) POL (c) POLY (d) O

2.15 Which command is used to access the AutoCAD Offset function?
(a) O (b) OF (c) OS (d) CTRL + O

2.16 A Polyline can be broken into individual lines and arcs using which of the following command?
(a) break (b) trim (c) explode (d) overkill

2.17 Plagiostomi angle can be created by
(a) chamfer (b) fillet (c) offset (d) mirror

2.18 The function of a polar array is to create object in

(a) grid pattern (b) straight line (c) circular pattern (d) all of them

2.19 What is the shortcut for Break command?

(a) B (b) BK (c) BR (d) BE

2.20 What is the shortcut for Join command?

(a) J (b) JJ (c) JN (d JO

Answers of Multiple-Choice Questions
2.1 (a), 2.2 (b), 2.3 (b), 2.4 (b), 2.5 (d), 2.6 (c), 2.7 (c), 2.8 (a), 2.9 (c), 2.10 (c), 2.11 (a), 2.12 (b), 2.13 (a), 2.14 (b), 2.15 (a), 2.16 (c), 2.17 (a), 2.18 (c), 2.19 (c), 2.20 (a)

B. Subjective Questions

2.1	Describe any five commands from draw menu.
2.2	List and explain all types of circle drawing command.
2.3	List the advantages of using polyline command.
2.4	List the uses of spline and divide command.
2.5	List the commands from modify menu and explain any five of them.
2.6	Illustrate the difference between stretch and lengthen command.
2.7	Explain the uses of extend and explode.
2.8	Discuss the needs of the array command.
2.9	Discuss the functions of mirror and rotate.
2.10	Elucidate the differences between polar array and rectangular array.

C. Problems for Practice

2.1 Draw the following figures in AutoCAD.

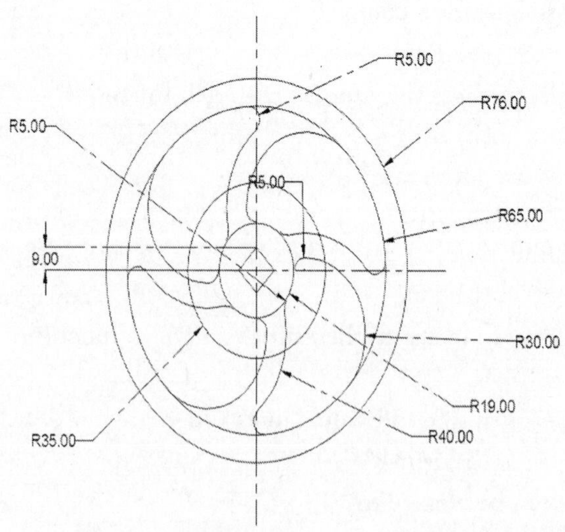

Fig. 2.1.1

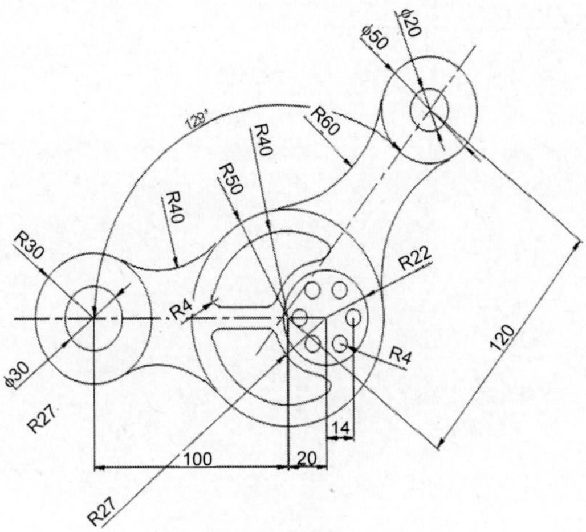

Fig. 2.1.2

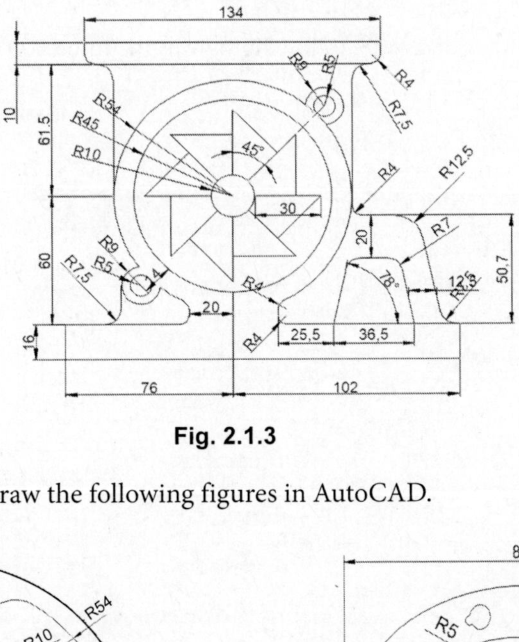

Fig. 2.1.3

2.2 Use array command to draw the following figures in AutoCAD.

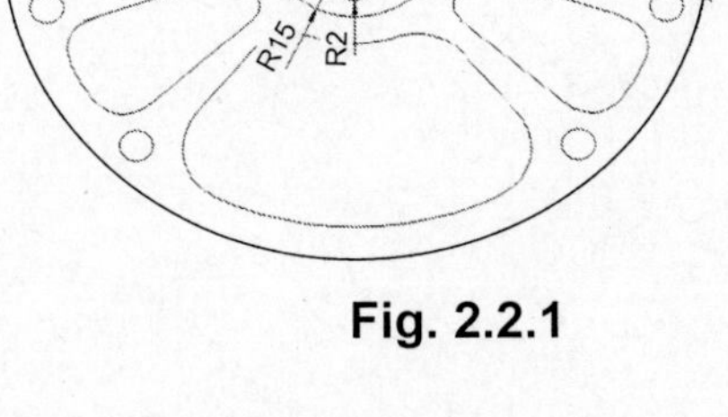

Fig. 2.2.1

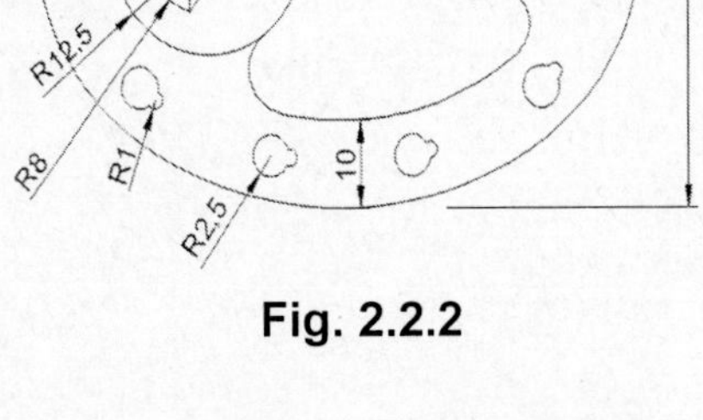

Fig. 2.2.2

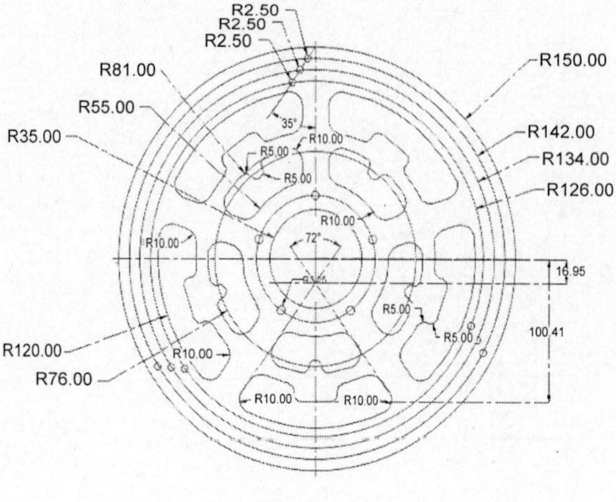

Fig. 2.2.3

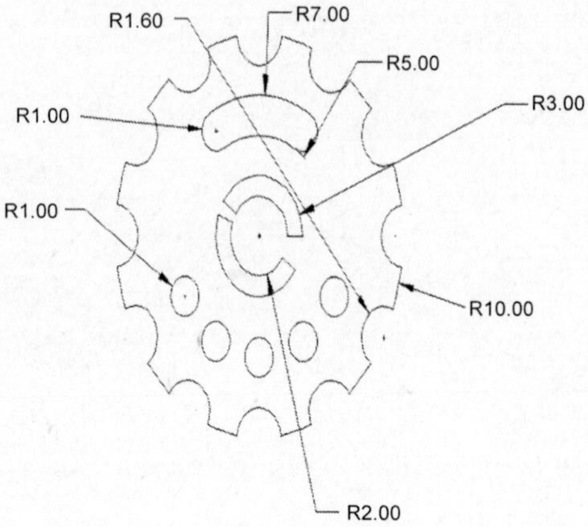

Fig. 2.2.4

2.3 Use mirror command to draw the following figures in AutoCAD.

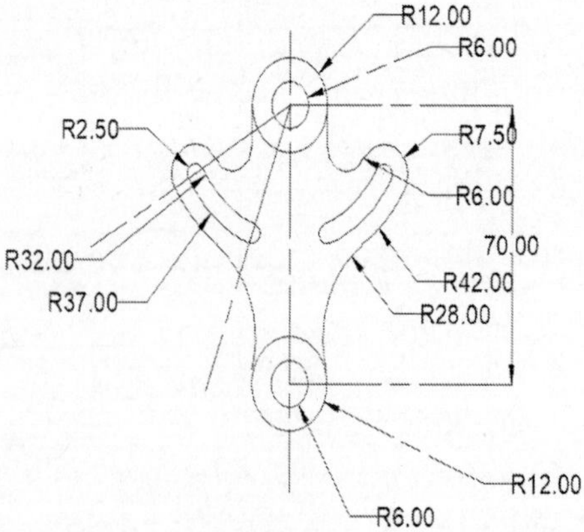

Fig. 2.3.1

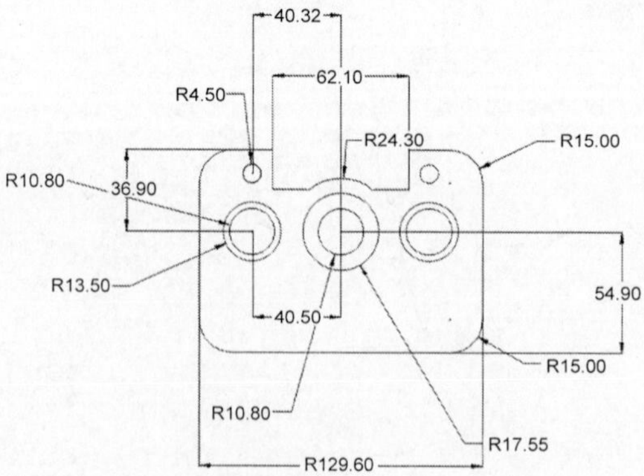

Fig. 2.3.2

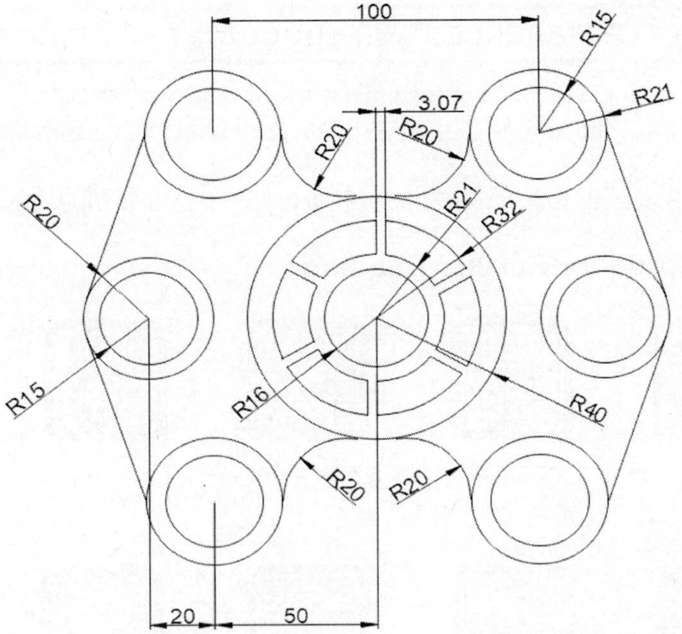

Fig. 2.3.3

KNOW MORE

- *Quick Select Tool*: The Quick Select tool is a valuable asset for efficiently selecting the objects you need to modify. With this tool, you can filter and select objects based on their properties, making it easier to manipulate specific elements within your drawing without manually picking them one by one.
- *Layer States Manager*: Managing your drawing's layers effectively is crucial for a streamlined working process. The Layer States Manager allows you to save, restore and manage different layer configurations, facilitating the visibility and editing process of various elements within your drawing.
- *Grouping and Ungrouping*: For complex drawings with numerous elements, the Group command can be a lifesaver. It allows you to combine different objects into a single group, making it easier to move, copy, or modify them collectively. Similarly, the Ungroup command lets you disband these groups when individual editing is required.
- *Quick Calculator*: AutoCAD comes with a built-in calculator that can be swiftly accessed with the 'CAL' command. This calculator is not just for arithmetic; it can also perform trigonometric, logarithmic and algebraic operations, proving to be an indispensable tool for engineers and designers.
- *DesignCenter*: The AutoCAD DesignCenter provides a centralized location where you can manipulate and insert content such as blocks, xrefs and hatch patterns from other drawings. This feature enhances efficiency, especially when working on large projects that require consistent use of standardized elements.

REFERENCES AND SUGGESTED READINGS

1. Cadfolks. AutoCAD 2020 For Beginners, 7th ed.; Kishore, 2019.
2. Palm, B.S. Introduction to AutoCAD 2020: 2D and 3D Design; Routledge: London, England, 2020.
3. Hawkes, B. Succeeding with AutoCAD; McGraw-Hill Publishing: London, England, 1995.

Dynamic QR Code for Further Reading

Circle command

Ellipse command

Arc command

Divide command

Spline command

Mirror command

Offset command

Trim command

Circle command

3 Basic Dimensioning and Annotations in AutoCAD

UNIT SPECIFICS

This unit presents information related to the following topics:

- dimensioning in AutoCAD, including quick, menu, command and toolbar dimensioning;
- different dimension types available in AutoCAD, such as linear, aligned, angular etc.;
- dimension editing tools and techniques;
- text writing in AutoCAD;

This unit is designed to provide readers with a comprehensive understanding of dimensioning and annotations in AutoCAD. This unit covers various dimensioning techniques and tools available in AutoCAD, such as quick dimensioning, menu dimensioning, command dimensioning and toolbar dimensioning. The different dimension types available in AutoCAD, such as linear, aligned, angular, radius, ordinate and jogged dimensions, are also discussed in detail. In addition, readers will learn how to use dimension editing tools and techniques to modify and adjust dimension lines, text and leaders. This unit also covers text writing in AutoCAD, including single-line text and multiline text tools and techniques.

The unit concludes with a summary of the key concepts covered in the unit, as well as multiple-choice and descriptive type questions with Bloom's taxonomy action verbs. Suggested readings and references are provided to help readers practice and improve their dimensioning and annotation skills. QR codes for supportive knowledge and video resources are also included to supplement the learning experience. Practical examples with detailed dimensions are given to help readers apply the concepts covered in the unit.

RATIONALE

In engineering and architecture, dimensioning and annotations are critical aspects of communicating design specifications and requirements. Effective dimensioning and annotation play a crucial role in ensuring that the drawings are clear, accurate and convey the necessary information to manufacturers, assemblers and other stakeholders. Therefore, it is essential for professionals to have a comprehensive understanding of dimensioning and annotation techniques to produce high-quality and functional technical drawings.

In today's digital era, computer-aided design programs like AutoCAD provide a variety of tools and techniques to help designers create accurate and precise dimensioned drawings. By understanding the different dimension types available in AutoCAD, such as linear, aligned, angular, radius, ordinate and jogged dimensions, readers can choose the most suitable dimensioning method for a given drawing. The unit also covers various dimension editing tools and techniques, enabling readers to modify and adjust dimension lines, text and leaders accurately. In addition, text writing tools and techniques in AutoCAD, such as single-line text and multiline text, are covered in detail, enabling readers to create annotations that are clear and legible. By the end of this unit, the readers will have developed the necessary skills to create accurate and precise dimensioned drawings using AutoCAD.

PRE-REQUISITES

Before reading this unit, the students are advised to revisit the following:

- Mathematics: Coordinate and Plane geometry (Class XII)
- Engineering Graphics (ES101)
- AutoCAD commands mentioned in Unit-1

UNIT OUTCOMES

After studying this unit students will be able to:

U3-O1: Identify and apply various dimensioning techniques and tools available in AutoCAD

U3-O2: Differentiate between different types of dimensions such as linear, aligned, angular, radius, ordinate and jogged dimensions and apply them correctly to drawings

U3-O3: Apply dimension editing tools and techniques to modify and adjust dimension lines, text and leaders

U3-O4: Create and edit text in AutoCAD, including single-line and multiline text

U3-O5: Apply dimensioning and annotation tools effectively to create professional drawings in AutoCAD.

Unit-3 Outcomes	EXPECTED MAPPING WITH COURSE OUTCOMES (1 – Weak Correlation; 2 – Medium correlation; 3 – Strong Correlation)				
	CO-1	CO-2	CO-3	CO-4	CO-5
U3-O1	3	1	1	1	2
U3-O2	3	1	1	1	2
U3-O3	3	1	1	1	2
U3-O4	3	1	1	1	2
U3-O5	3	1	1	1	2

3.1 INTRODUCTION

In the field of engineering, machine drawings play a vital role in communicating design concepts and specifications between designers, manufacturers and end-users. A machine drawing is essentially a graphical representation of a machine or its components, showing its shape, size and geometric features in detail. However, without accurate dimensioning and annotations, machine drawings are incomplete and useless. Dimensioning and annotations are essential components of any machine drawing, providing crucial information about the size, shape and other characteristics of the machine or its parts.

In this unit, we will explore the fundamental concepts of dimensioning and annotations in machine drawing. We will discuss the various types of dimensioning, the techniques for adding dimensions and the rules and conventions for annotating a drawing. Additionally, we will cover how to use Computer-

Aided Design (CAD) tools to create accurate and professional-looking dimensioned drawings quickly and efficiently.

3.2 DIMENSIONING IN AUTOCAD

Dimensioning in AutoCAD is the process of adding accurate measurements and annotations to a drawing. It involves indicating the size, shape and other characteristics of an object or component. It is a critical aspect of engineering and technical design as it ensures that the manufactured product or part will meet the intended specifications and fit correctly with other components. Dimensioning involves the use of various symbols and conventions to communicate information about the object or component's size, location and orientation, such as lines, arrows and text. Proper dimensioning requires adherence to standards and guidelines to ensure clarity, consistency and accuracy in communicating design information.

3.2.1 Quick Dimensioning

Quick Dimensioning in AutoCAD is a tool that allows you to quickly add dimensions to your drawing without having to go through the Dimension Style Manager. Here are the steps to use Quick Dimensioning:

Step 1: Select the Quick Dimension tool from the toolbar or type QDIM in the command line and press Enter.

Step 2: Click on the first point of the object you want to dimension. This can be an endpoint, midpoint, or any other reference point.

Step 3: Move your cursor to the second point and click on it. This will add the dimension to your drawing.

Step 4: If you want to add additional dimensions, simply repeat steps 2 and 3.

Step 5: To exit Quick Dimensioning, press Enter or right-click and select Exit.

Note that Quick Dimensioning uses the current dimension style, so make sure you have the correct style selected before using the tool. You can also modify the dimension text and other settings by right-clicking on the dimension and selecting Properties.

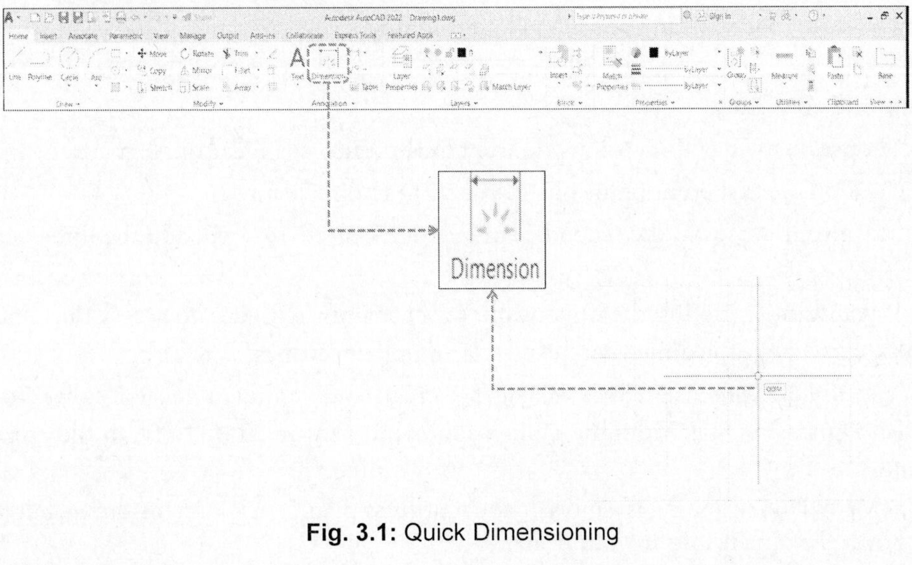

Fig. 3.1: Quick Dimensioning

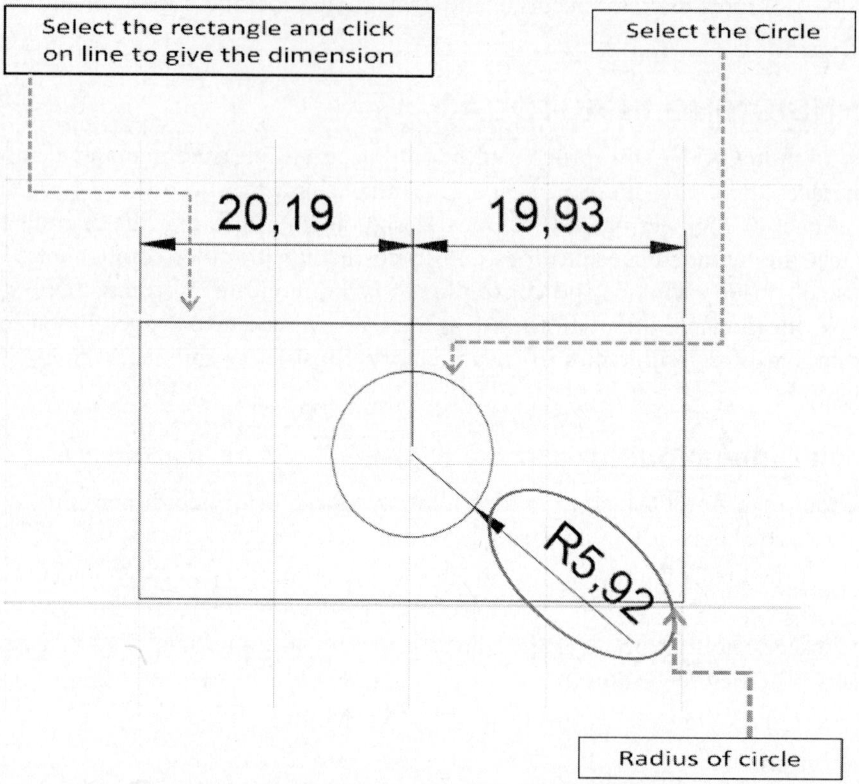

Fig. 3.2: Example of Quick Dimensioning

3.2.2 Menu Dimensioning

Menu dimensioning in AutoCAD refers to using the Dimension menu to add and modify dimensions in your drawing. Here are the steps to use Menu Dimensioning:

Step 1: Select the Dimension menu from the menu bar or type DIMENSION (or DIM) in the command line and press Enter.

Step 2: Choose the type of dimension you want to add, such as Linear or Aligned.

Step 3: Click on the first point of the object you want to dimension.

Step 4: Move your cursor to the second point and click on it. This will add the dimension to your drawing.

Step 5: If you want to modify the dimension, select it and right-click to access the context menu. From here, you can change the dimension text, style and other settings.

You can also use the Dimension Style Manager to create and modify dimension styles. To access the manager, select Dimension Style from the Dimension menu or type DIMSTYLE in the command line and press Enter.

Note that Menu Dimensioning is a more manual process than Quick Dimensioning, but it gives you more control over the dimensions in your drawing.

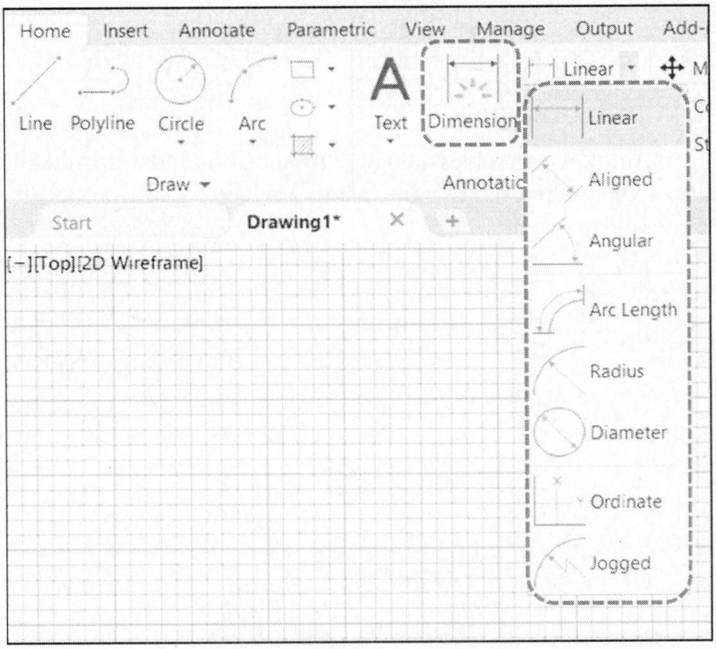

Fig. 3.3: Menu Dimensioning

3.2.3 Command Dimensioning

Command dimensioning in AutoCAD involves using various dimensioning commands to add and modify dimensions in your drawing. Here are some of the most commonly used dimensioning commands:

DIMLINEAR: Adds a linear dimension between two horizontal or vertical points.

DIMALIGNED: Adds dimension between two inclined points.

DIMANGULAR: Adds an angular dimension between two lines or edges.

DIMORDINATE: Adds co-ordinate position or location.

DIMDIAMETER: Adds a diameter dimension to a circle or arc.

DIMRADIUS: Adds a radius dimension to a circle or arc.

DIMCENTER: Adds the centre position in drawing object.

DIMARC: Adds the dimension of arcs.

DIMEDIT: To edit the dimension.

DIMTEDIT: To edit the text of dimension.

DIMCONTINUE: Adds a dimension line that extends from an existing dimension line.

DIMBASELINE: Adds a series of dimensions that are aligned with a baseline.

DIMSTYLE: Opens the Dimension Style Manager, which allows you to create and modify dimension styles.

To use these commands, simply type the command name in the command line and press Enter. You will then be prompted to select the objects you want to dimension and specify other options such as the dimension text and style.

Note that Command Dimensioning gives you more control over the dimension settings than Menu Dimensioning or Quick Dimensioning, but it can be more time-consuming.

3.2.4 Toolbar Dimensioning

Toolbar dimensioning in AutoCAD involves using the dimension toolbar to add and modify dimensions in your drawing. The toolbar can be used to edit various attributes of dimensions.

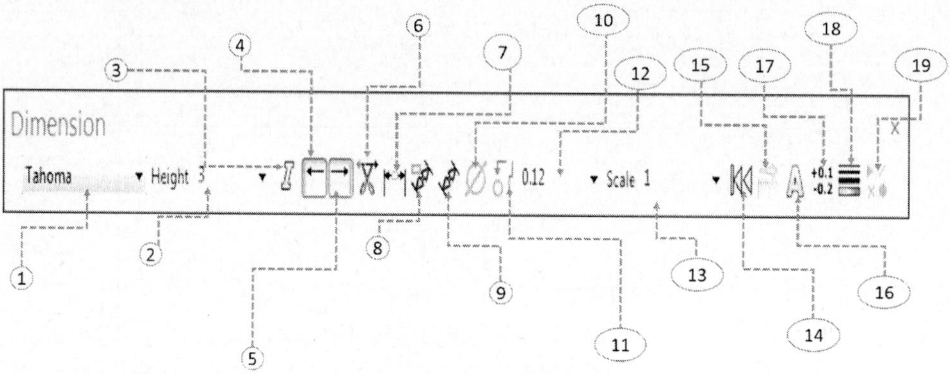

Fig. 3.4: Dimension Toolbar

The various elements of the dimension toolbar shown in Fig. 3.4 are discussed in Table 3.1.

Table 3.1: Description of the various elements in the dimension toolbar

Sl. No.	Element	Symbol	Utility
1.	Font Style		Select various types of fonts for dimensioning.
2.	Font Height	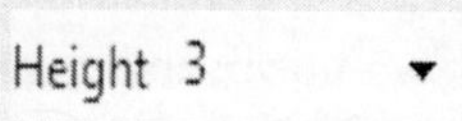	Assign the height of fonts.
3.	Italics		Toggle between italics and italics texts.
4.	Removing First witness		Deselect this button to remove the first witness line from the dimension. 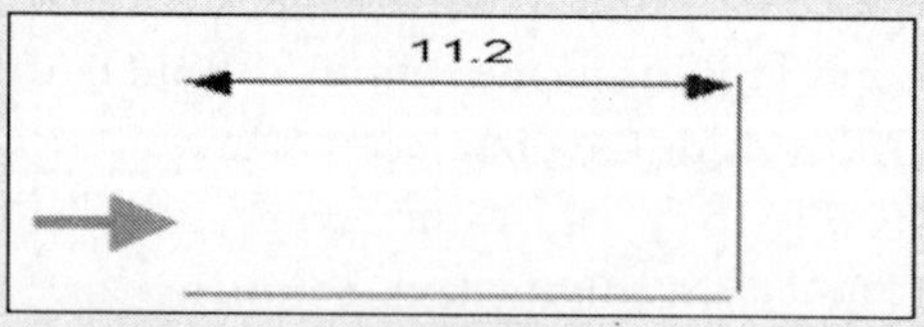
5.	Removing second witness		Deselect this button to remove the second witness line from the dimension. 

<table>
<tr><td>6.</td><td>Trim Dimension</td><td></td><td>Trims the leader line to the position of the dimension annotation. This can be used when the end of the leader is obvious, for example, in base-point and chain dimensions. Trimming leaders is only possible for linear dimensions.
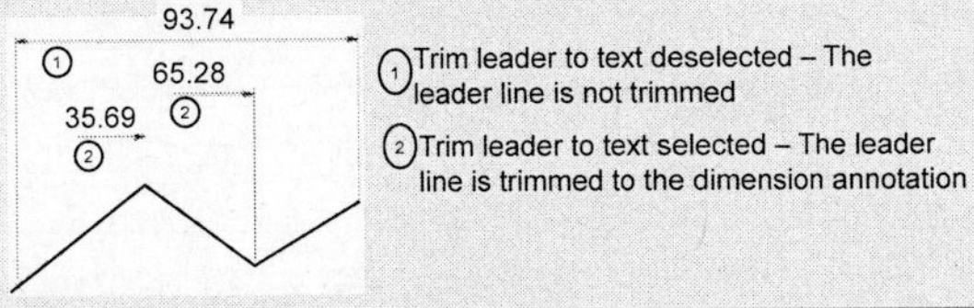
</td></tr>
<tr><td>7.</td><td>Embeds the text in the leader line</td><td></td><td>Embedding the text in the leader line
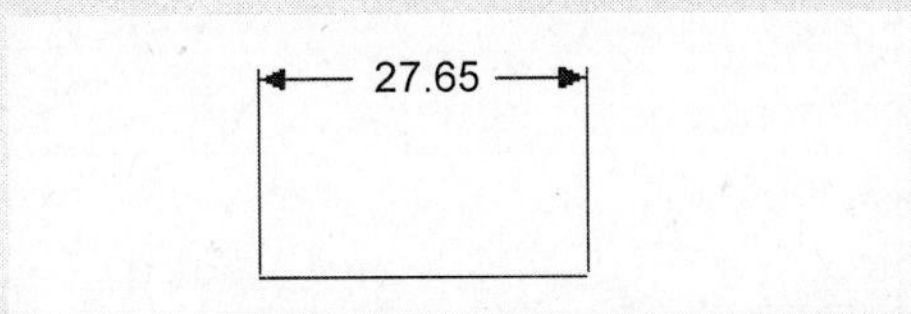
</td></tr>
<tr><td>8.</td><td>Text Positioning</td><td></td><td>This changes all text so that it is positioned horizontally to the principal plane</td></tr>
<tr><td></td><td colspan="3">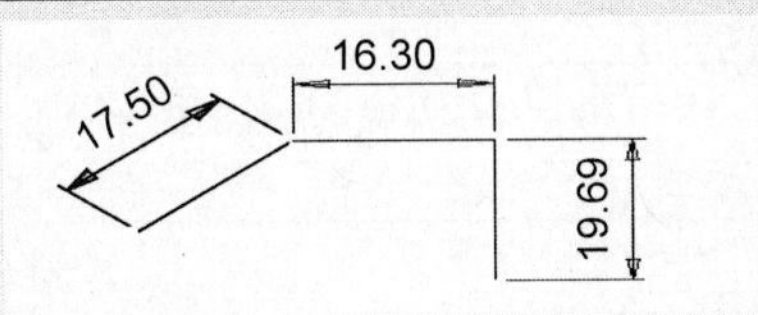

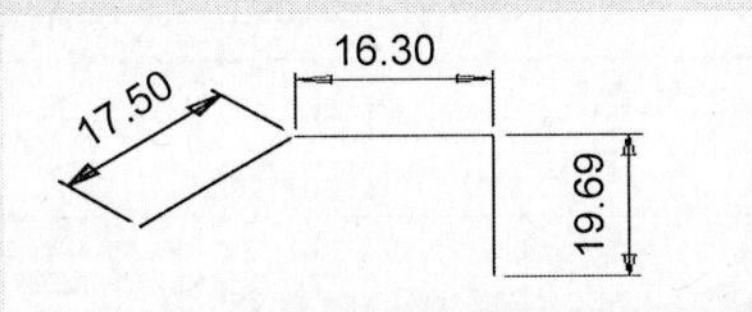
</td></tr>
<tr><td>9.</td><td>True dimensioning</td><td></td><td>Select this button to display the true measurements for linear dimensions.</td></tr>
<tr><td></td><td colspan="3">Projected dimension
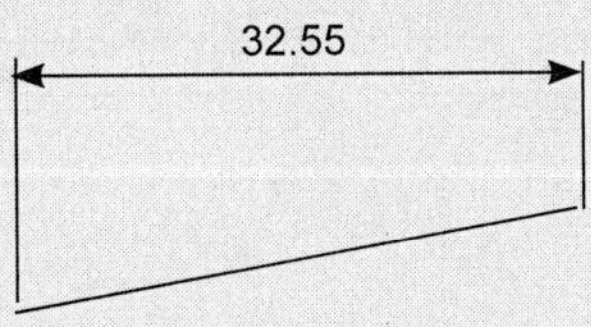

True dimension

</td></tr>
<tr><td>10.</td><td>Diameter</td><td></td><td>Select this button to display the diameter symbol before the dimension value. If the dimension is the radius of an arc, the value changes to its diameter.</td></tr>
<tr><td>11.</td><td>Dog-leg leader line</td><td></td><td>Use this button to switch between straight and dog-leg leader lines for radial and ordinate dimensions
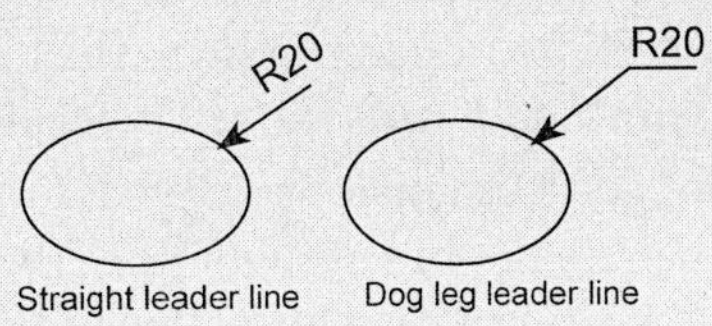
</td></tr>
</table>

12.	Decimal Places	0.12	List to set the number of decimal places or denominator of the fractional part of the dimension
13.	Scale	Scale 1	Scales the value of the dimension. In your model, you can create objects using a different scale. To create dimensions of the 'true' values, enter the value of the scale in the Scale box.
14.	Reset		Resets the attributes to the settings on the Dimension pages on the Options dialog.
15.	Restart base-point, chain and ordinate dimensions		Restart creating base-point, chain and ordinate dimensions.
16.	Annotation		Displays the Annotation tab of the Dimension dialog.
17.	Tolerance tab	+0.1 -0.2	Displays the Tolerance tab of the Dimension dialog.
18.	Style Tab		Displays the Style tab of the Dimension dialog.
19.	Mark Tab		Displays the Mark tab of the Dimension dialog.

3.3 DIMENSION TYPES

In AutoCAD, there are several types of dimensions that you can use to annotate your drawings. Here are some of the most commonly used dimension types:

- **Linear Dimensions:** Linear dimensions are used to measure the distance between two points on a drawing. They are typically shown as a horizontal line with an arrow at each end, along with a text value indicating the distance between the two points.
- **Radial Dimensions:** Radial dimensions are used to measure the radius of a circle or arc. They are shown as a horizontal line with a single arrowhead pointing towards the center of the circle, along with a text value indicating the radius.
- **Angular Dimensions:** Angular dimensions are used to measure angles between two lines or edges. They are shown as an arc with a center point and two lines extending from it, along with a text value indicating the angle between the two lines.
- **Diameter Dimensions:** Diameter dimensions are used to measure the diameter of a circle or arc. They are shown as a horizontal line with two arrowheads pointing towards the center of the circle, along with a text value indicating the diameter.
- **Ordinate Dimensions:** Ordinate dimensions are used to measure the distance of a point from a specified reference line. They are shown as a horizontal or vertical line with an arrowhead pointing towards the reference line, along with a text value indicating the distance.
- **Jogged Dimensions:** Jogged dimensions are used to measure the distance between two points that are not in a straight line. They are shown as a series of connected lines with arrows at each end, along with a text value indicating the distance.

Each of these dimension types can be customized in terms of their appearance and behaviour, allowing you to create precise and accurate drawings. By using a combination of these dimension types, you can create detailed and comprehensive drawings that accurately communicate your design intent.

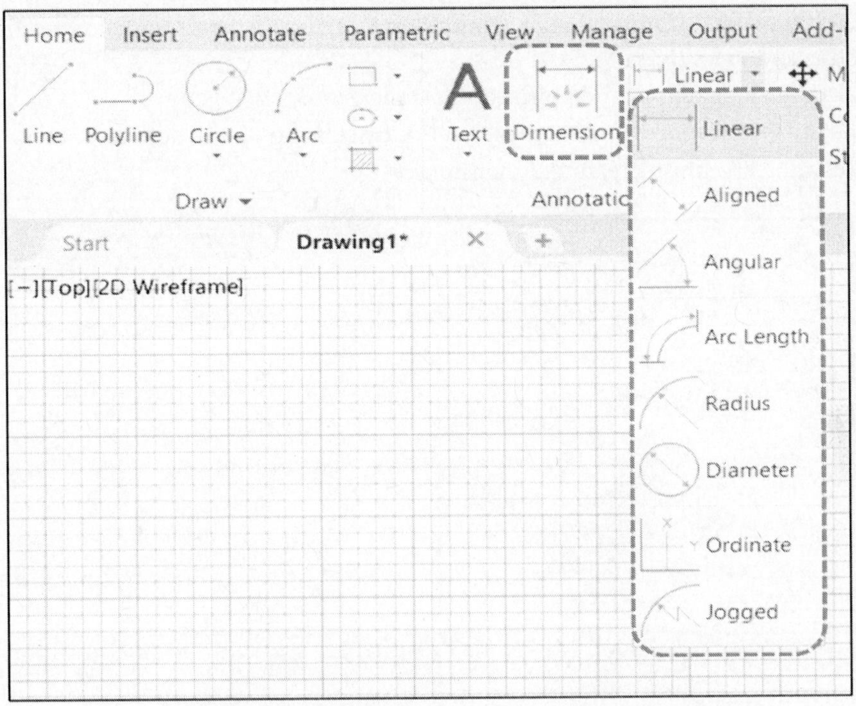

Fig. 3.5: Different types of Dimension Tools

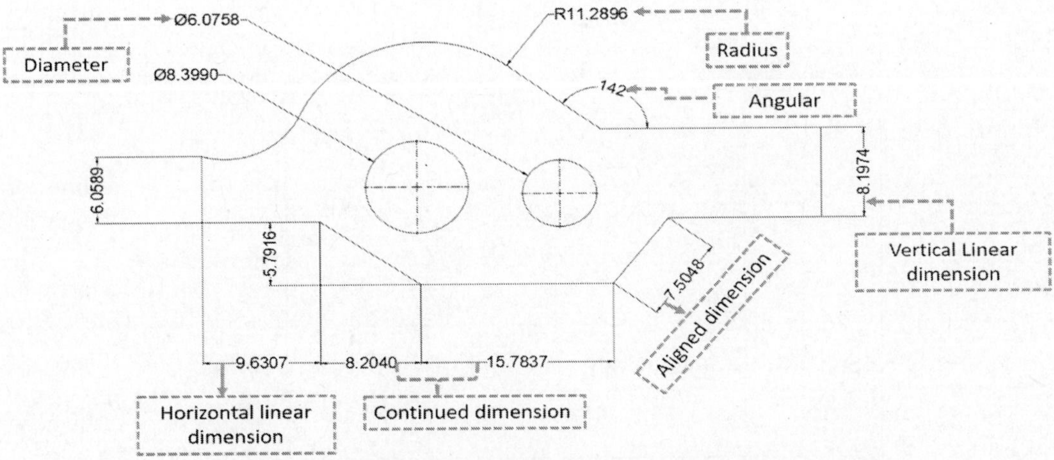

Fig. 3.6: Example of a part drawing with various dimensions

The various types of dimensions that can be used in AutoCAD are depicted below.

3.3.1 Linear

Linear dimensions are used to measure the distance between two points on a drawing. To add a linear dimension in AutoCAD, you can follow these steps:

Step 1: Click Annotate > Dimensions > Dimension > Linear on the ribbon or type DIMLINEAR in the command line and press Enter.

Step 2: Select the first and second points of the dimension.

Step 3: Specify the location for the dimension line by clicking on the desired point.

Step 4: Press Enter to finish creating dimensions.

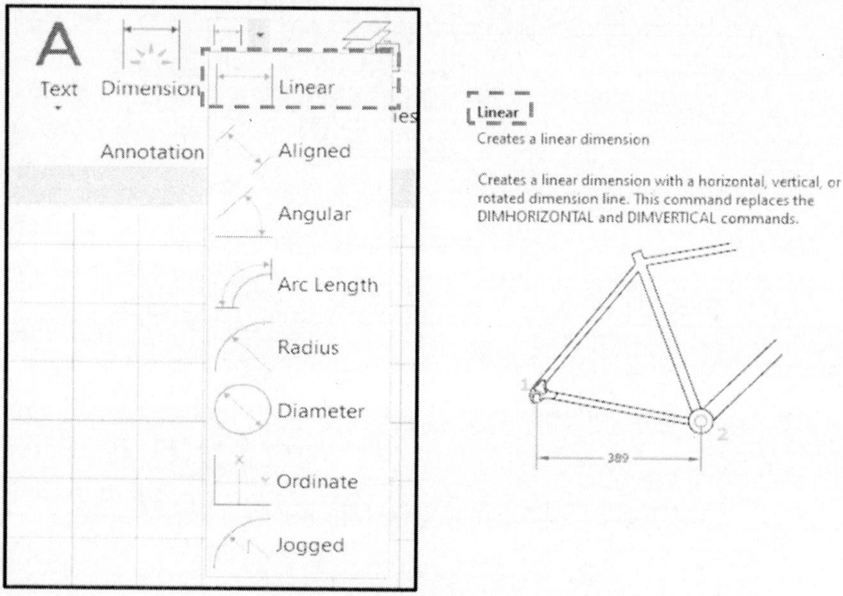

Fig. 3.7: Linear Dimension Tool

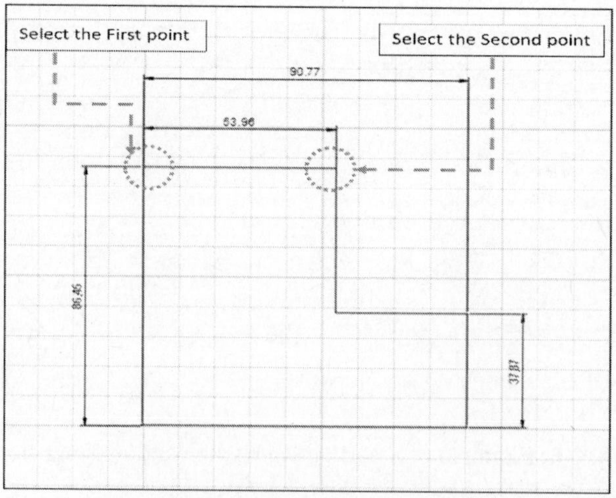

Fig. 3.8: Example of linear dimensioning

3.3.2 Aligned

Aligned dimension is a type of dimensioning in which the dimension lines are parallel to the object being dimensioned. This type of dimensioning is commonly used for linear features that are not perpendicular to the drawing sheet, such as angled or slanted lines. To add an aligned dimension in AutoCAD, you can follow these steps:

Step 1: Click Annotate > Dimensions > Dimension > Aligned on the ribbon or type DIMALIGNED in the command line and press Enter.

Step 2: Select the first and second points of the dimension.

Step 3: Specify the location for the dimension line by clicking on the desired point.

Step 4: Press Enter to finish creating dimensions.

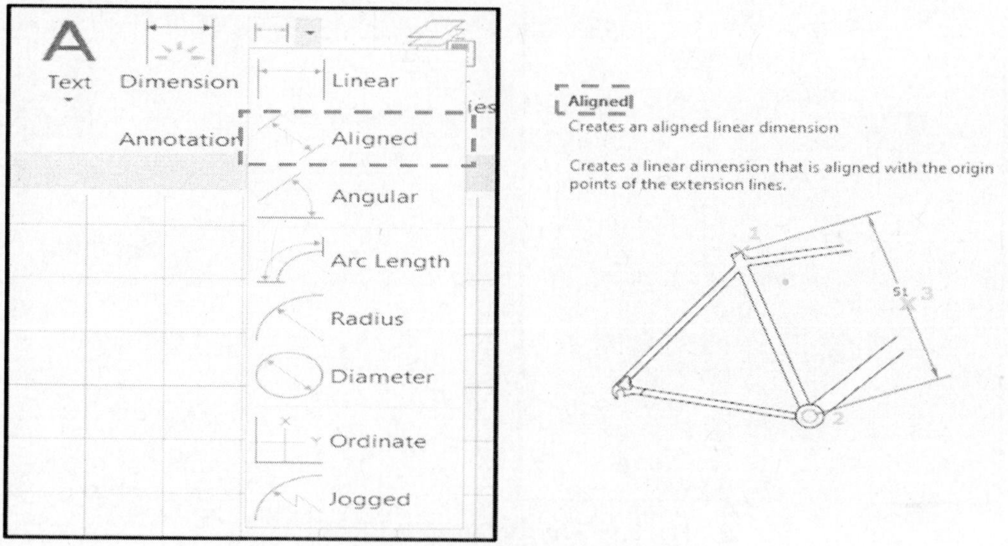

Fig. 3.9: Aligned Dimension Tool

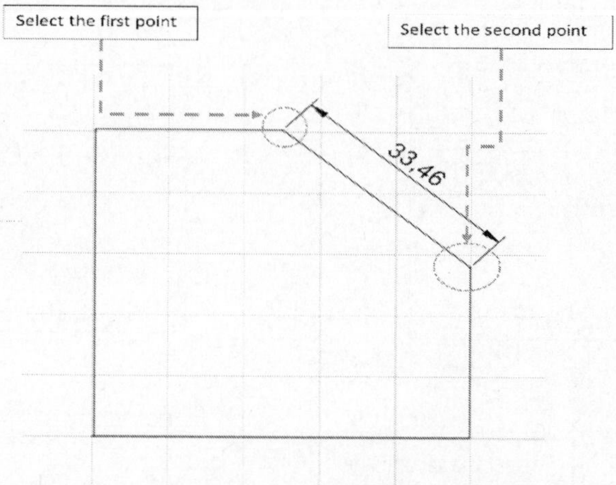

Fig. 3.10: Example of aligned dimensioning

3.3.3 Angular

The Angular Dimension tool in AutoCAD allows you to measure the angle between two lines or a line and a plane. To add an angular dimension in AutoCAD, you can follow these steps:

Step 1: Click Annotate > Dimensions > Dimension > Angular on the ribbon or type DIMANGULAR in the command line and press Enter.

Step 2: Click on the first line or edge of the object that you want to measure. Then click on the second line or edge to define the angle.

Step 3: Specify the location for the dimension line by clicking on the desired point.

Step 4: Press Enter to finish creating dimensions.

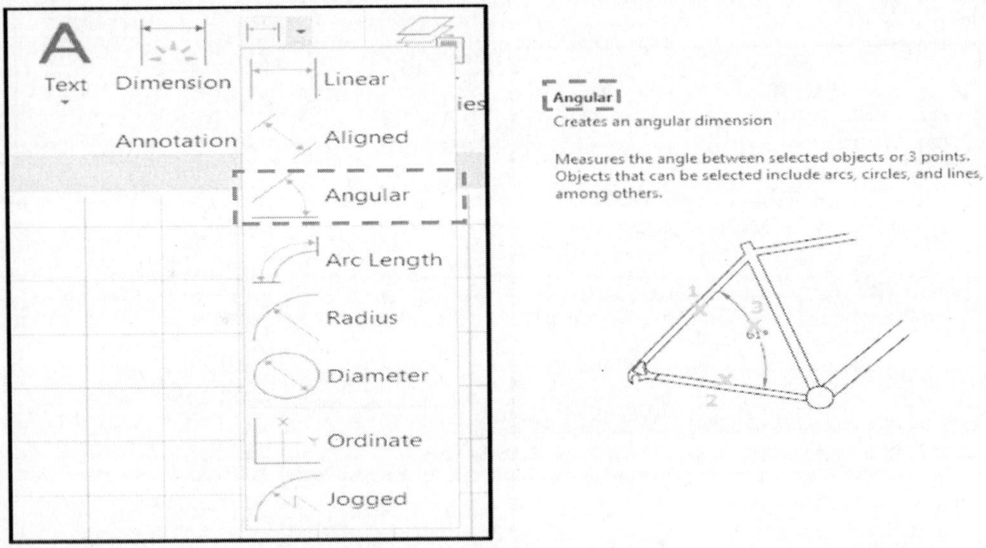

Fig. 3.11: Angular Dimension Tool

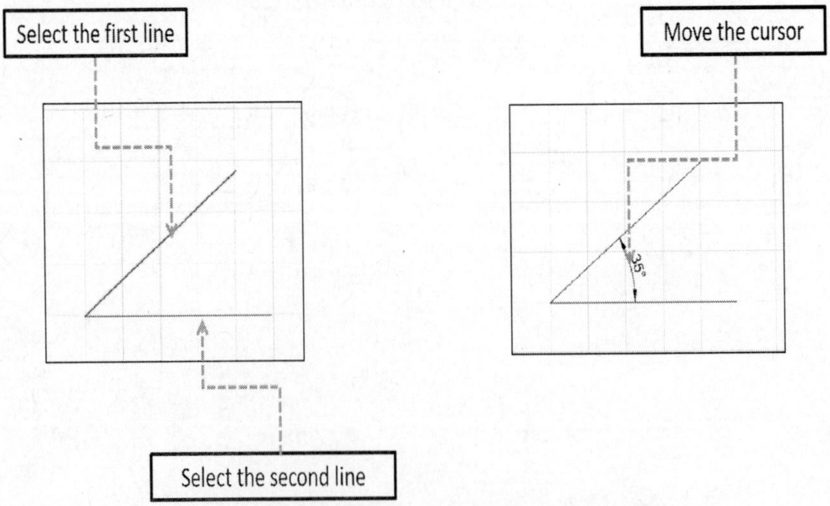

Fig. 3.12: Example of angular dimensioning

3.3.4 Arc length

The Arc Length Dimension tool in AutoCAD allows you to measure the length of an arc. To add an arc length dimension in AutoCAD, you can follow these steps:

Step 1: Click Annotate > Dimensions > Dimension > Arc Length on the ribbon or type DIMARC in the command line and press Enter.

Step 2: Click on the arc that you want to measure the length. The arc length dimension will be automatically created.

Step 3: Specify the location for the dimension line by clicking on the desired point.

Step 4: Press Enter to finish creating dimensions.

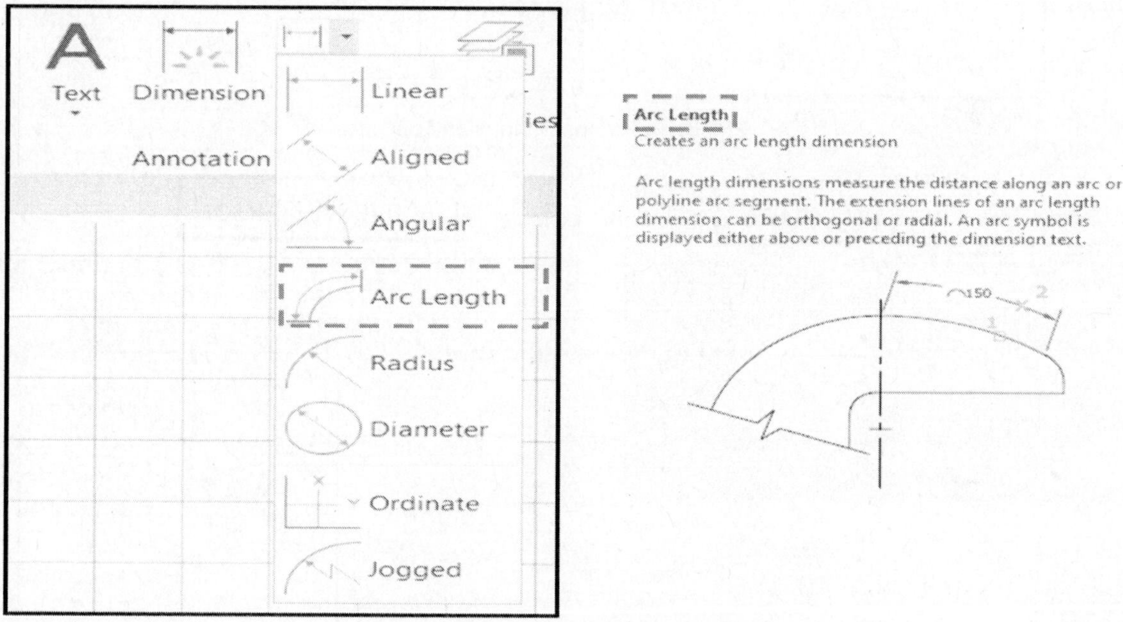

Fig. 3.13: Arc Length Dimension Tool

3.3.5 Radius

The Radius Dimension tool in AutoCAD allows you to measure the radius of a circle or an arc. To add a radius dimension in AutoCAD, you can follow these steps:

Step 1: Click Annotate > Dimensions > Dimension > Radius on the ribbon or type DIMRADIUS in the command line and press Enter.

Step 2: Click on the arc or circle that you want to measure the radius of. The radius dimension will be automatically created.

Step 3: Specify the location for the dimension line by clicking on the desired point.

Step 4: Press Enter to finish creating dimensions.

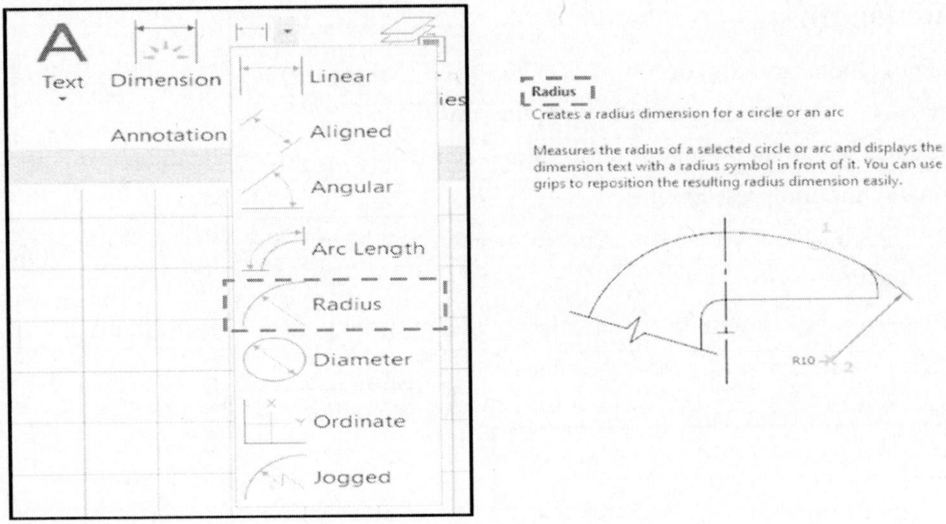

Fig. 3.14: Radius Dimension Tool

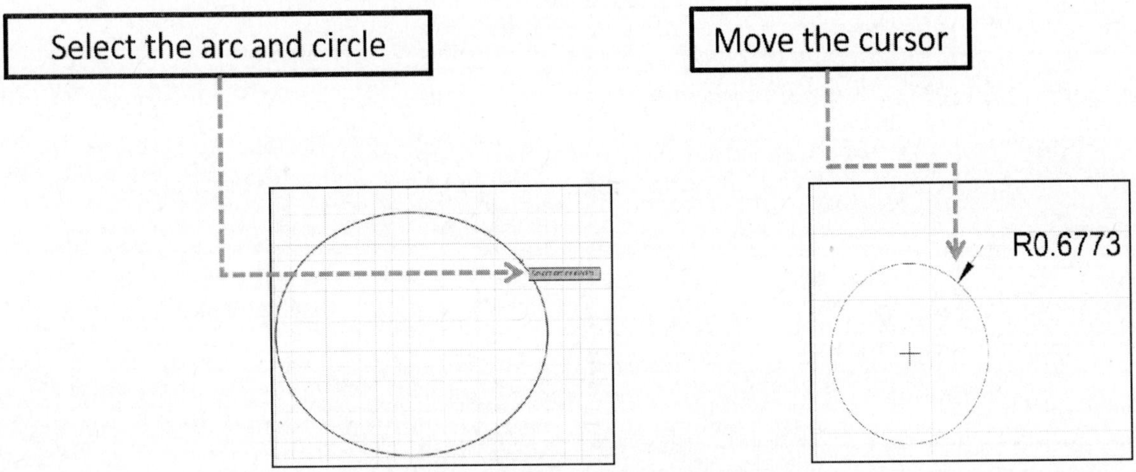

Fig. 3.15: Example of radius dimensioning

3.3.6 Diameter

The Diameter Dimension tool in AutoCAD allows you to measure the diameter of a circle or an arc. To add a diameter dimension in AutoCAD, you can follow these steps:

Step 1: Click Annotate > Dimensions > Dimension > Diameter on the ribbon or type DIMDIAMETER in the command line and press Enter.

Step 2: Click on the circle or arc that you want to measure the diameter of. The diameter dimension will be automatically created.

Step 3: Specify the location for the dimension line by clicking on the desired point.

Step 4: Press Enter to finish creating dimensions.

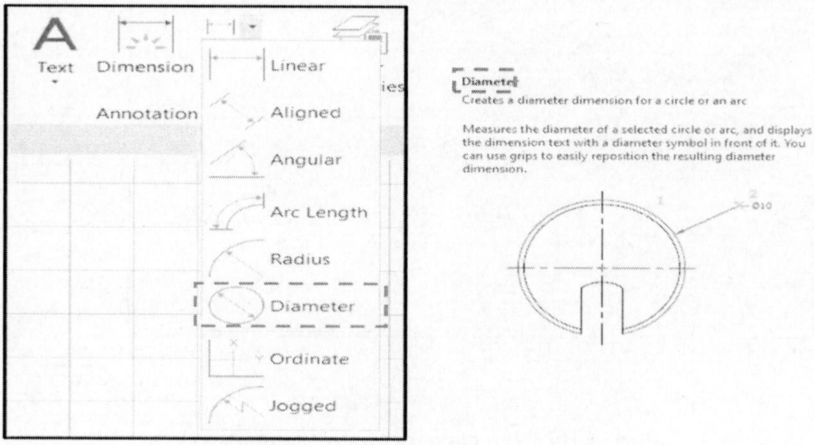

Fig. 3.16: Diameter Dimension Tool

3.3.7 Ordinate

The Ordinate Dimension tool in AutoCAD allows you to measure the X or Y distance between two points, relative to a specified base point. To add an ordinate dimension in AutoCAD, you can follow these steps:

Step 1: Click Annotate > Dimensions > Dimension > Ordinate on the ribbon.

Step 2: Click on the base point, which is the point that you want the X or Y coordinates to be measured from.

Step 3: Click on the first point that you want to measure. Then click on the second point to define the direction of the dimension line

Step 4: Specify the location for the dimension line by clicking on the desired point.

Step 5: Press Enter to finish creating dimensions.

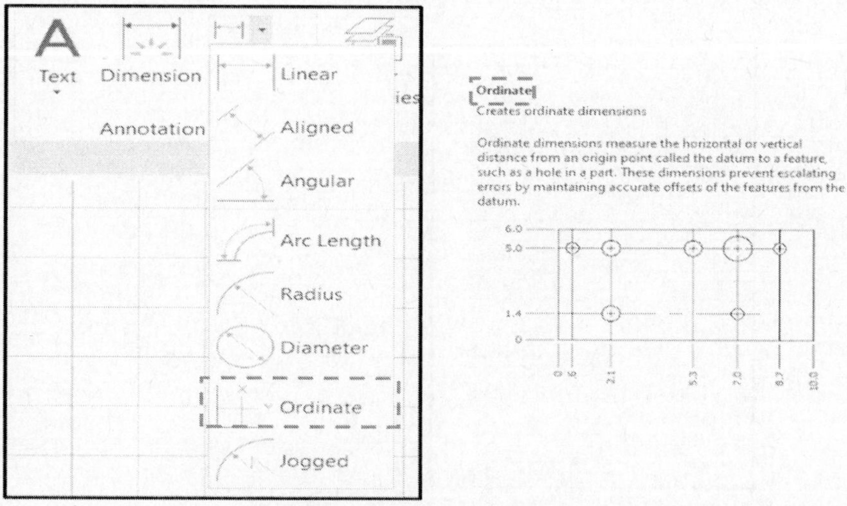

Fig. 3.17: Ordinate Dimension Tool

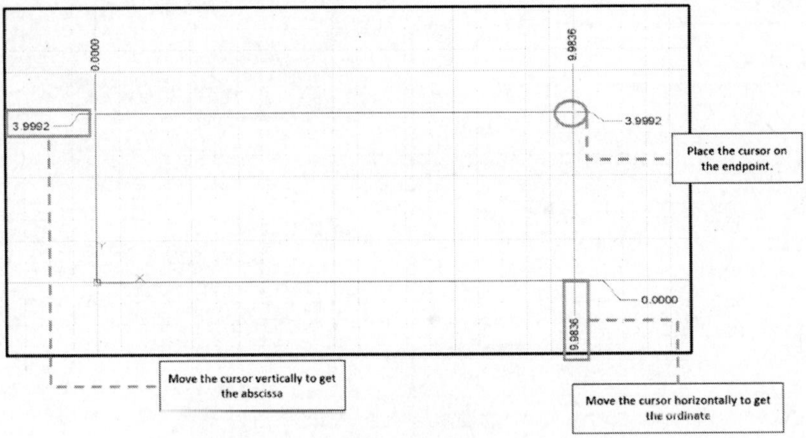

Fig. 3.18: Example of ordinate dimensioning

3.3.8 Jogged

The Jogged Dimension tool in AutoCAD allows you to create a series of connected dimensions with jogged lines, which are useful when you need to dimension objects that have irregular shapes or complex geometries. To add a jogged dimension in AutoCAD, you can follow these steps:

Step 1: Click Annotate > Dimensions > Dimension > Jogged on the ribbon.

Step 2: Click on the first point where you want the dimension line to start.

Step 3: Move the cursor to the location where you want the jog point to be and then click to define the jog point.

Step 4: Move the cursor to the location where you want the next dimension line to start and then click to define the second point.

Step 5: Repeat steps 3 and 4 for each additional jog point and dimension line that you want to create.

Step 6: Specify the location for the dimension line by clicking on the desired point.

Step 7: Press Enter to finish creating dimensions.

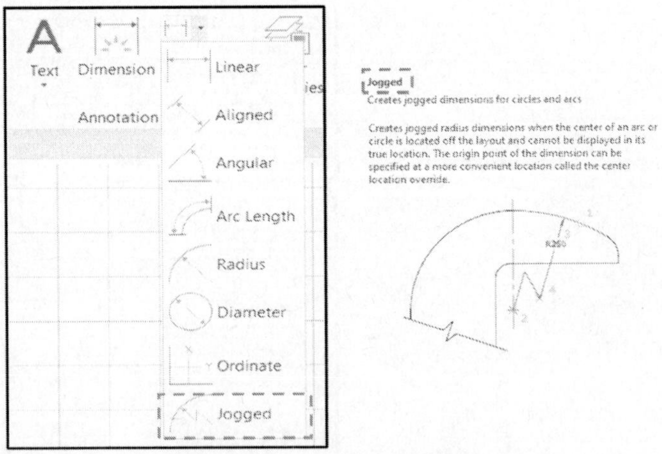

Fig. 3.19: Jogged Dimension Tool

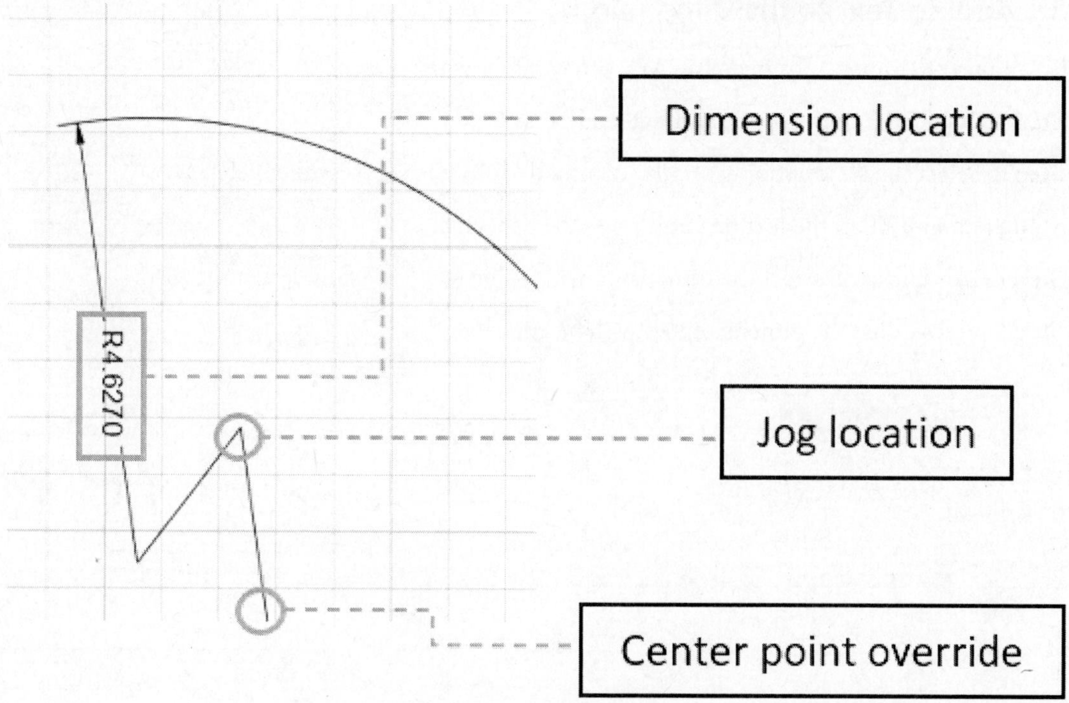

Fig. 3.20: Example of jogged dimensioning

3.4. DIMENSION EDITING

Dimension editing in AutoCAD allows you to modify the size, location and appearance of dimensions in your drawing. To edit dimensions in AutoCAD, you can follow these steps:

Step 1: Open the drawing that contains the dimension you want to edit.

Step 2: Click Annotate > Dimensions > Dimension on the ribbon.

Step 3: Select the dimension you want to edit by clicking on it with the mouse.

Step 4: You can modify the dimension text by double-clicking on the dimension text to activate the text editor. You can then edit the text as you would in any text editor.

Step 5: You can modify the dimension line by clicking on the dimension line to activate the grip points. You can then move, stretch, or rotate the dimension line as needed.

Step 6: You can modify the dimension style by clicking Annotate > Dimensions > Dimension > Dimension Style. You can then modify the style settings, such as text height, arrow size and leader length.

Step 7: Once you have made the desired changes, click on the "Save" button to save your changes.

By following these steps, you can easily edit dimensions in AutoCAD and ensure that your drawing is accurate and precise.

The editing of the dimensions can also be done by DIMEDIT command. The DIMEDIT command can be used to modify dimension. Using this command, you can add text to a dimension, rotate the dimension text and extension lines or reset the position of the dimension text.

3.4.1 Adding Text to the dimension

To add text to dimensions in AutoCAD, you can follow these steps:

Step 1: Type DED in the command line and press Enter.

Step 2: Select the New option from the command line; a text box will appear.

Step 3: Enter TYP in the text box and press the Spacebar.

Step 4: Left-click and select the dimension with value (say for example 10)

Step 5: Press Enter, the dimension text will be changed.

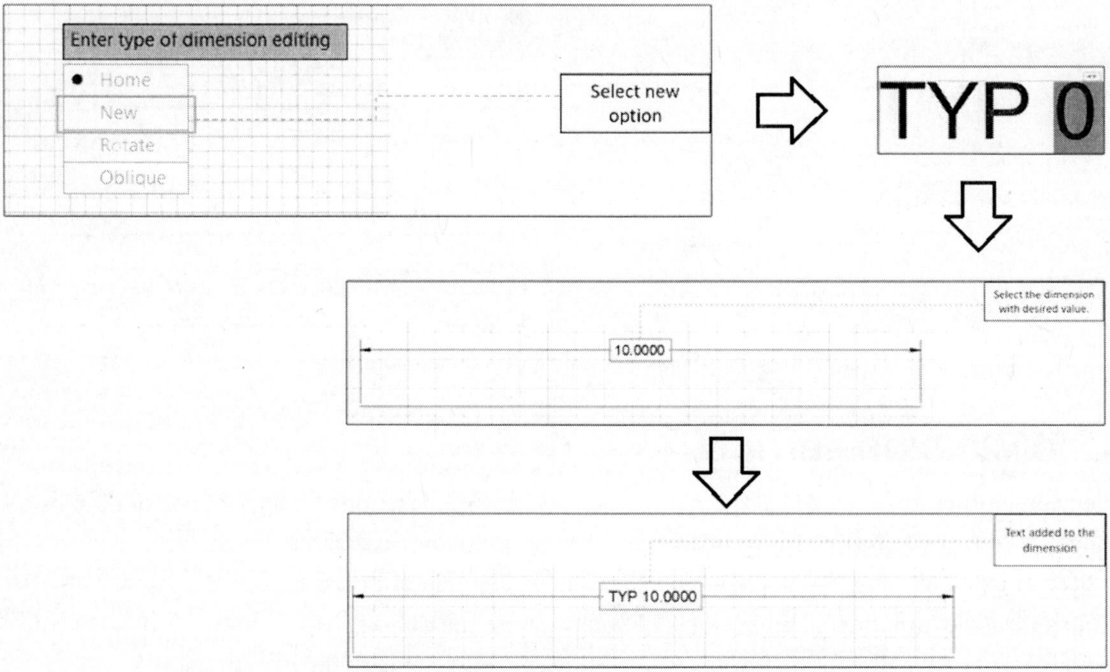

Fig. 3.21: Example of addition of text to the Dimension

3.4.2 Rotating the dimension text

To rotate the text of dimensions in AutoCAD, you can follow these steps:

Step 1: Type DED in the command line and select the Rotate option; the message "Specify angle for dimension text" appears in the command line.

Step 2: Type angle value (say for example 45) and press Enter.

Step 3: Select the dimension with the value TYP 50 and right click; the angle of the dimension text is changed to 45 degrees. Note that the angle is measured from the horizontal axis (X-axis).

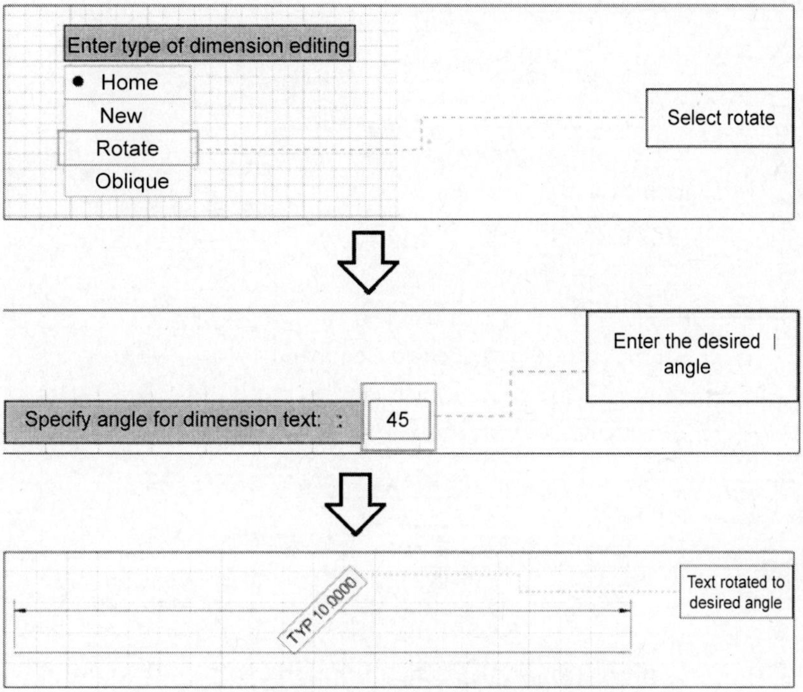

Fig. 3.22: Example of text rotated by a given angle

3.5 TEXT WRITING IN AUTOCAD

One of the essential features of AutoCAD is its ability to create and edit text in your drawings. Whether you need to label dimensions, add notes, or annotate your designs, AutoCAD provides a variety of text tools to help you accomplish your task. In this section, we will cover the basics of text writing in AutoCAD.

To create text in AutoCAD, you can use the DT command or the MT command. The DT command allows you to create single-line text, while the MT command allows you to create multi-line text. Both commands can be accessed from the ribbon or the command line.

3.5.1 Single Line Text

To create single line text using the Text command in AutoCAD, you can follow these steps:

Step 1: Type DT in the command line or click on the single line text icon in the ribbon.

Step 2: Specify the insertion point for the text.

Step 3: Specify the height of the text.

Step 4: Specify the rotation angle of the text. To type your text horizontally, enter the value of angle as 0.

Step 5: Type the text you want to create.

Step 6: Press Enter to end the command.

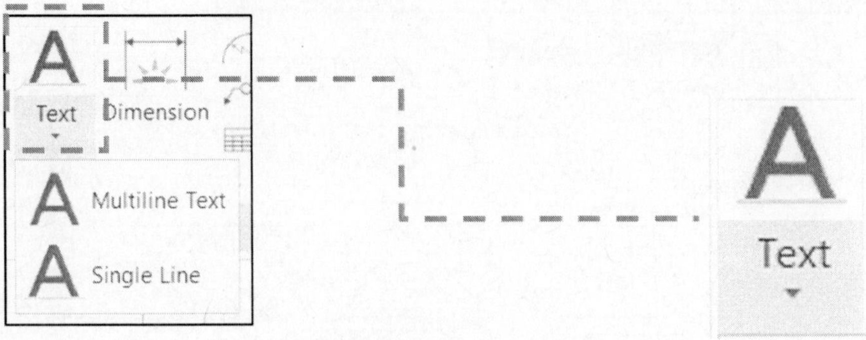

Fig. 3.23: Text Command

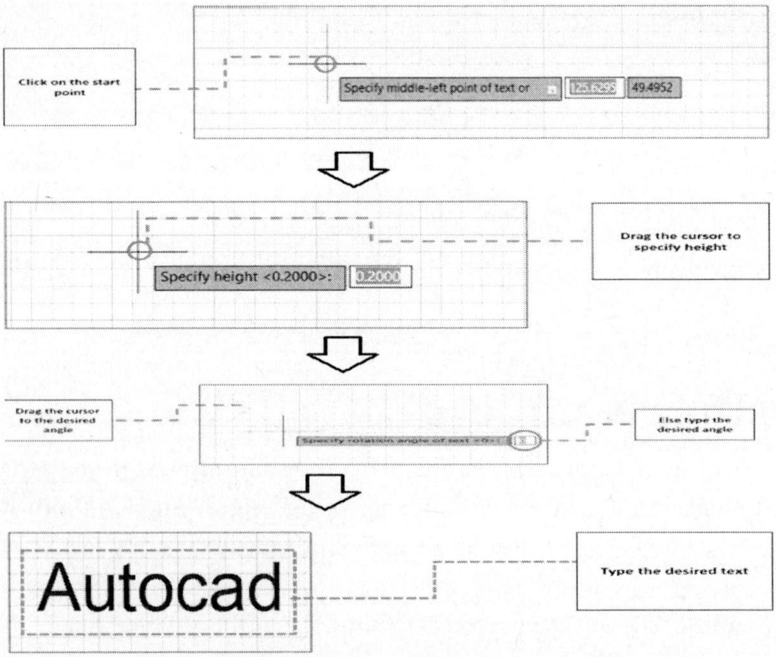

Fig. 3.24: Single line text command

3.5.2 Multiline Text

To create multiline text using the Text command in AutoCAD, you can follow these steps:

Step 1: Type MT in the command line or click on the multiline text icon in the ribbon.

Step 2: Specify the first corner of the text box.

Step 3: Specify the opposite corner of the text box.

Step 4: Type the text you want to create.

Step 5: The height of the box will depend on the height of the text. For setting height, press the H button, then press Enter button of the keyboard. Specify the value of height and press Enter.

Step 6: Press Enter to end the command.

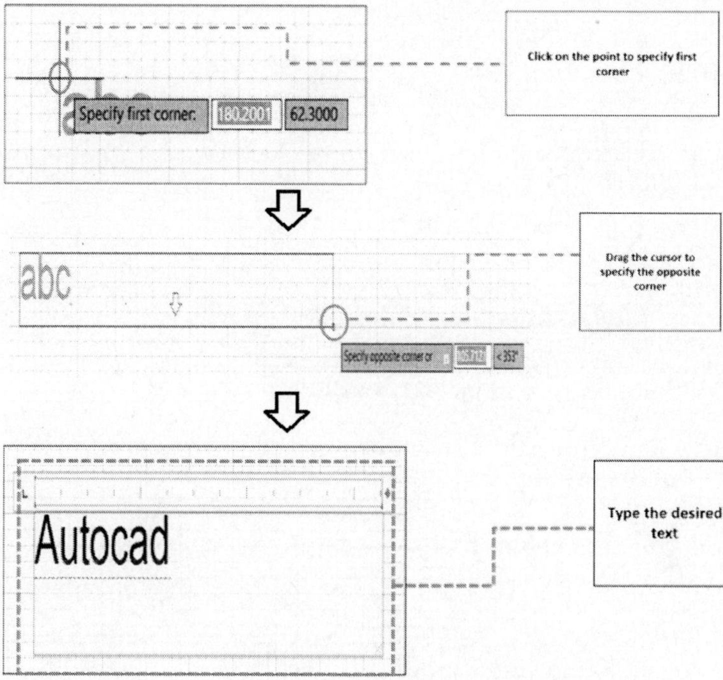

Fig. 3.25: Multiline text command

You can also change the font style of your text. For changing the font, select your desired text, then click on the drop-down arrow option and choose your desired font style from here.

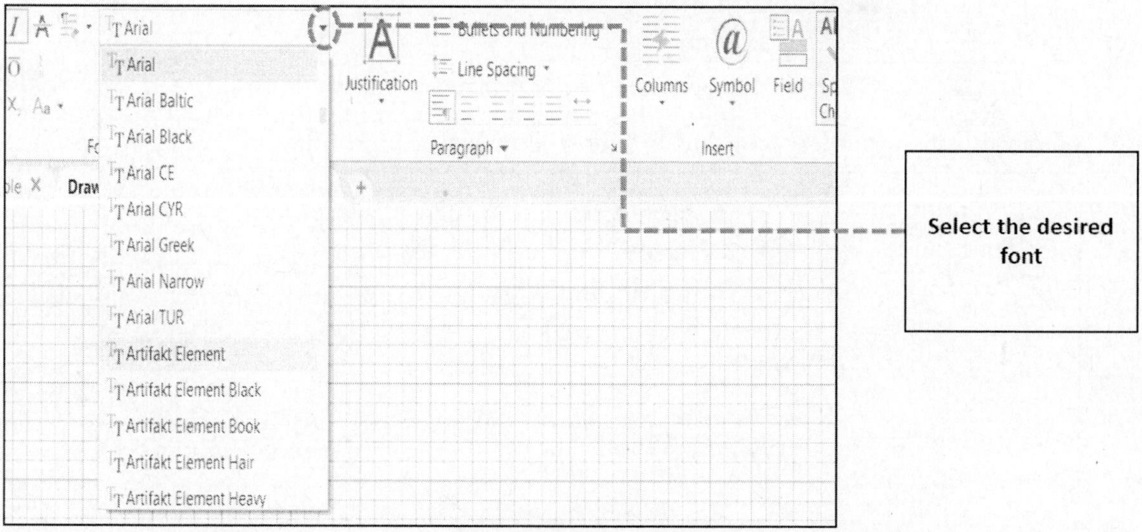

Fig. 3.26: Font style

You can also change the colour of your text. For changing the colour, select your desired text, then click on the drop-down arrow option and choose your desired colour from the colour box by clicking on it.

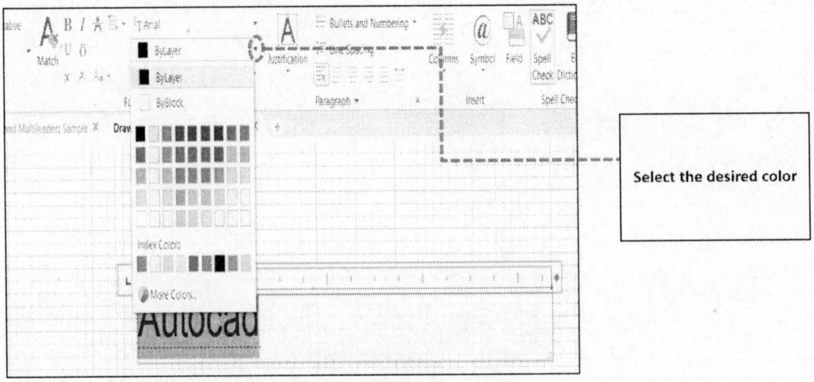

Fig. 3.27: Font Colour

You can also increase or decrease the distance between the characters of the text.

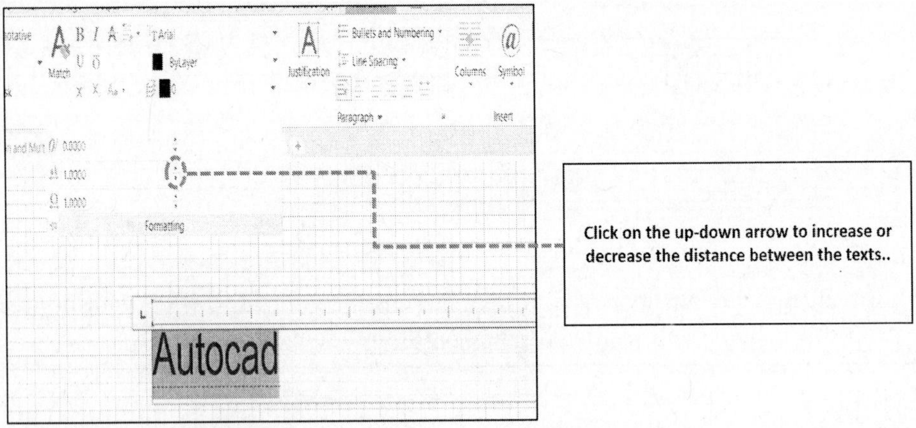

Fig. 3.28: Space between characters of the text

You can also align your text according to your requirement.

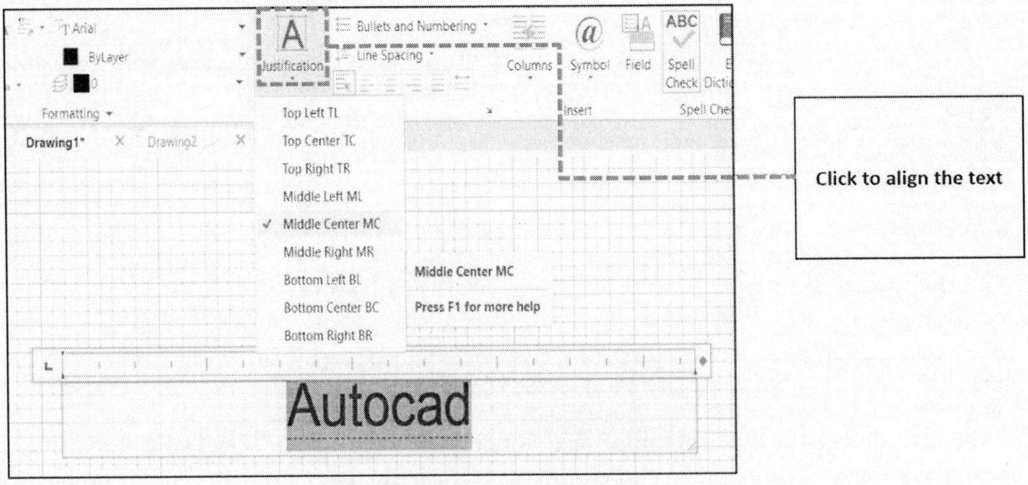

Fig. 3.29: Text alignment

UNIT SUMMARY

In this unit you have learned about:

- Different methods of dimensioning in AutoCAD such as quick dimensioning, menu dimensioning, command dimensioning and toolbar dimensioning.
- Various dimension types such as linear, aligned, angular, radius, ordinate and jogged dimensions.
- Various dimensions allow for precise and accurate measurement annotations on drawings.
- Techniques for editing dimensions to ensure clarity and consistency in design documentation.
- Text writing in AutoCAD using single line text and multiline text to add informative annotations to drawings.

EXERCISES

A. Multiple Choice Questions

3.1 In AutoCAD, the LIMITS command allows you to determine the
(a) size of drawing space (b) size of text
(c) amount of time needed to draw (d) none

3.2 A uniform pattern of dots/lines on the CAD screen is called a/an
(a) snap pattern (b) reference dot pattern
(c) grid (d) none

3.3 Which of the following lines are used to show that the object is cut and then viewed?
(a) centre lines (b) hatching lines (c) leader lines (d) hidden

3.4 Scaling the object makes it
(a) smaller (b) bigger
(c) either smaller or bigger (d) non

3.5 One of the below options is wrong about type of AutoCAD drawing units, which is it?
(a) architectural (b) engineering (c) decimal (d) millimeter

3.6 Which of the following is NOT a property of an object?
(a) line weight (b) measure (c) hyperlink (d) elevation

3.7 Which state grid is used to design perspective?
(a) parametric (b) isometric (c) pro-optic (d) rectangular

3.8 In which dimensioning, the dimension text is aligned by default with the object being dimensioned.
(a) aligned (b) linear (c) angular (d) ordinate

3.9 When in absolute Cartesian coordinates have points A (10.8) and B (6.5), then to make a line from A->B with relative polar coordinates will write
(a) @-5<36.88 (b) @4<30 (c) @5<216.88 (d) @3<60

3.10 Ordinate dimension measures

(a) horizontal distance (b) vertical dimension

(c) both a and b (d) none

3.11 With multiline text editor, you can do all of the following except:

(a) insert numbered and bulleted lists

(b) insert specific drafting and engineering symbols

(c) Insert pictures

(d) create fields such as date, time and author

3.12 The following are some of the properties and settings that you can save in a layout, except

(a) plot scale (b) paper size

(c) line weight (d) drawing orientation

3.13 You can set viewport scale factor by typing the following in command line

(a) VSCALE (b) VPSCALE (c) VIEWPORTSCALE (d) none

3.14 Which of the following dimensions are automatically updated when the object to which they are assigned is modified?

(a) linear (b) angular

(c) ordinate (d) true associative dimensions

3.15 Extra leaders can be added or removed from existing multileaders by using the following tool.

(a) angular (b) ordinate

(c) Remove Leader (d) Align multileaders

Answers of Multiple-Choice Questions
3.1 (a), 3.2 (c), 3.3 (b), 3.4 (c), 3.5 (b), 3.6 (b), 3.7 (b), 3.8 (a), 3.9 (c), 3.10 (c), 3.11 (c), 3.12 (c), 3.13 (d), 3.14 (d), 3.15 (c)

B. Subjective Questions

3.1 List the different methods of dimensioning in AutoCAD? Briefly explain all.

3.2 List the different types of dimensions and explain them.

3.3 Write the procedures for dimension editing and its benefits.

3.4 Elucidate the differences between Ordinate dimension and Jogged dimension.

3.5 Differentiate between a Dimension line and an Extension line.

3.6 Describe how text writing in AutoCAD is beneficial.

3.7 Discuss the utility of TEXT command by an example.

3.8 Illustrate the process of setting up a new text style in AutoCAD.

3.9 Define justification. Describe the justifications available in TEXT command?

3.10 Elucidate the process of making corrections in text using AutoCAD?

C. Problems for Practice

Draw the following figures in AutoCAD with proper dimensions.

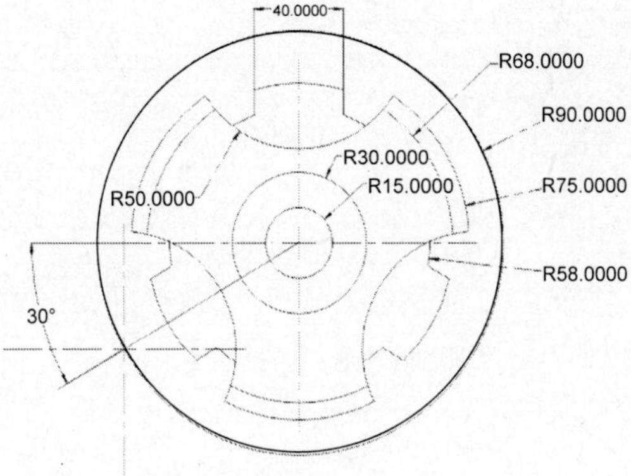

Fig. 3.1

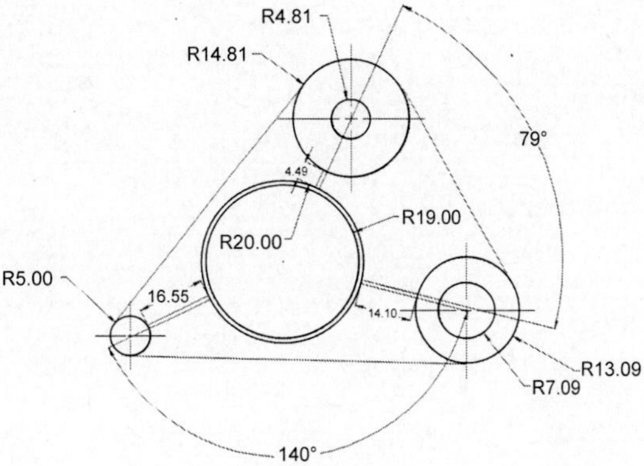

Fig. 3.2

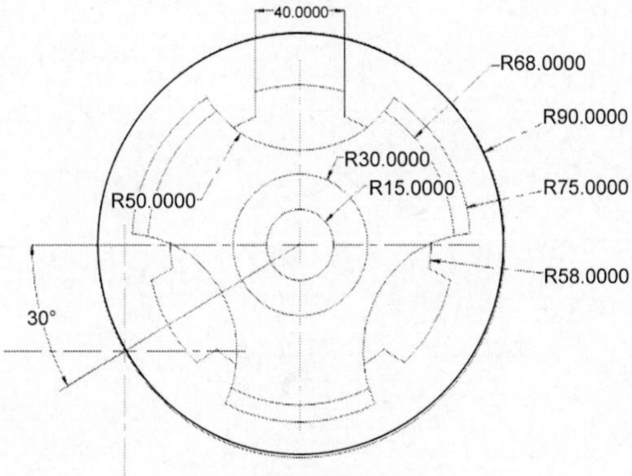

Fig. 3.3

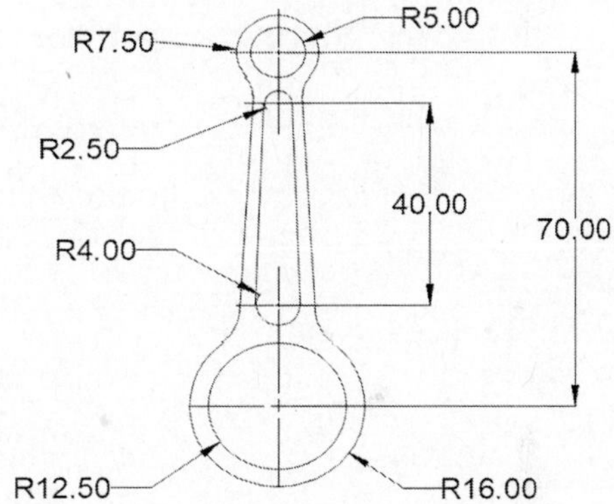

Fig. 3.4

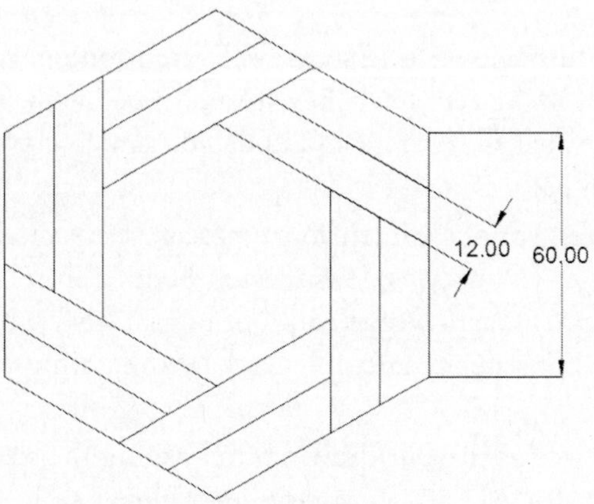

Fig. 3.5

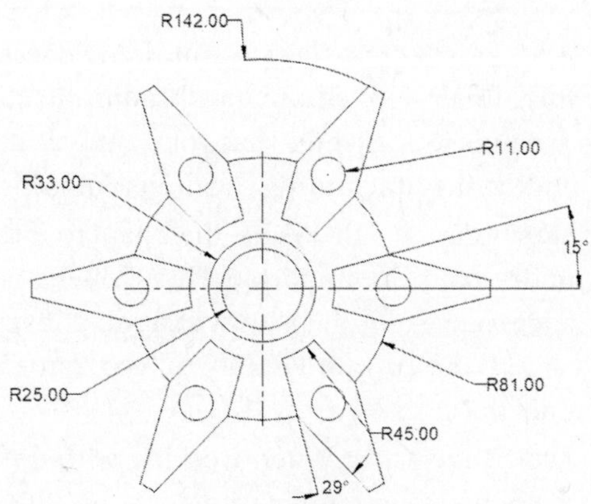

Fig. 3.6

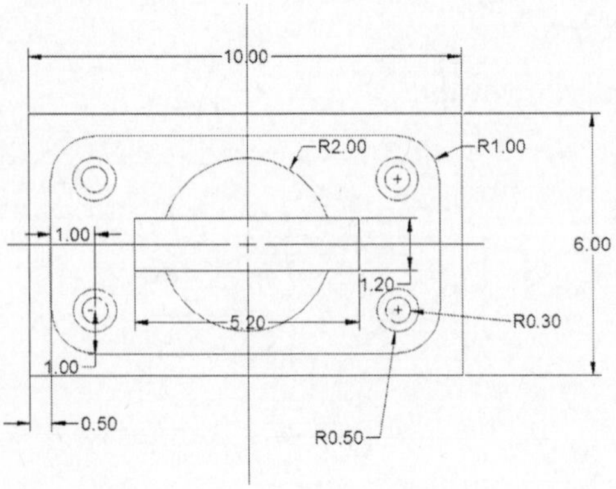

Fig. 3.7

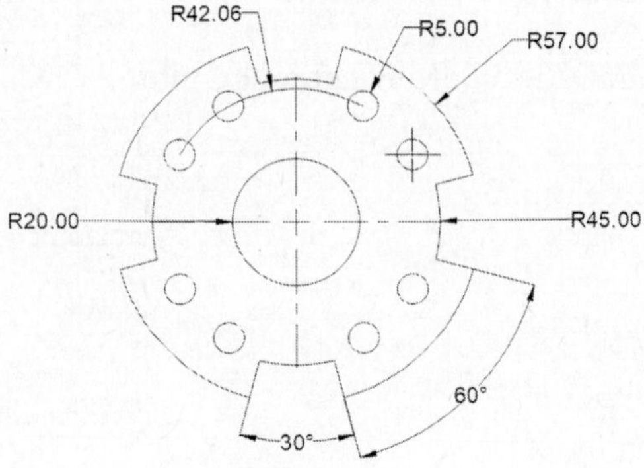

Fig. 3.8

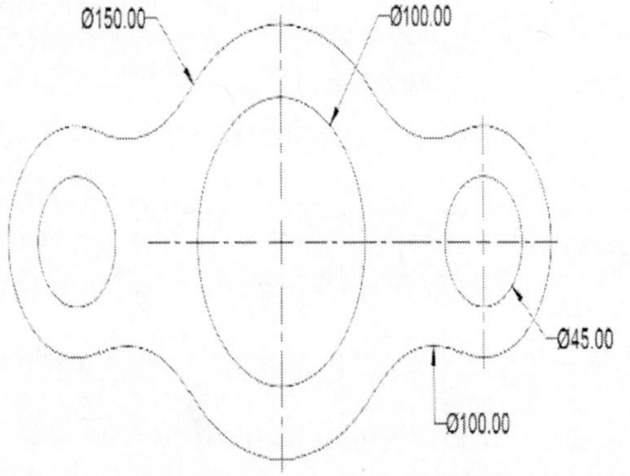

Fig. 3.9

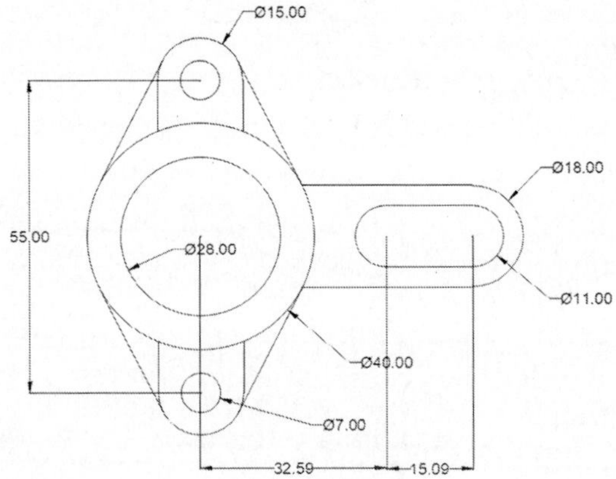

Fig. 3.10

KNOW MORE

- *Associative Dimensioning*: AutoCAD offers a feature known as Associative Dimensioning. With this, dimensions automatically adjust to changes in the geometry they are associated with. This dynamic link ensures that your annotations are always accurate and updated, reducing errors and saving time during the design modification process.
- *Dimension Styles*: While the chapter introduces various dimension types, it is essential to understand the importance of Dimension Styles (DimStyles). Dimension Styles control the appearance of dimensions, including text placement, arrow type and unit format. By creating and saving custom DimStyles, you can ensure consistency and efficiency in annotating your drawings.
- *Dimension Scale*: When working with different scales in your drawings, managing the size of your dimension annotations can be crucial. AutoCAD allows you to set a global scale that adjusts the size of dimension arrows, text and other elements to match the scale of your viewport or drawing, facilitating readability and precision.

- *Quick Leader*: The Quick Leader command is a valuable tool for adding leader lines that point to specific objects or areas in your drawing, accompanied by descriptive text. Leaders are especially useful in technical drawings where you need to annotate and clarify specific components.
- *Text Formatting*: Beyond single and multiline text, AutoCAD provides extensive text formatting options. You can adjust the font, size, style and alignment of your text and even incorporate fields that display data that can automatically update as your drawing changes.
- *Text Styles*: Just like Dimension Styles, Text Styles in AutoCAD allow you to save and apply preset text formatting settings. This feature is crucial for maintaining consistent annotation appearances across different drawings and projects.
- *In-Place Text Editing*: AutoCAD offers an In-Place Text Editor that facilitates the process of editing text. This feature provides a word processor-like interface within the drawing area, allowing you to see the text formatting changes in real-time as you make them.
- *Spell Check*: To ensure the professionalism and accuracy of your annotated drawings, AutoCAD incorporates a Spell Check feature. This tool works similarly to those in word processors, helping you identify and correct spelling mistakes in your annotations.
- *Hyperlinks*: You can embed hyperlinks within your text in AutoCAD. This feature is particularly useful when you need to reference external documents, web pages, or other relevant resources directly from your drawing, providing additional context or information to colleagues and clients.

REFERENCES AND SUGGESTED READINGS

1. Fane, Bill. AutoCAD for dummies. John Wiley & Sons, 2016.
2. Shoukry, Yasser and Jaiprakash Pandey. Practical Autodesk AutoCAD 2021 and AutoCAD LT 2021: A no-nonsense, beginner's guide to drafting and 3D modeling with Autodesk AutoCAD. Packt Publishing Ltd, 2020.
3. Nyemba, Wilson R. Computer Aided Design: Engineering Design and Modeling Using AutoCAD. CRC Press, 2022.

Dynamic QR Code for Further Reading

Linear dimensioning

Aligned dimensioning

Angular dimensioning

Radius dimensioning

Diameter dimensioning

Arc length dimensioning

Jogged dimensioning

Ordinate dimensioning

Text addition

4 Hatching, Blocks and Views in AutoCAD

UNIT SPECIFICS

This unit presents information related to the following topics:

- Hatching commands, including the hatch tool, properties of hatch lines, island detection tools, hatching with text and hatch line edits
- Block making, including creating block definitions, inserting blocks, working with attributes, exploding blocks and purging unused block definitions;
- Layers command and its uses;

This unit introduces the readers to hatching, block making and using layers in AutoCAD. First, the readers are presented with hatching commands. Hatching is when a certain area of a drawing is filled with lines or patterns. This helps in understand the drawing better. Next, block creation and their manipulation are covered. Blocks are a way to group different objects into one. This is useful when the same group of objects is used many times in a drawing. Lastly, the unit deals with layers. Layers are like transparent sheets that can be put on top of each other in AutoCAD drawings. This helps in organizing the drawing better. All these topics are discussed with reference to their procedural steps in AutoCAD.

Apart from this at the end of the unit, a succinct recapitulation of the overall broad concepts is provided in form of a unit summary. Besides, a large number of multiple-choice questions as well as descriptive type questions with Bloom's taxonomy action verbs are included. A list of references and suggested readings are given in the unit so that one can go through them for practice. It is important to note that for getting more information on various topics of interest some QR codes have been provided in different sections which can be scanned for relevant supportive knowledge. Video resources along with QR codes are mentioned for getting more information on various topics of interest which can be surfed or scanned through mobile phones for viewing.

RATIONALE

Hatching, blocks and layers are essential features in CAD software like AutoCAD, allowing users to create more efficient, organized and visually appealing drawings. Hatching adds textures and patterns to drawings, enhancing their readability and representation. Blocks enable the reuse of frequently used objects, streamlining the drawing process and reducing file size. Layers help organize and manage complex drawings by separating different components, making it easier to edit and navigate the drawing. By mastering these features, students will be able to create more efficient, organized and visually appealing drawings in AutoCAD, enhancing their drafting skills and productivity. The objectives of this unit are performance-based, so once you have finished the unit, you will be able to carry out all of the tasks it specifies. In this lesson, students will learn the fundamentals of hatching, block and layer creation and their applications. This unit includes detailed explanations of how to execute each command, as well as exercises to hone your skills.

PRE-REQUISITES

Before reading this unit, the students are advised to revisit the following:

- Mathematics: Coordinate and Plane geometry (Class XII)
- Engineering Graphics (ES101)
- AutoCAD commands mentioned in Unit-1

UNIT OUTCOMES

After studying this unit students will be able to:

U4-O1: Understand and apply hatching commands in AutoCAD to create visually appealing drawings with textures and patterns

U4-O2: Create, insert and manage blocks in AutoCAD to streamline the drawing process and reduce file size

U4-O3: Work with attributes to create fill-in-the-blank blocks for customizable objects

U4-O4: Use the layers command in AutoCAD to effectively organize and manage complex drawings

U4-O5: Apply the skills learned in this unit to create efficient, organized and visually appealing drawings in AutoCAD

Unit-4 Outcomes	EXPECTED MAPPING WITH COURSE OUTCOMES (1 – Weak Correlation; 2 – Medium correlation; 3 – Strong Correlation)				
	CO-1	**CO-2**	**CO-3**	**CO-4**	**CO-5**
U4-O1	3	1	1	1	2
U4-O2	1	-	1	-	-
U4-O3	1	-	1	-	-
U4-O4	2	-	1	-	2
U4-O5	2	1	1	1	1

4.1 INTRODUCTION

Hatching is the process of using repeated lines, patterns, or filled areas to indicate the material of an object, differentiate between different sections, or highlight specific areas in a drawing. It provides an additional level of information that can not be conveyed by contours and lines alone. In the complex arena of Computer-Aided Machine Drawing, these seemingly subtle elements can significantly enhance the comprehensibility of your designs, making them more effective communication tools.

Whether it is to indicate the composition of a particular section in a mechanical component, or to differentiate between different parts in an assembly drawing, hatching brings in a level of clarity that is indispensable. Furthermore, hatching can also be used to create a sense of depth in a two-dimensional drawing, adding an element of realism to the designs.

In essence, hatching, when properly used, can enrich your drawings, making them more detailed, understandable and aesthetically pleasing. It is a powerful tool that, when understood and utilized correctly, can greatly enhance the quality and effectiveness of your machine drawings in AutoCAD.

4.2 HATCHING COMMANDS

The Hatch command is a dynamic feature in AutoCAD, offering users the opportunity to add depth, detail and clarity to their drawings. This command is used to fill an enclosed area with a range of patterns – hatched, gradient, or solid fill.

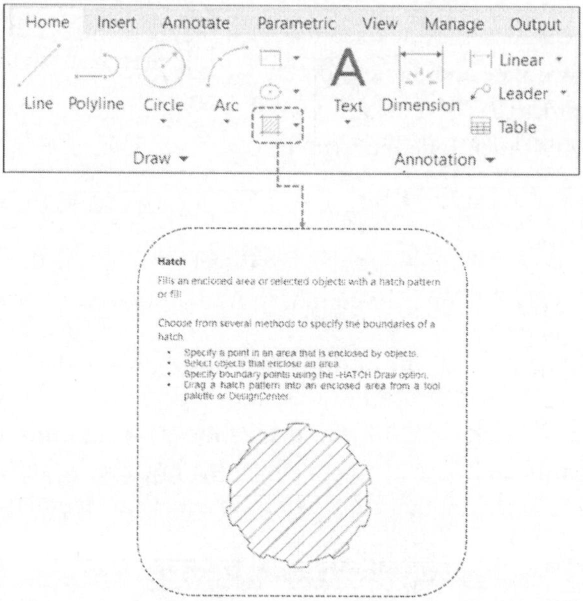

Fig. 4.1: Hatch command

Hatched patterns are characterized by a series of lines, providing texture and differentiation to the area they fill. The Hatch Pattern Scale, an important feature in this process, allows you to adjust the spacing between these lines, thereby modifying the pattern's overall appearance.

Gradient patterns, on the other hand, offer a smooth transition between two colours, creating a sense of depth and volume in a two-dimensional space.

In addition, AutoCAD provides a wide array of hatch patterns to choose from. These patterns cater to diverse requirements and design contexts, making the Hatch command a versatile and essential tool in creating intricate and comprehensive drawings.

This section will guide you through the various commands, tools and properties associated with hatching in AutoCAD. It will equip you with the knowledge and skills needed to effectively use the Hatch command, enhancing the visual distinction and overall quality of your CAD drawings.

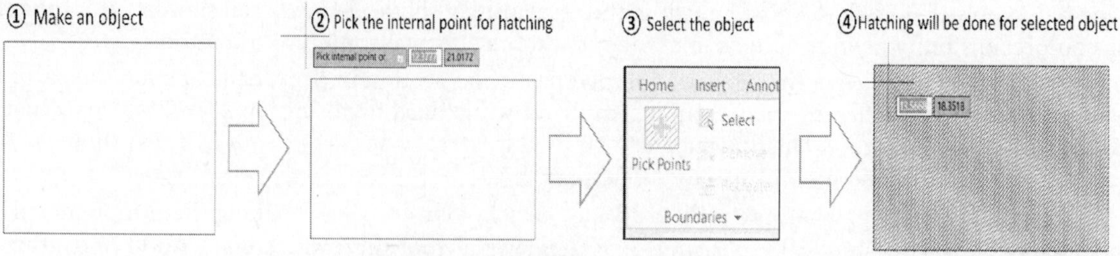

Fig. 4.2: Example of Hatching

4.2.1 Hatch Tool

The Hatch tool is the primary means of creating hatches in AutoCAD. It is a versatile tool that lets you select a pattern or solid fill, specify the area to be hatched, adjust the scale, angle and other properties of the hatch.

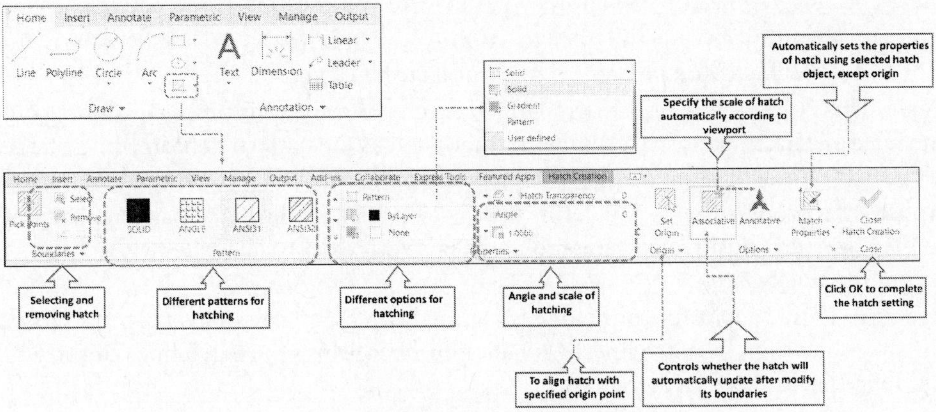

Fig. 4.3: Hatch tool

To use the Hatch tool, click **Home > Draw > Hatch** on the ribbon.

You can also use the Hatch tool from the command line by typing **HATCH** or **H** and press Enter.

4.2.2 Properties of Hatch Lines

The properties of hatch lines play a critical role in how effectively your hatches convey information. This includes factors like the type and scale of the pattern, the colour and layer of the hatch and the angle of the hatch lines. Understanding and effectively manipulating these properties can significantly enhance the effectiveness of your hatches.

Fig. 4.4: Properties of Hatch Lines

These properties can be accessed and edited through the **Pattern Properties** tab in the **Boundary Hatch** dialog box. The various hatch line properties are —

1. **Pattern Type**: There are two types you can choose from— *Predefined* and *User Defined*. The *Predefined* option allows you to select a pattern from the standard or ISO hatch pattern library files. The *User Defined* option allows you to create your custom hatch pattern.
2. **Color**: This property enables you to choose a colour for your hatch pattern. It could be *BYLAYER, BYBLOCK, a predefined color*, or *a custom color*.
3. **Scale**: This is applicable only to predefined hatch patterns. This property allows you to specify the scale of a hatch pattern. A smaller number makes the pattern denser, while a larger number reduces the amount of hatching within a boundary.
4. **Angle**: This property helps you define the rotation angle of the hatch pattern in degrees, relative to the boundary. Positive numbers rotate the pattern clockwise, while negative numbers rotate it counter-clockwise.
5. **Spacing**: This is applicable only to user-defined hatch patterns. This property lets you specify the space between the hatch lines. A smaller number brings the hatch lines closer, while a larger number, spaces them farther apart.
6. **ISO Pen Width**: This is applicable to predefined and ISO hatch patterns. This property lets you define the hatch-line width for ISO hatch patterns.

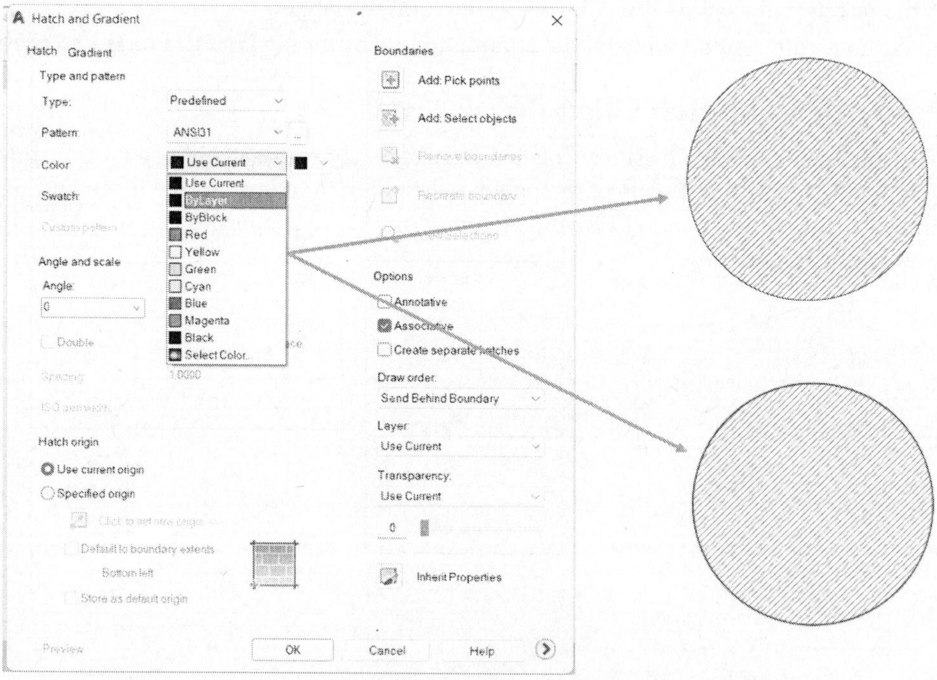

Fig. 4.5: Hatching with different colours

7. **Cross-Hatched**: This property is available only for user-defined hatch patterns. It superimposes another copy of the pattern at a 90-degree angle over the original when checked.
8. **Hatch Attributes**: Checking the *Associative* option links the hatch pattern to its boundaries, enabling automatic updates whenever the extents are modified.

9. **Gap Tolerance**: This shows the permissible gap between non-touching entities while still being used to create the hatch pattern. AutoCAD calculates this hatch tolerance automatically, depending on the size of the software's window.
10. **Hatch Transparency**: This property lets you control the transparency of your hatch pattern. There are three options—
    Bylayer— The hatch pattern assumes the transparency defined in the currently selected layer.
    Byblock— The hatch pattern takes on the properties of a block, if the pattern is included in one.
    Specify value— you can set a specific transparency value.

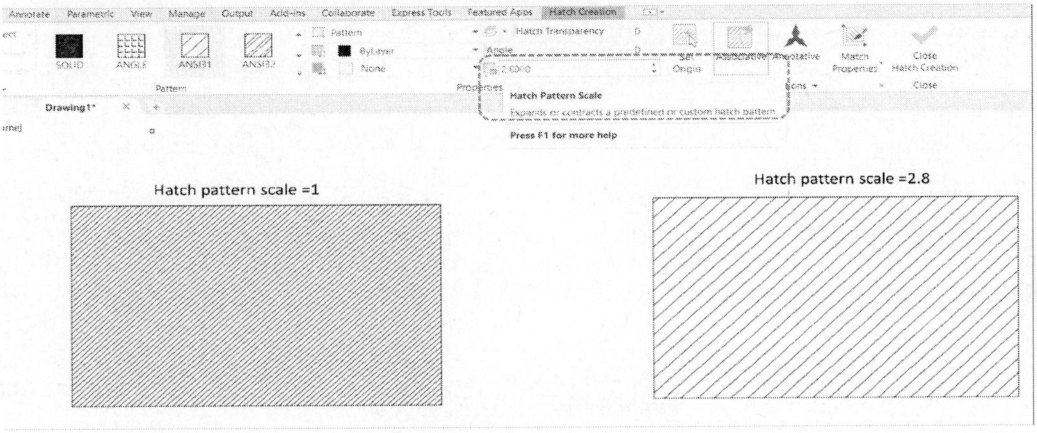

Fig. 4.6: Hatch pattern scale

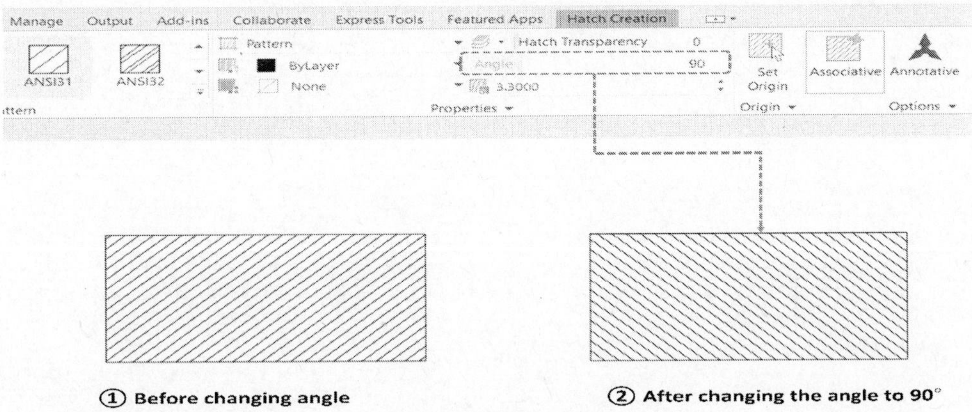

Fig. 4.7: Hatch angle

Copy Hatch Properties function allows you to duplicate the properties of an existing hatch pattern in the project to a new one. This comes in handy when you want to maintain consistency in your drawings.

4.2.3 Island Detection Tools

Island detection tools in AutoCAD allow you to control how hatches interact with other objects in the drawing. They enable you to specify whether a hatch should fill an entire enclosed area or leave space around certain objects within that area, known as "islands".

The Island Detection Tools can be used in AutoCAD by using the following steps —

Step 1: Click on Hatch tool

Step 2: Click on Hatch Creation

Step 3: Click on option tool

Step 4: Click on island detection

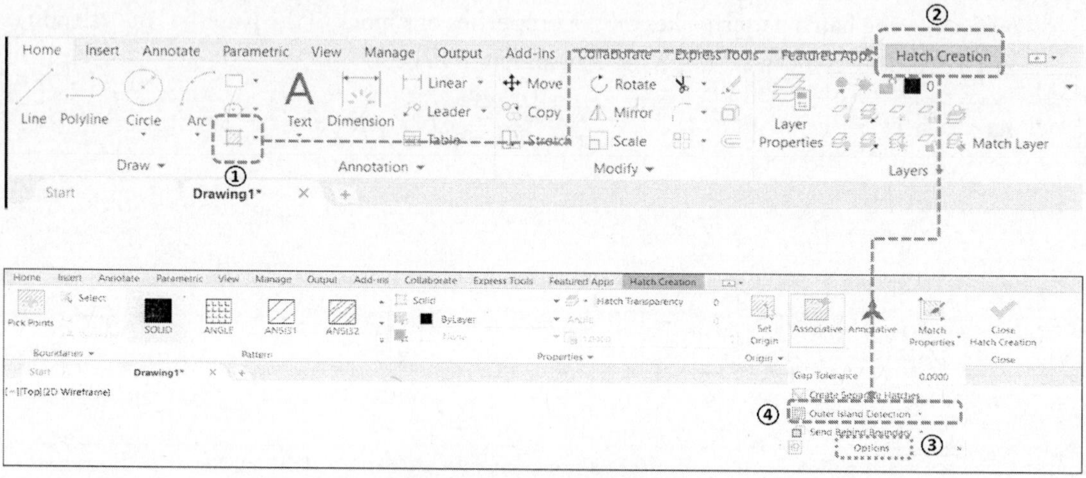

Fig. 4.8: Island detection tool

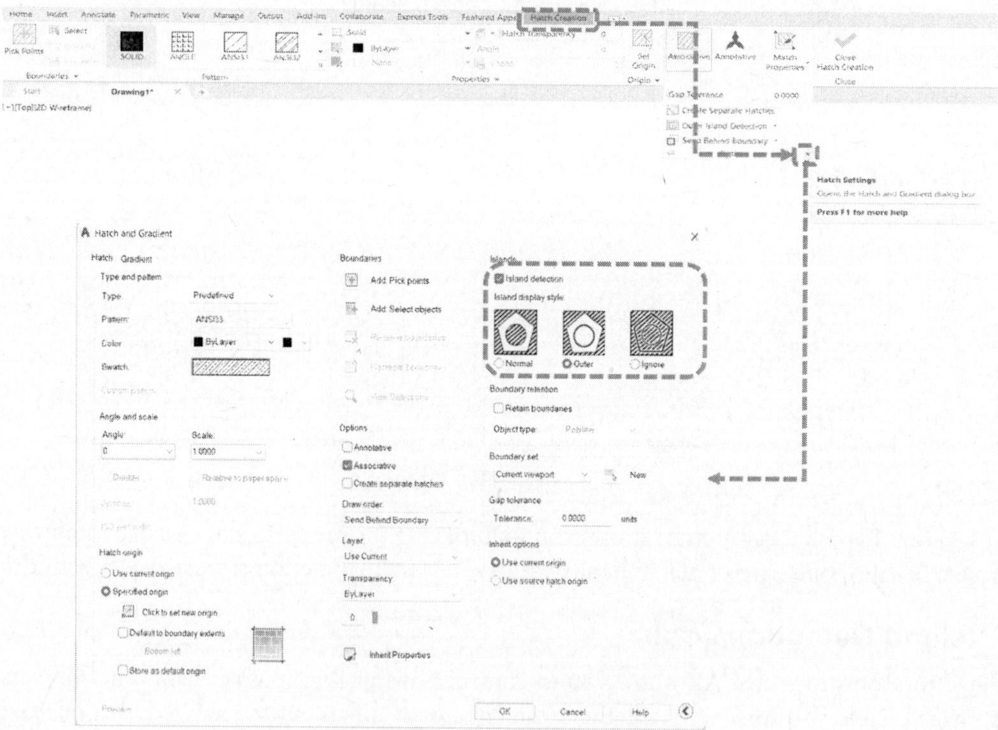

Fig. 4.9: Island detection tool options

There are four types of island detection hatching method

1. **Normal Island Detection**: It automatically hatches islands inwards from the area specified by the hatch pick point.
2. **Outer Island Detection**: It hatches only the area between the outer hatch boundary and any interior islands relative to the location of the hatch pick point.
3. **Ignore Island Detection**: Hatches Inward from the outermost hatch boundary and ignore the interior objects.
4. **No Island Detection**: Turn off to use the legacy island detection method.

Fig. 4.10: Types of Island detection tool

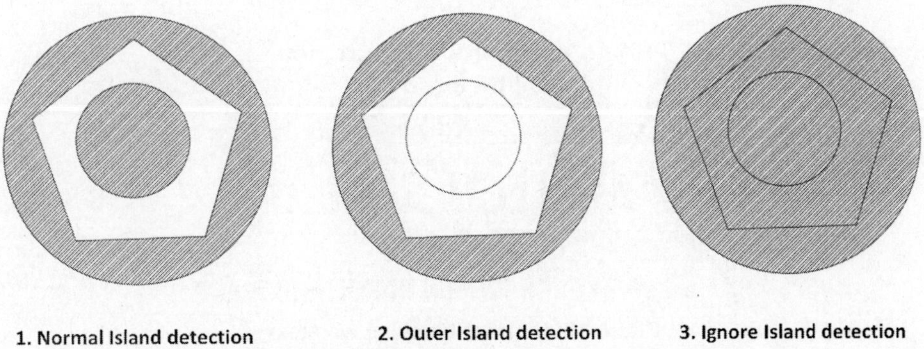

Fig. 4.11: Example of Island detection tool

The following steps can be followed to hatch an object using the island detection tool —

Step 1: Draw the object

Step 2: Click on Home > Draw > Hatch > Hatch creation > select island detection type (normal/ outer /ignore)

Step 3: Select the object to be hatched

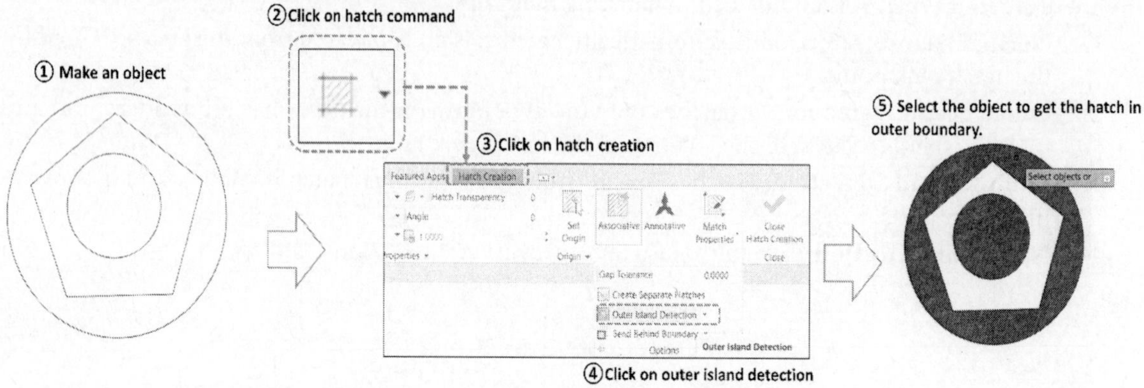

Fig. 4.12: Steps in hatching an object with Island detection tool

4.2.4 Hatching and Text

When working with hatches, it is essential to understand how they can interact with text in your drawings. This section will cover how to ensure that your text remains clear and legible when it overlaps with a hatch and how to use hatches to highlight or emphasize text.

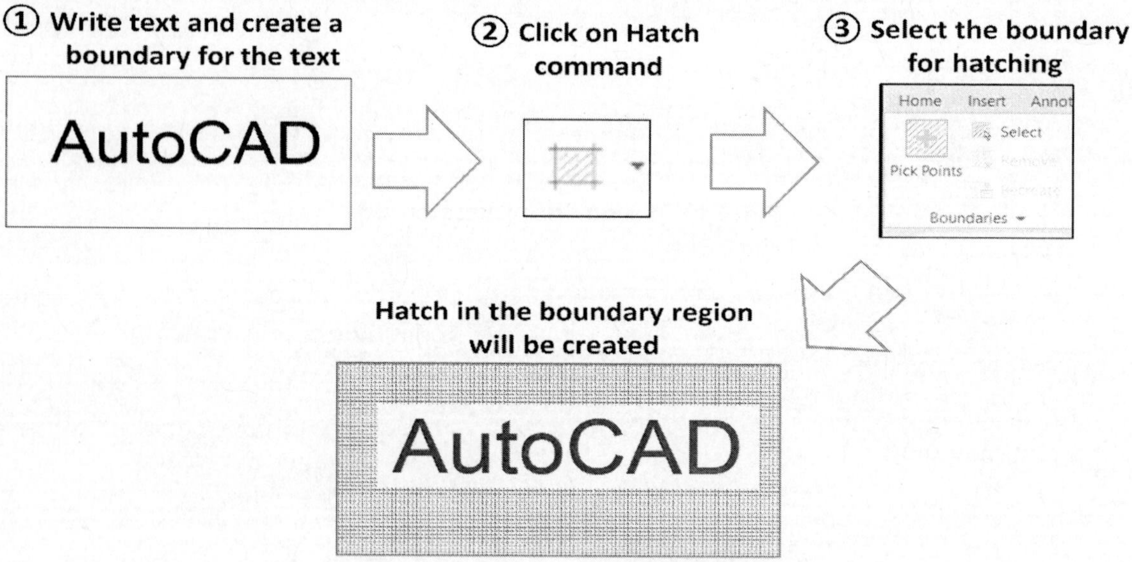

Fig. 4.13: Steps in hatching with text

The following steps can be followed to hatch an object with text —

Step 1: Write the text and create a boundary for the text

Step 2: Click on Home > Draw > Hatch

Step 3: Select the boundary for hatching

In case, you want to remove the text boundary and hatch the entire text, Right click on hatch > hatch edit > remove boundaries.

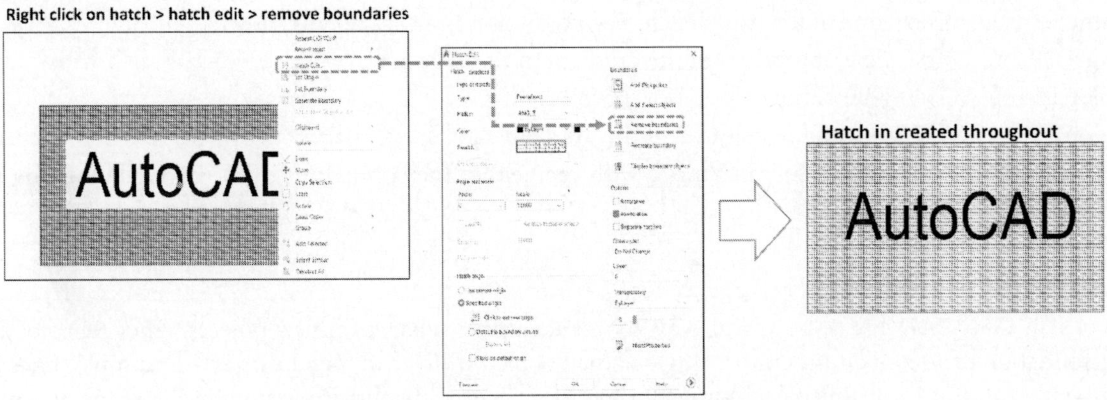

Fig. 4.14: Remove boundary of text while hatching

4.2.5 Hatch Line Edits

Sometimes, you may need to modify or edit your hatches after they have been created. This section will cover the commands and tools you can use to edit hatch lines, including changing the hatch pattern, scale, or angle, adjusting the hatch boundaries, or even removing the hatch entirely.

A hatch has a set of properties which can be viewed and edited. Some parameters can be changed in the Properties panel, some others – only in the dialog box Boundary hatch and Gradient, which is called with the help of a double mouse click over the created hatch. The contour of the hatch can not be changed. It can be moved, copied etc.

Hatch properties

Color	Defines the colour of the lines of the hatch and solid fill.
Layer	This property assigns the layer to which the hatch belongs. It comprises a list of all layers available in the drawing, offering the option to relocate the hatch to a different layer.
Line Weight	This property sets the line weight of the hatch, selectable from a range of standard values. You have the option to enable or disable the feature that displays lines according to their respective line weights. Each line weight value corresponds to a specific number of pixels on the screen, a number that remains constant regardless of scaling within the program window. The line weight display scale on the screen is configurable within the Visualization tab of the Options window.
Name	The name of the sample of the used hatch. To change it call the dialog box Boundary hatch and Gradient.
Angle	The angle of rotation of the hatch in the XY plane.
Scale	Sets the rate of hatch lines. The higher the scale is the lower the frequency of the hatch lines is.

Gradient properties

Color	Defines the colour of the lines of the hatch and solid fill.
Layer	Defines the layer to which the gradient belongs. The property contains a list of all available in the drawing layers and allows to move the gradient to another layer.

Name	The name of the used hatch. For gradient it is always GRADIENT.
Angle	The angle of rotation of the gradient in the XY plane.
Color 1	The value of the colour 1 of the gradient.
Color 2	The value of the colour 2 of the gradient.

The order of the hatch can be changed with the help of the commands of the dialog box **Bring to Front** and **Send to Back**.

4.3 BLOCK MAKING

Blocks in AutoCAD are a powerful way to manage and reuse complex objects. They enable the consolidation of a set of entities into a single named object, which can then be inserted multiple times in a drawing without recreating it each time. This not only saves time but also maintains consistency across the design.

4.3.1 CREATING BLOCK DEFINITIONS

Creating a block begins with defining it. Block definitions include the objects that make up the block, as well as a base point and a name for the block. Once created, the block definition is added to the block library and can be used in any drawing.

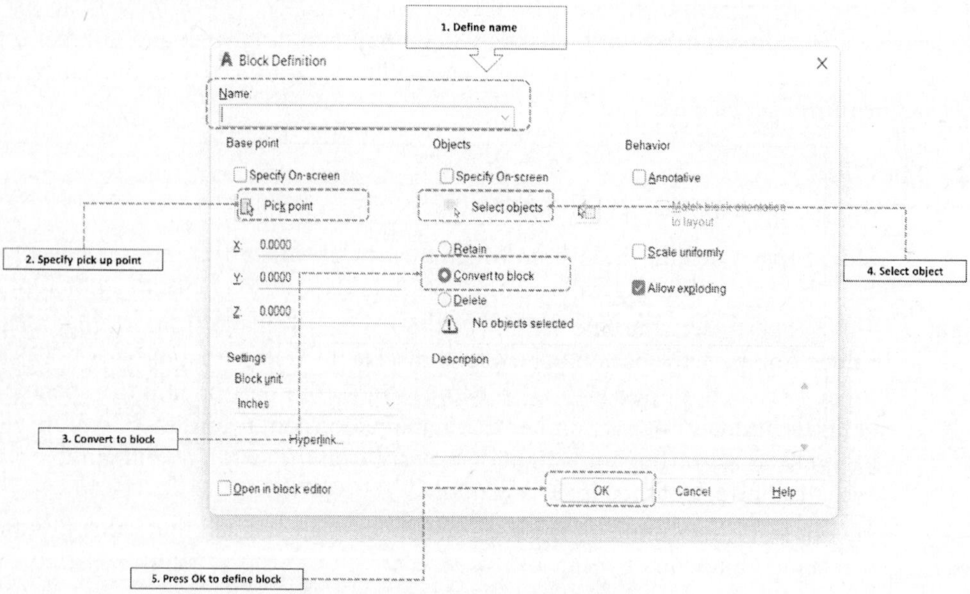

Fig. 4.15: Block definition

To create a block, follow these steps:

Step 1: Initiate your process by sketching the components you desire to include within the block.

Step 2: Navigate to the Home tab, select the Block panel and click Create to commence the BLOCK command. This action will open the Block Definition dialog box.

Step 3: In the Name text box, input a unique identifier for your block. Feel free to use spaces in the name.

Step 4: The next step involves designating a base point for your block. This serves as the insertion point when employing the block. In the Base Point section, select the Pick Point option, using an object snap for precision. Upon completion, you will be redirected back to the dialog box.

Step 5: Within the Objects section, activate the Select Objects button, choose the required objects and press Enter to revisit the dialog box.

Step 6: Adjacent to this, you have the option to either Retain, Convert to Block, or Delete. These selections dictate the outcome post-block creation.

Step 7: In the Behavior section, you are given the flexibility to make the block Annotative, mandate uniform scaling and decide on the exploding option.

Step 8: Progressing to the Settings area, you are required to assign the block unit. Opting for Unitless is viable, but selecting a unit will prompt AutoCAD to automatically scale the block when it is introduced into a different drawing. An additional feature is the ability to append a hyperlink if desired.

Step 9: For improved organization, a description can be appended in the Description box. This is beneficial when you aim to import the block from another drawing through the Design Center.

Step 10: To finalize the process, click OK. If you previously chose the Delete option, the objects will vanish. However, they can be retrieved using the OOPS command.

4.3.2 Inserting Blocks

Inserting blocks into a drawing is a straightforward process. From the Block Library, you can choose the desired block and insert it at a specified location. Additional attributes like rotation, scale and angle can also be specified while inserting a block.

To insert a block, follow these steps:

Step 1: Make an object i.e., Block

Step 2: Go to insert > create block > block definition > name > select the object

Block will be created and can be utilised for future drawings wherever it will be required.

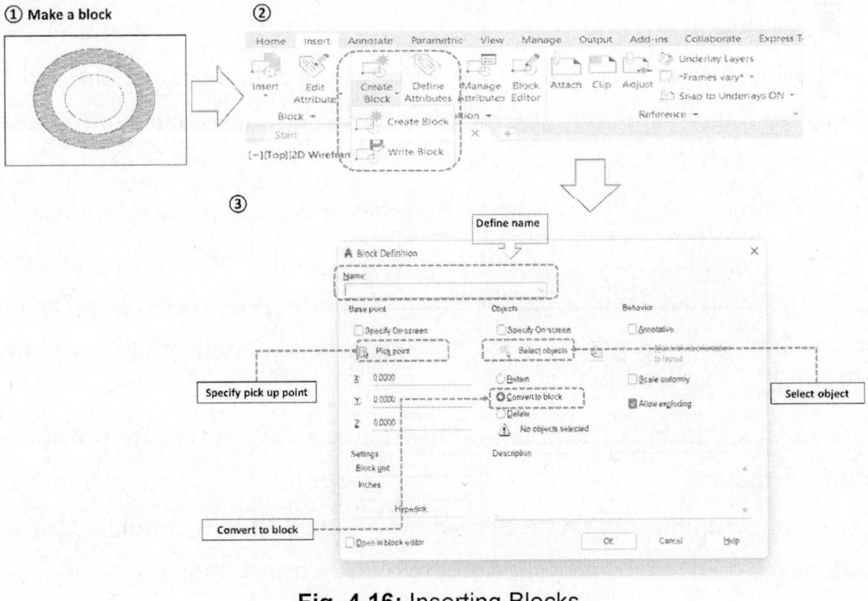

Fig. 4.16: Inserting Blocks

4.3.3 Attributes: Fill-in-the-Blank Blocks

Fill-in-the-blank blocks, often called attribute blocks, are a type of block with variable content. The process to create these blocks is straightforward and can greatly enhance your AutoCAD projects. The following is a step-by-step guide to creating fill-in-the-blank blocks using the Attribute Definition tool (ATTDEF command):

Step 1: Selecting the Layer— Firstly, you need to switch to the layer where you plan to create the attribute definition.

Step 2: Accessing the Attribute Definition Tool— On the Home tab, locate the Block panel slideout and select Define Attributes. This will open the *Attribute Definition* dialog box. The first four mode settings (*Invisible*, *Constant*, *Verify* or *Preset*) are usually left deselected for most use cases. If you need more information on these modes, hover over the items or click on the dialog box Help button.

Step 3: Locking the Position— The Lock Position checkbox decides whether the attributes within the block reference can be relocated. If selected, the block is treated as a single entity. Deselecting it allows for individual attributes to be moved by dragging their grips.

Step 4: Enabling Multiple Lines— If you wish to enter multiple lines of text, check the Multiple Lines box in the Mode area. This will disable the Default text box and offer a button to open the Multiline Editor, which provides several formatting options. If the system variable ATTIPE is set to 1, all formatting options in the In-Place Text Editor will be enabled.

Step 5: Filling out Attribute Details— Under the Attribute area, provide details for the tag (the unique identifier), the user prompt and the default value. The Tag text box should not contain any spaces, while the Prompt and Default text boxes can.

Step 6: Working with Multiple Lines— If the Multiple Lines checkbox was selected in Step 4, you can click the Multiline Editor button to enter multiline default attribute values and apply formatting, if needed.

Step 7: Text Settings— Specify the Justification, Text Style, Annotative property, Text Height, Rotation and Boundary Width (only for multiline attributes) in the Text Settings area.

Step 8: Specifying Insertion Point— Use the Specify On-Screen option to choose an insertion point for the attribute definition. Remember to use precision tools such as snap or object snap for precise positioning of attribute values.

Step 9: Creating the Attribute Definition— Once all necessary details are provided, click OK to create the attribute definition.

Step 10: Creating Additional Attribute Definitions— Repeat Steps 1 through 9 for any additional attribute definitions you need to create. To align multiple attribute definitions neatly, use the Align Below Previous Attribute Definition checkbox.

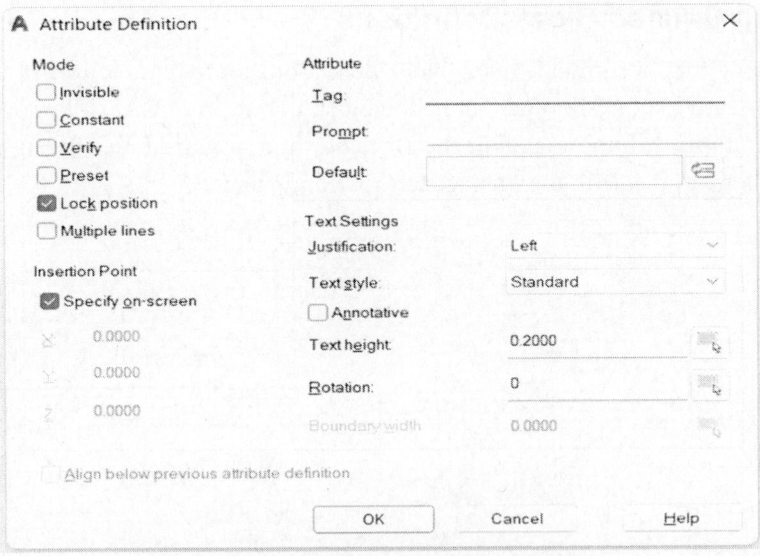

Fig. 4.17: The Attribute Definition dialog box

To generate nonadjacent attributes, duplicate the initial attribute definition and then utilize the Properties palette to modify the copied version. You can ensure that attributes remain fixed within the block by selecting the Lock Position option within the Attribute Definition dialog box.

4.3.4 Exploding Blocks

While blocks are incredibly useful, there may be instances where you need to break them back into their constituent entities, for example, to make edits to a specific part of the block. This can be done using the EXPLODE command, which disassembles the block back into individual objects.

To explode a block, follow these steps:

Step 1: Insert a block from recent block or Select the Block you want to explode

Step 2: Click on explode command and press enter

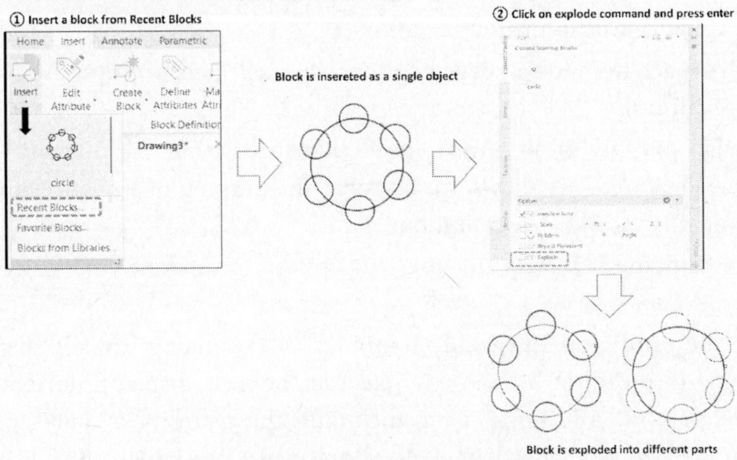

Fig. 4.18: Exploding Blocks

4.3.5 Purging Unused Block Definitions

Over time, a drawing may accumulate many block definitions that are no longer in use. These unused definitions can increase the file size and clutter the block library. AutoCAD allows you to purge these unused block definitions, helping maintain the efficiency and organization of your drawing files. The various Purge tools are available in the Manage tab > Cleanup panel.

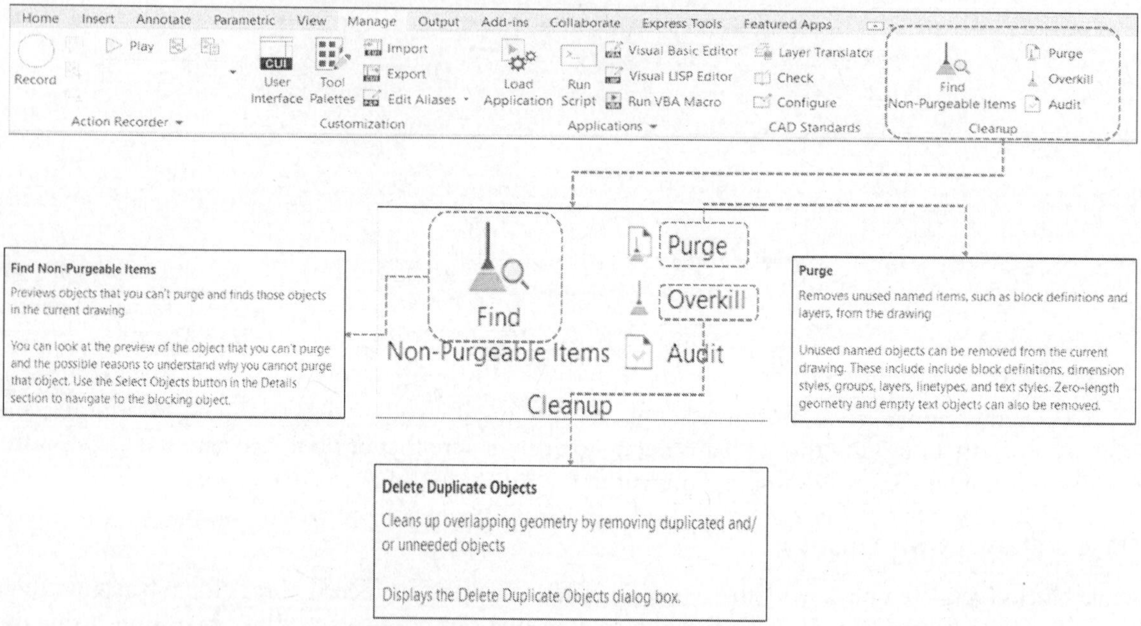

Fig. 4.19: Purge tools in Cleanup panel

The Purge tool initiates the Purge dialog box, enabling you to select the category of the item to be purged, such as Blocks or Layers, amongst others. Each category list can be expanded to select individual items for purging. The dialog box's Preview area exhibits an image of the prospective item for purging. Several options are available within the Purge tool—

- The **Confirm each item to be purged** option, when selected, prompts user verification for each item before purging.
- For thorough purging of all unreferenced elements within the drawing, opt for the **Purge nested items** feature. This allows, for instance, the purging of any unreferenced layers nested within an unreferenced block definition.
- The **Purge Unnamed Objects** area presents options to separately purge *Zero-length geometry* and *Empty text objects*.
- Choosing the **Find Non-Purgeable Items** tab in the dialog box reveals a list of currently in-use items that cannot be purged. Select an item to display information explaining its unpurgeable status. Additional data, including the quantity of items per layer and their impact on file size, is also provided. To zoom into a particular non-purgeable object, click the *Select Objects* button.

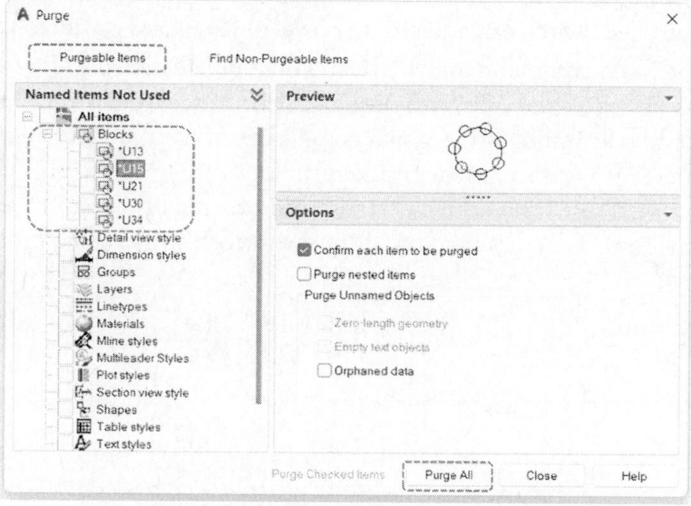

Fig. 4.20: Purgeable Items option in Purge tools

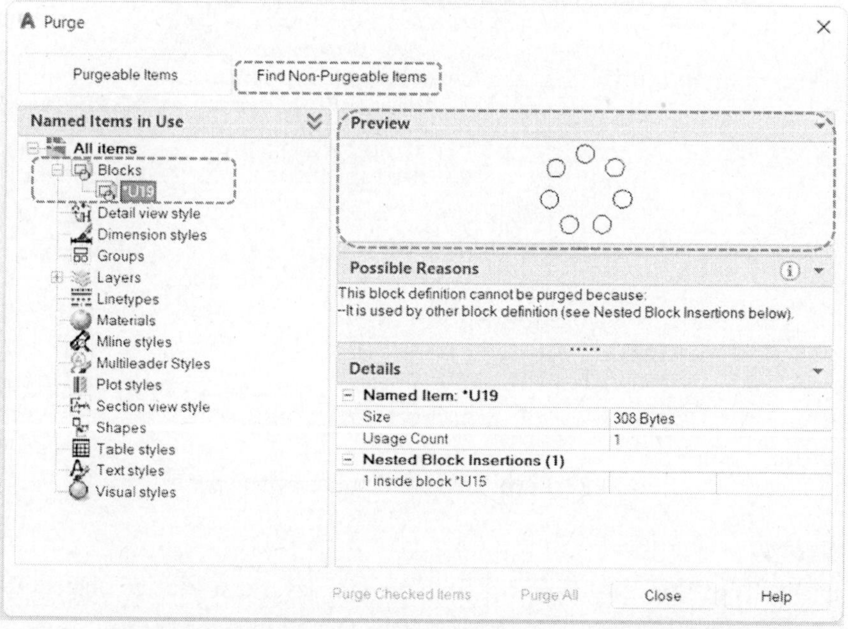

Fig. 4.21: Find Non-Purgeable Items option in Purge tools

4.4 LAYERS COMMAND

Layers are integral to managing complex AutoCAD drawings. They work as transparent overlays, allowing you to organize your drawing into logical sections, control the visibility of certain parts and efficiently edit your work.

Working with layers involves creating new layers, setting a current layer, freezing or hiding layers and changing the properties of layers. Understanding and mastering these commands can greatly increase your efficiency in managing complex drawings. Layer properties like colour, line type and line weight can also be controlled, which can add another level of detail and clarity to your drawings.

AutoCAD uses the Layer command to manage and manipulate drawings for various purposes. This command significantly enhances the software's display performance by allowing parts of the drawing to be hidden when necessary. It simplifies the visual complexity of the drawing by creating a set of layers with different properties. For instance, when designing a floor plan or house plan, we can create distinct layers for doors, walls, etc. These layers can be created and ordered as required, each assigned a unique name for identification. The Layer command can be quickly accessed by typing **LA** in the command line.

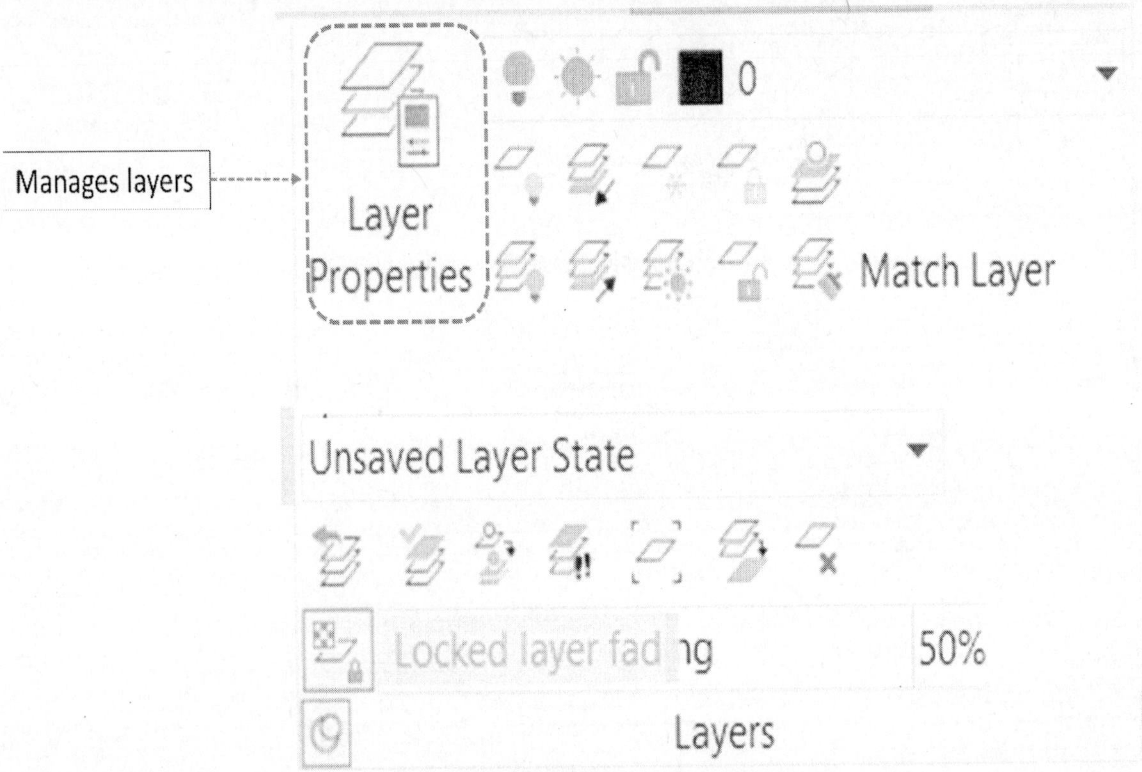

Fig. 4.22: Layer command on the ribbon pane

The various layer properties are—

- **Layer ON/OFF**: This command is used to turn the layer of a selected object ON or OFF. It can be accessed using the shortcut commands LayON and LayOFF.
- **Layer Isolate/Unisolate**: The Isolate command is used to hide or lock layers. Only the layers isolated are visible and remain locked except for selected objects. The Unisolate command, on the other hand, unlocks the layers.
- **Layer Freeze/Thaw**: Freeze is used to immobilize the layer of a selected object, whereas Thaw is used to remove this freeze. Objects become invisible after applying the Layer Freeze command.
- **Layer LOCK/UNLOCK**: This command is used to either lock or unlock the layer of a selected object. The Layer Lock command specifically prevents the object from being modified.

- **Current Layer**: This command sets the selected layer as the active layer on the viewport. All new objects created will be based on this current layer.
- **Match Layer**: The Match Layer command is used to change the current layer to match the target layer.

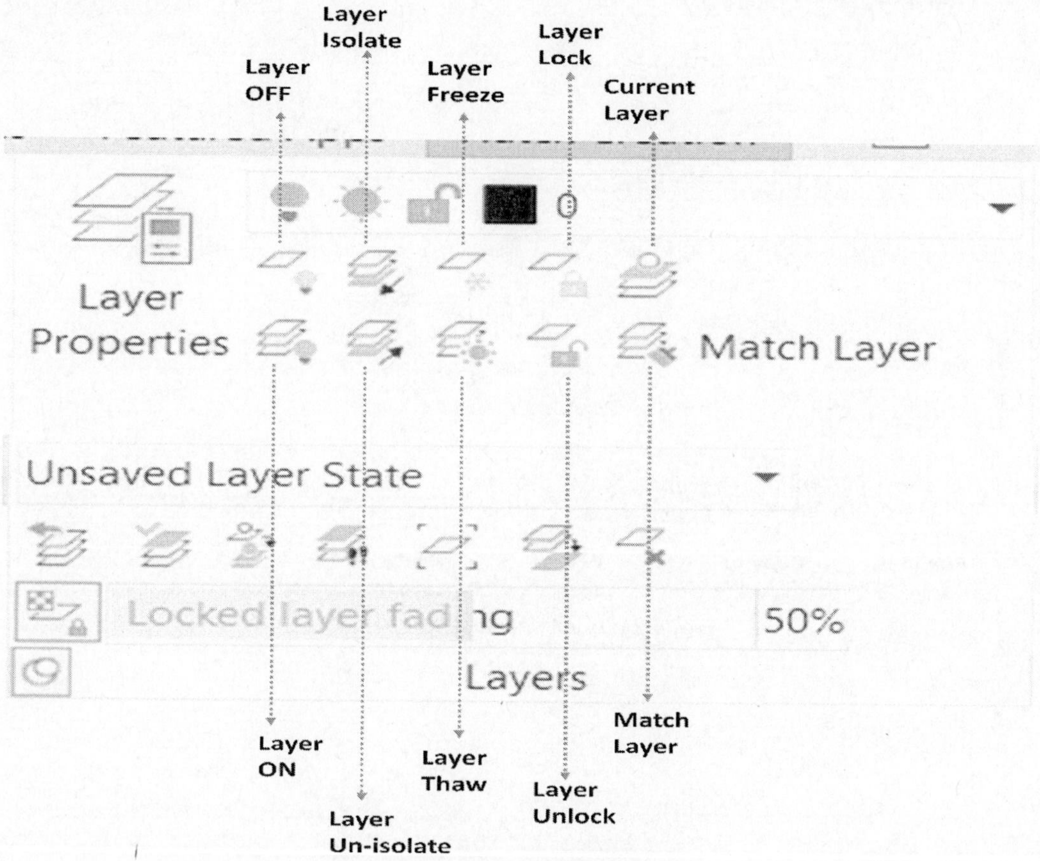

Fig. 4.23: Various layer properties

Some additional layer-related commands are —

- **Previous Layer**: This command undoes the set of changes made to the layer settings.
- **Copy to Layer**: This command is used to copy one or more objects to another layer.
- **Walk Layer**: This command is used to display the objects of the selected layer while hiding objects on other layers.
- **VPI Layer:** This command freezes layers in all layout viewports, except the current one.
- **Merge Layer**: This command merges the selected layers into a target layer, removing previous layers from the drawing.
- **Delete Layer**: This command deletes the selected layer from the layer settings or the list of layers, also removing all objects on that layer.

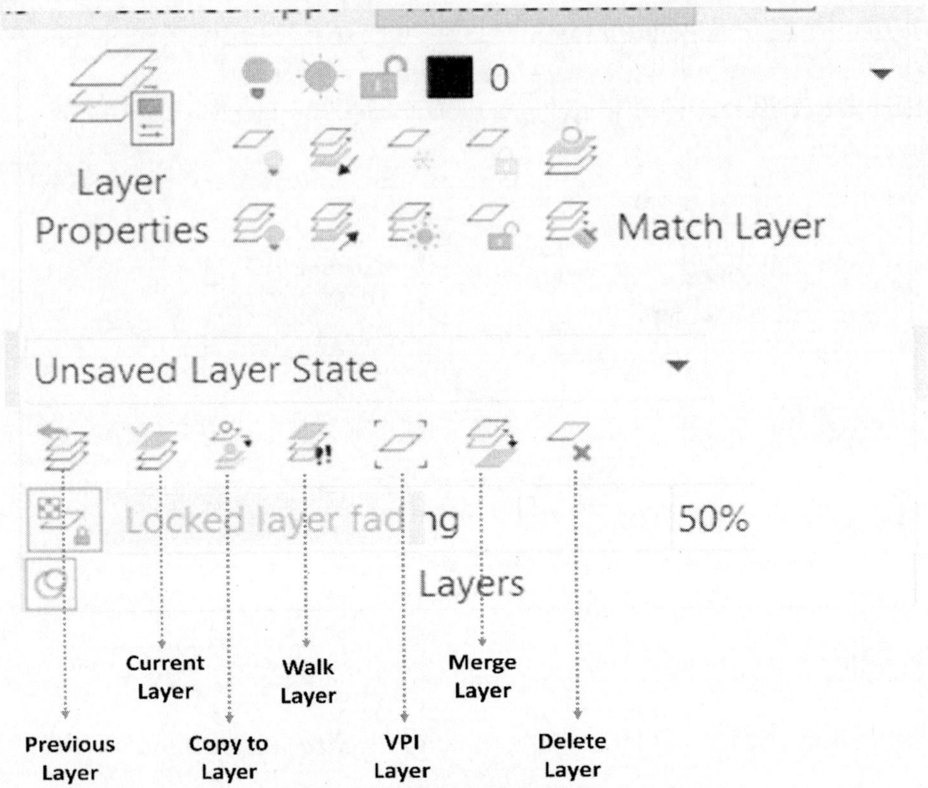

Fig. 4.24: Additional layer properties

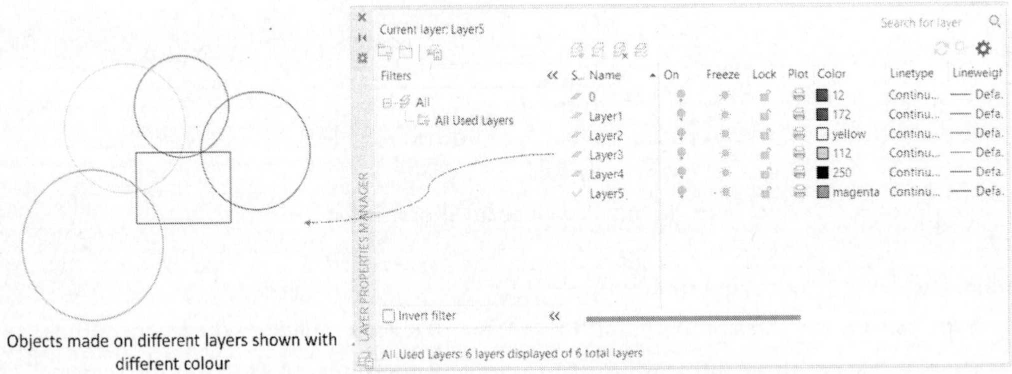

Fig. 4.25: Example of objects drawn on different layers

The steps to create several objects in different layers are given below —

Step 1: Type **LA** on the command line and press Enter. The Layer Properties Manager dialog box will appear.

Step 2: Click on the New Layer icon at the top of the dialog box. A new layer will appear where you can specify the name, color, Linetype, Transparency, Layer Freeze and other settings as per your requirements.

Step 3: Repeat these steps to create as many layers as you need.

Step 4: To create an object on a particular layer, set that layer as the current layer.

Step 5: Keep changing the current layer according to the object you are creating to effectively use all layers in the drawing.

UNIT SUMMARY

Based on the contents covered in this unit you should remember that:

- Hatching in AutoCAD involves filling specific areas of a drawing with patterns or solid fills, providing visual distinction between different parts.
- The Hatch tool is the primary means of creating hatches, allowing selection of patterns or solid fills and adjustment of their properties.
- Properties of hatch lines, such as pattern type, scale, colour and layer, play a critical role in effectively conveying information.
- Island detection tools control how hatches interact with other objects within an enclosed area, allowing for a more nuanced representation.
- Hatches can interact with text, which can be used to ensure text remains clear when overlapping with a hatch or to highlight specific sections.
- Block making involves creating, inserting and modifying defined objects, facilitating efficient management of complex patterns or objects.
- Blocks are created through block definitions, which include a set of entities, a base point and a name.
- Fill-in-the-blank blocks, or blocks with attribute definitions, allow for varying information within the same block across different instances.
- Exploding blocks disassembles a block back into individual objects, useful for making edits to specific parts of a block.
- Purging unused block definitions helps maintain an organized block library and efficient file sizes.
- Layers in AutoCAD are like transparent overlays that help organize drawings into logical sections and control the visibility of certain parts.
- Working with layers involves creating new layers, setting a current layer, freezing or hiding layers and changing layer properties.

EXERCISES

A. Multiple Choice Questions

4.1 The Hatch and Gradient dialogue can be accessed by

(a) selecting the Hatch icon in the DRAW toolbar (b) typing 'h' in the command line

(c) typing 'hatch' in the command line (d) any of the above methods

4.2 The term 'associative hatch' refers to

(a) hatch patterns that can be connected
(b) hatching linked to the color of the hatched area
(c) hatching connected to objects that are moved within a hatched area
(d) hatching associated with a hatched area from another section of the drawing

4.3 When hatching a sectional view in an engineering drawing

(a) every detail intersected by a section plane should be included
(b) the parts to be hatched are at the discretion of the draftsperson
(c) there is no requirement to hatch any part of the drawing
(d) a bolt within a sectional view is typically not hatched

4.4 When text is placed in a hatched area

(a) a non-hatch zone surrounds the text only if Advanced Normal is enabled
(b) a non-hatch zone surrounds the text only if Advanced Normal is disabled
(c) a non-hatch zone surrounds the text only if Advanced Ignore is enabled
(d) a non-hatch zone surrounds the text if Advanced settings are activated

4.5 When using the Section tool on a 3D solid model drawing

(a) the sectional view is automatically hatched
(b) the section outline should be removed from the 3D solid model drawing
(c) only an outline of the sectional view is created
(d) the sectional view appears as a distinct view in the drawing area

4.6 Which of the following statement is correct?

(a) the Purge tool is only functional when inserting blocks
(b) the Purge tool can be used to remove unnecessary details from any drawing
(c) the Purge tool is solely for deleting exploded blocks from drawings
(d) the Purge tool can only be invoked by typing 'purge' at the command line

4.7 Dependent views allow you to

(a) observe the same object from different perspectives
(b) alter all your views simultaneously
(c) add various views of the same model
(d) partition the same view onto separate sheets

4.8 If a layer is locked

(a) elements can be added to a locked layer
(b) elements can be erased from a locked layer
(c) elements cannot be added or erased from a locked layer
(d) elements can be added to a locked layer, but they vanish when the drawing file is saved

4.9 When a layer is turned off

(a) file space is conserved upon saving the file
(b) it does not affect the layer significantly, elements can still be added
(c) elements cannot be erased from the layer
(d) elements on the layer are invisible

4.10 The Design Center palette can be launched from the:

(a) layer toolbar (b) properties toolbar

(c) modify toolbar (d) standard toolbar

Answers to Multiple-Choice Questions
4.1 (d), 4.2 (c), 4.3 (d), 4.4 (a), 4.5 (c), 4.6 (b), 4.7 (b), 4.8 (a), 4.9 (d), 4.10 (b)

B. Subjective Questions

4.1 Define the Hatch tool and provide a succinct explanation of its purpose and function.

4.2 Enumerate the characteristics of hatch lines.

4.3 Analyse the functionality of Island detection tools.

4.4 Illustrate the process of block making.

4.5 Interpret the mechanism of exploding blocks and rationalize its applications.

4.6 Outline the steps involved in block making.

4.7 Elucidate the concept of purging and demonstrate its application in block making.

4.8 Identify what layers are and expound on their uses.

KNOW MORE

- *Advanced Hatching*: AutoCAD also supports gradient hatching, which smoothly transitions between two colours. This can be particularly useful in certain design contexts, such as architectural drawings or 3D rendering. The HATCH command includes a variety of gradient patterns to choose from, offering designers a broad palette for visual expression.
- *Dynamic Blocks*: Beyond simple block creation and manipulation, AutoCAD offers a feature known as Dynamic Blocks. Dynamic Blocks are more versatile than regular blocks because they can be manipulated (like stretched, rotated, or mirrored) after insertion, without exploding them. This functionality enables designers to create flexible designs that can adapt to specific requirements while maintaining consistency and efficiency.
- *Layer States*: In addition to basic layer management, AutoCAD provides a feature called Layer States. Layer States allow you to save and restore configurations of layer properties and states. This can be incredibly helpful when working on complex drawings that require different layer configurations at different stages of the design process.
- *Annotation Scaling*: An aspect linked with Blocks and Layers is Annotation Scaling. This feature in AutoCAD allows annotations (like text, dimensions and blocks) to automatically adjust their size based on the scale of the viewport, ensuring that they appear at the appropriate size on different scales of the drawing.
- *Interoperability*: It is also important to understand the interoperability of AutoCAD with other design and drafting tools. Blocks, Hatches and Layers, while being intrinsic to AutoCAD, can also be exported to and imported from another CAD software. This facilitates seamless collaboration and transition across different software environments.

- *The Evolution of AutoCAD*: Lastly, it is worth appreciating the evolution of AutoCAD over the years. The features discussed in this unit have undergone significant improvements since their initial versions, thanks to Autodesk's commitment to continuous development. The software continues to evolve, offering increasingly sophisticated and user-friendly tools to aid the work of designers and engineers worldwide.

REFERENCES AND SUGGESTED READINGS

1. Omura, George. Mastering AutoCAD 2010 and AutoCAD LT 2010. John Wiley & Sons, 2011.
2. Yarwood, Alfred. Introduction to AutoCAD 2012. Routledge, 2011.
3. Onstott, Scott. AutoCAD 2017 and AutoCAD LT 2017: Essentials. John Wiley & Sons, 2016.

Dynamic QR Code for Further Reading

Hatching Commands

Hatch Line Edits

Hatching and Text

Island Detection Tools

Block Creation

Inserting Blocks

Exploding Blocks

Purging Unused Block

Layers Toolbar

5 Viewport and Isomeric Drawing in AutoCAD

UNIT SPECIFICS

This unit presents information related to the following topics:

- Viewports creation;
- Clipping existing viewports;
- Scaling and maximizing viewports;
- Freezing layers and layer override in viewports;
- Isometric drawing;

This unit guides readers through the intricate but essential features of AutoCAD, namely viewports and isometric drawings. Beginning with transforming existing entities into viewports and further altering them through clipping, this unit will acquaint you with the versatile use of viewports in handling complex designs. A focus on scaling and maximizing viewports follows, elucidating its role in maintaining accuracy during the transition between model space and paper space. Thereafter, freezing layers in viewports and overriding layer properties is dealt in detail, offering readers an effective means to control design element visibility across multiple viewports. Isometric drawings are introduced in this unit as a potent tool for representing three-dimensional designs in two dimensions, thereby broadening your understanding of spatial designs. Each topic in the unit is explained with detailed steps, making the implementation of these techniques in AutoCAD a straightforward task.

Apart from this at the end of the unit, a succinct recapitulation of the overall broad concepts is provided in form of a unit summary. Besides, a large number of multiple-choice questions as well as descriptive type questions with Bloom's taxonomy action verbs are included. A list of references and suggested readings are given in the unit so that one can go through them for practice. It is important to note that for getting more information on various topics of interest some QR codes have been provided in different sections which can be scanned for relevant supportive knowledge. Video resources along with QR codes are mentioned for getting more information on various topics of interest which can be surfed or scanned through mobile phones for viewing.

RATIONALE

Viewports and isometric drawing are essential tools in CAD software like AutoCAD, allowing users to present complex drawings in a more understandable and visually appealing manner. Viewports enable the display of different views, scales and layer settings of a drawing within a single layout, enhancing the readability and presentation quality. Isometric drawing allows the representation of three-dimensional objects in two dimensions, providing a clearer understanding of the object's geometry and proportions.

By mastering these techniques, students will be able to create more effective and visually appealing presentations of their drawings, improving their overall drafting skills and productivity. In this lesson, students will learn the fundamentals of viewports and isometric drawing and their applications. This unit includes detailed explanations of how to execute each command, as well as exercises to hone your skills.

PRE-REQUISITES

Before reading this unit, the students are advised to revisit the following:

- Mathematics: Coordinate and Plane geometry (Class XII)
- Engineering Graphics (ES101)
- AutoCAD commands mentioned in Unit-1

UNIT OUTCOMES

After studying this unit students will be able to:

U5-O1: Create and manage various types of viewports in AutoCAD to display different views, scales and layer settings within a single layout

U5-O2: Scale and maximize viewports in AutoCAD to improve drawing presentation and readability

U5-O3: Freeze layers in viewports to better organize and manage drawings

U5-O4: Apply layer override in viewports to customize the appearance of specific viewport elements

U5-O5: Create isometric drawings in AutoCAD to effectively represent three-dimensional objects in two dimensions

Unit-5 Outcomes	EXPECTED MAPPING WITH COURSE OUTCOMES (1 – Weak Correlation; 2 – Medium correlation; 3 – Strong Correlation)				
	CO-1	CO-2	CO-3	CO-4	CO-5
U5-O1	-	1	1	1	1
U5-O2	-	1	1	1	-
U5-O3	-	1	1	1	-
U5-O4	1	-	1	1	1
U5-O5	1	-	3	3	1

5.1 INTRODUCTION

The concepts of viewports and isometric drawing are integral elements in the domain of computer-aided design (CAD) practice. They fundamentally alter the way engineers, architects and designers visualize and interact with their creations. The ability to generate and manipulate designs digitally has significantly expanded the bounds of what is achievable in the realm of design and drafting and this unit intends to delve deeper into two essential aspects that form the foundation of this potential.

A viewport in AutoCAD is, essentially, a window into the model space, a uniquely defined viewing area that allows the user to work on different sections of a design simultaneously. Viewports serve as a portal to manage complex designs effectively. They permit various perspectives, angles and scales of the same design to be visible at the same time on a single layout, enhancing precision and productivity. Whether creating single or multiple rectangular viewports, forming polygonal viewports, or manipulating existing ones, this functionality is a cornerstone of efficient design practice.

On the other hand, isometric drawing, a method that depicts three-dimensional objects in two dimensions, plays a crucial role in accurately conveying the depth and spatial relationships within a design. Using AutoCAD to create isometric drawings offers designers a method to envision, plan and represent complex, multi-dimensional designs on a two-dimensional plane. This practice enables a clearer understanding of the design from different viewpoints, thus making it easier to construct, assemble, or interpret.

5.2 VIEWPORTS CREATION

AutoCAD consists of two distinct workspaces: Model space and Paper space.

Model space is where the creation process takes place. Here, you construct both 2D drawings and 3D models. Although plotting drawings directly from the model space is possible, it becomes complicated when dealing with drawings that require specific scales or consist of multiple views arranged at different scales.

To accommodate these complex plotting requirements, we use Paper space, also known as Layouts. This space allows you to work on additional details such as notes and annotations and conduct the plotting or publishing operations more effectively.

Within these Layouts, Viewports come into play. These are specialized windows that enable you to organize and display one or multiple views of a drawing, or even multiple drawings at specific scales. While viewports are typically rectangular, AutoCAD also allows for the creation of circular and polygonal viewports.

Throughout this unit, we will delve deeper into the concept and application of viewports, along with various annotative objects, providing you with a comprehensive understanding of their role in computer-aided design.

5.2.1 Layout creation

Layouts in AutoCAD symbolize conventional drawing sheets. They are created for the purpose of plotting drawings, whether in paper or electronic form. A single drawing can possess multiple layouts, allowing you to print in various sheet formats. AutoCAD comes with two default layouts: Layout 1 and Layout 2.

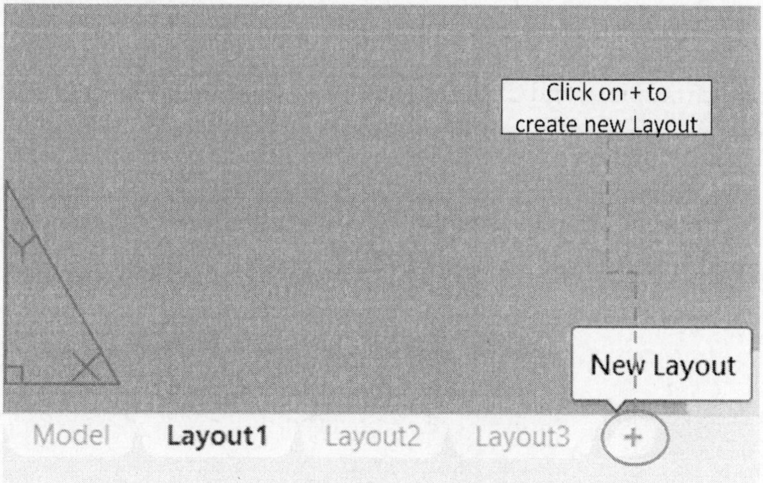

Fig. 5.1: Creating a new layout

The steps for creating and modifying a layout are as follows:

Step 1: Begin by clicking on the "Layout 1" tab, located at the bottom of the drawing window (refer to Fig. 5.2). Upon doing this, you will observe a white paper, accompanied by an automatically created viewport. The various components of this layout are shown in Fig. 5.3.

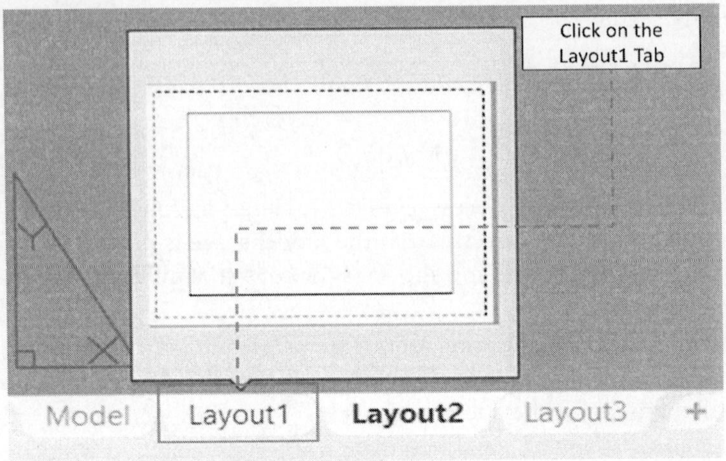

Fig. 5.2: Opening the created layout

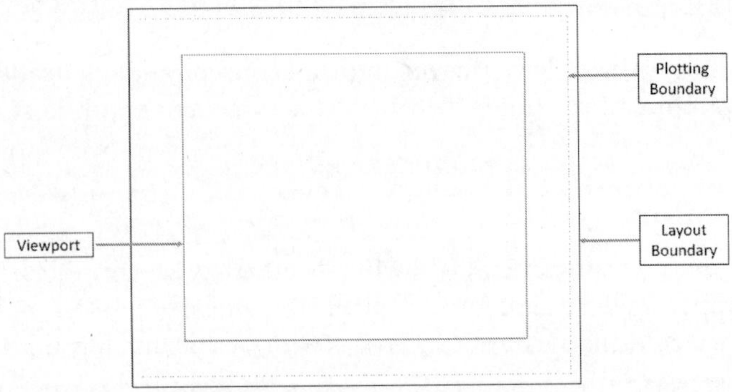

Fig. 5.3: Components of a layout

Step 2: Navigate to the ribbon and Click Output > Plot > Page Setup Manager (refer to Fig. 5.4). This will prompt the Page Setup Manager dialog box to appear on your screen.

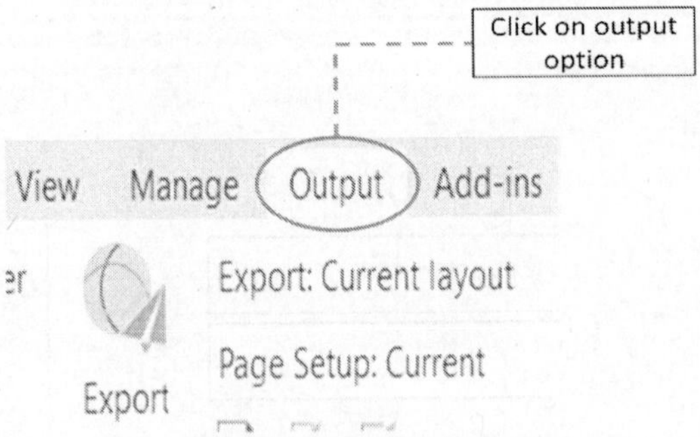

Fig. 5.4: Output option

Step 3: Inside the Page Setup Manager dialog box, click the "Modify" button (refer to Fig. 5.5). This will bring up the Page Setup – Layout1 dialog box (see Fig. 5.6).

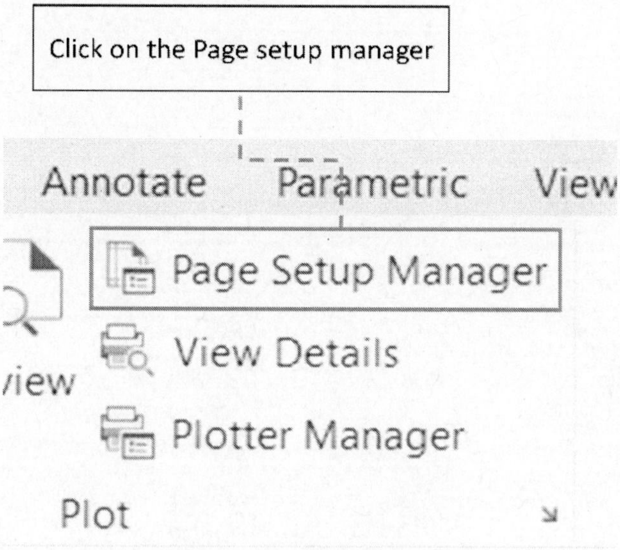

Fig. 5.5: Page setup manager

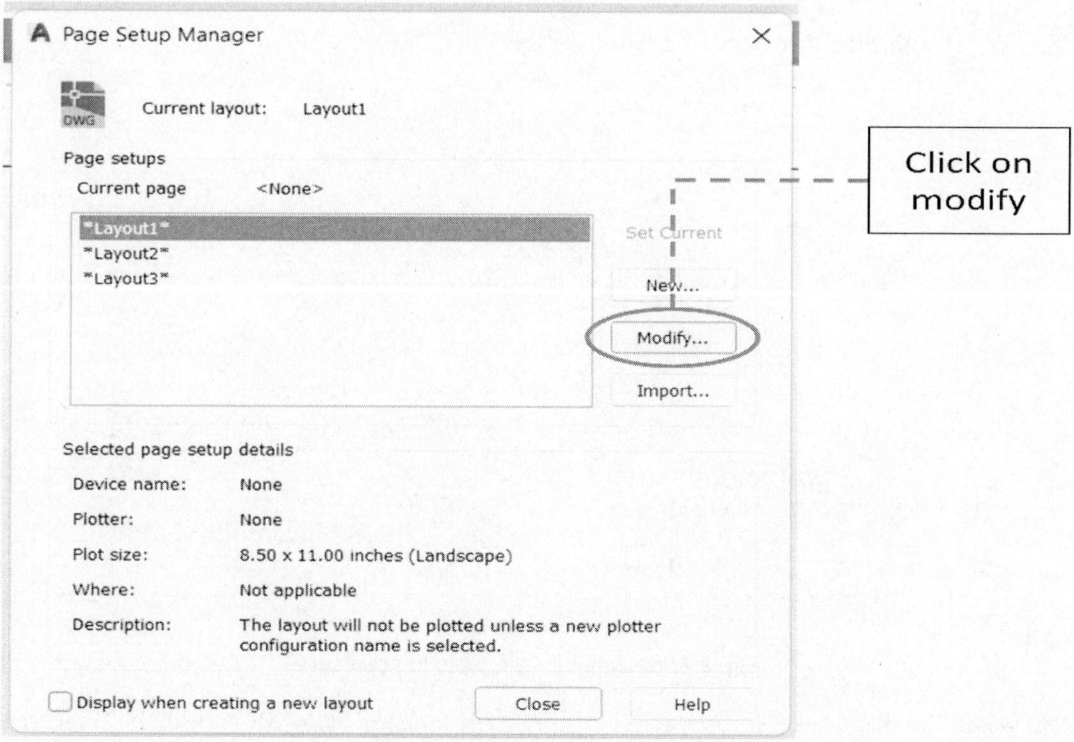

Fig. 5.6: Modifying the layout

Step 4: In the Page Setup dialog box, find the Printer/Plotter group and select "DWG to PDF.pc3" from the "Name" drop-down menu (refer to Fig. 5.7).

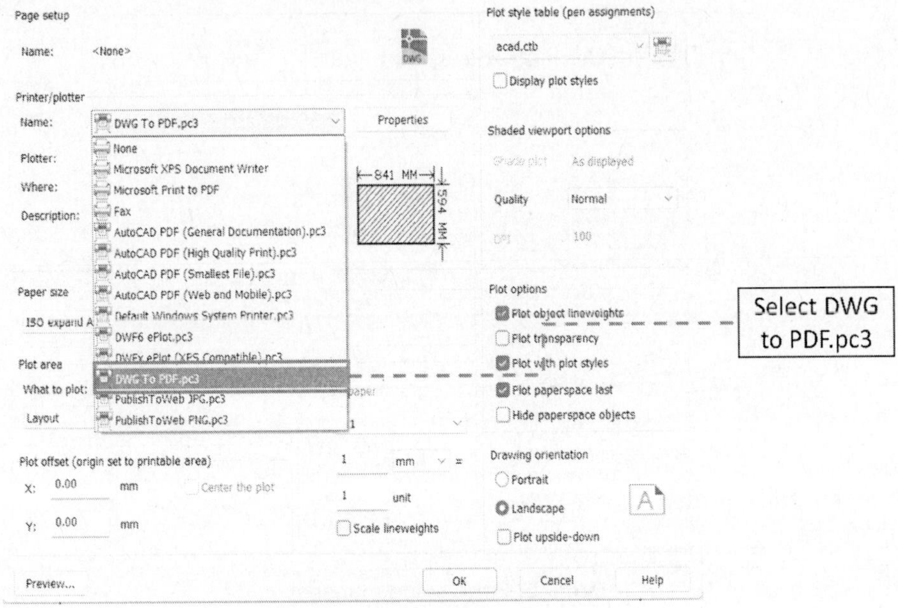

Fig. 5.7: Select DWG to PDF.pc3

Step 5: Set your Plot Style table to "acad.stb" (refer to Fig. 5.8).

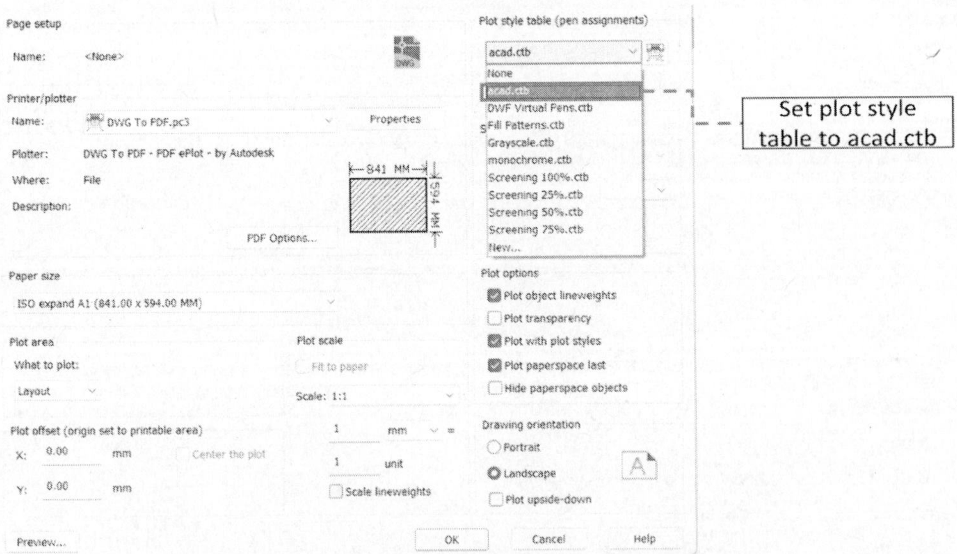

Fig. 5.8: Select plot style table to acad.std

Step 6: In the "Paper size" section, select "ISO A1 (841.00 x 594.00 MM)" and set the "Plot scale" to 1:1 (refer to Fig. 5.9 and Fig. 5.10).

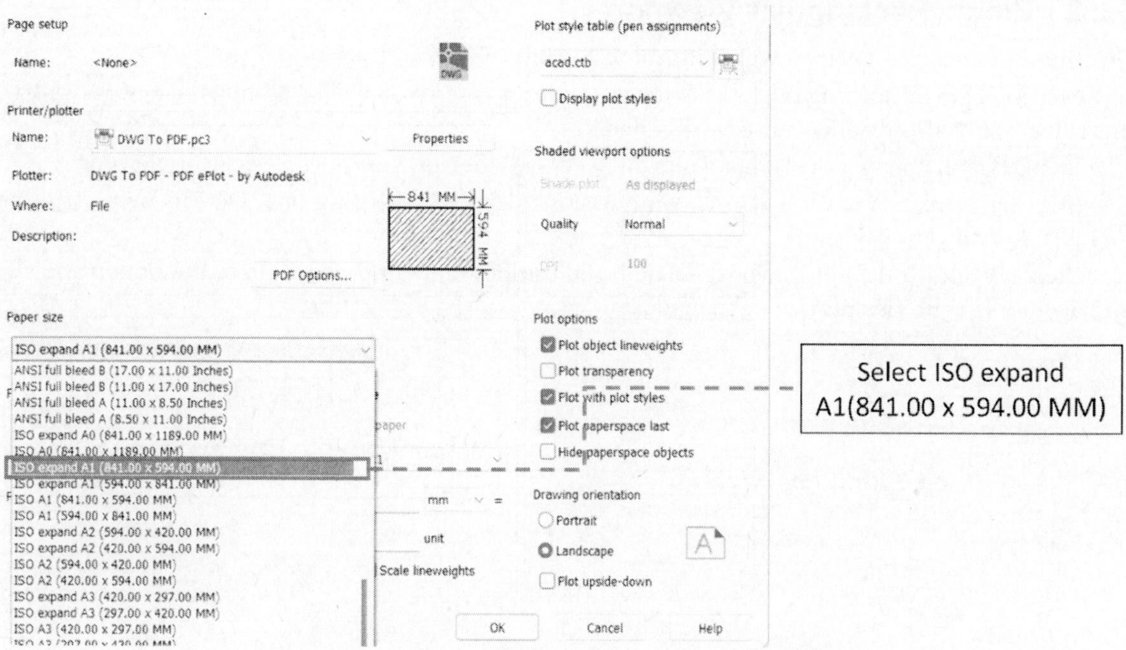

Fig. 5.9: Select ISO expand A1(841.00 x 594.00 MM)

Step 7: To apply these settings, click "OK" and then click "Close" on the Page Setup Manager dialog box.

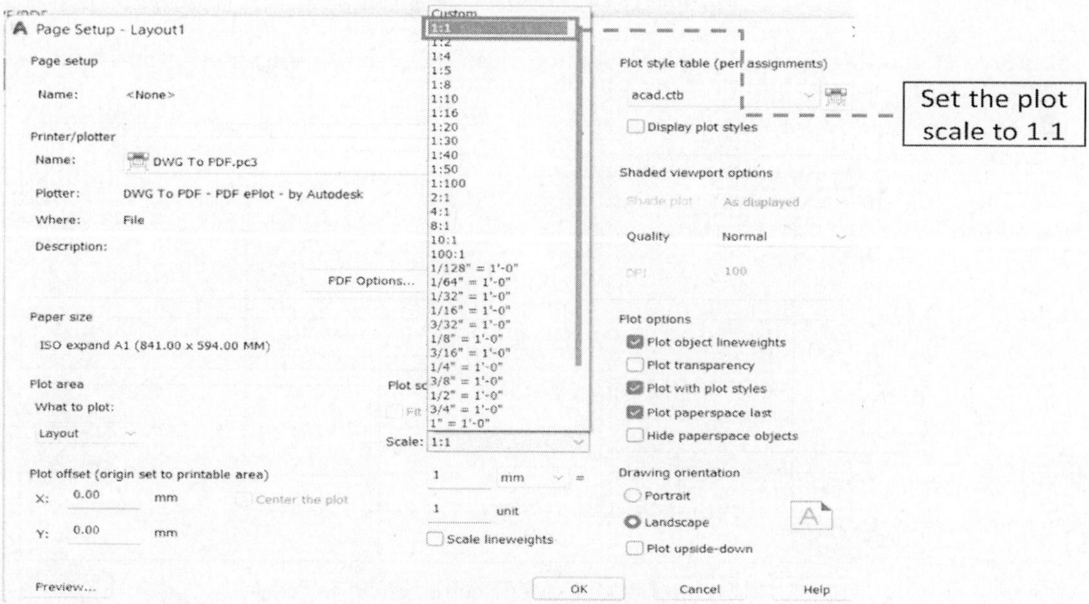

Fig. 5.10: Select plot scale to 1.1

These steps will help you create and modify layouts efficiently, enabling you to present your drawings in a variety of formats.

5.2.2 Single Rectangular Viewports

Creating a single rectangular viewport in AutoCAD involves several steps:

Step 1: Begin by opening the ISO A4 layout, which is where you will be adding the new viewport. If this layout is not already open, ensure you open it.

Step 2: Upon opening the layout, you will likely see a default viewport. Select this viewport.

Step 3: To create a new viewport, you must first delete the existing one. Do this by pressing the DELETE key on your keyboard.

Step 4: With the default viewport deleted, you can now add a new one. Go to the ribbon and click on Layout > Layout Viewports > Rectangular.

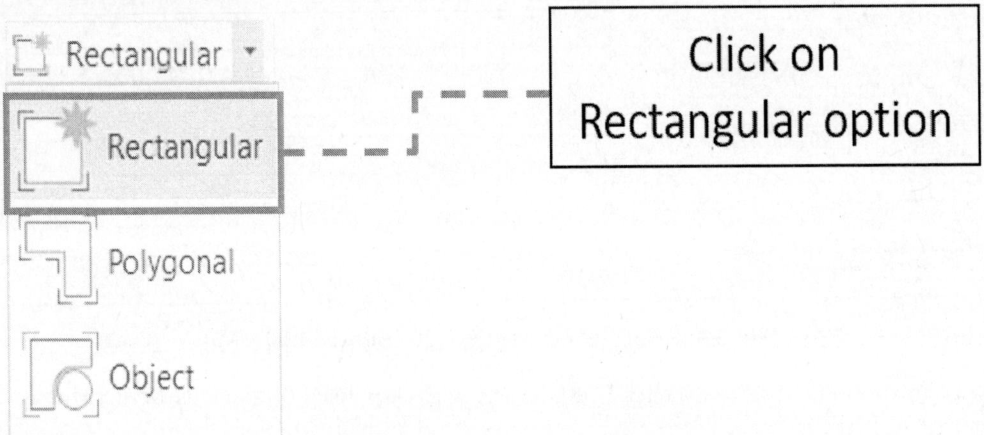

Fig. 5.11: Select rectangular option

Step 5: After selecting the 'Rectangular' option, you can create the rectangular viewport by choosing the first and second corner points.

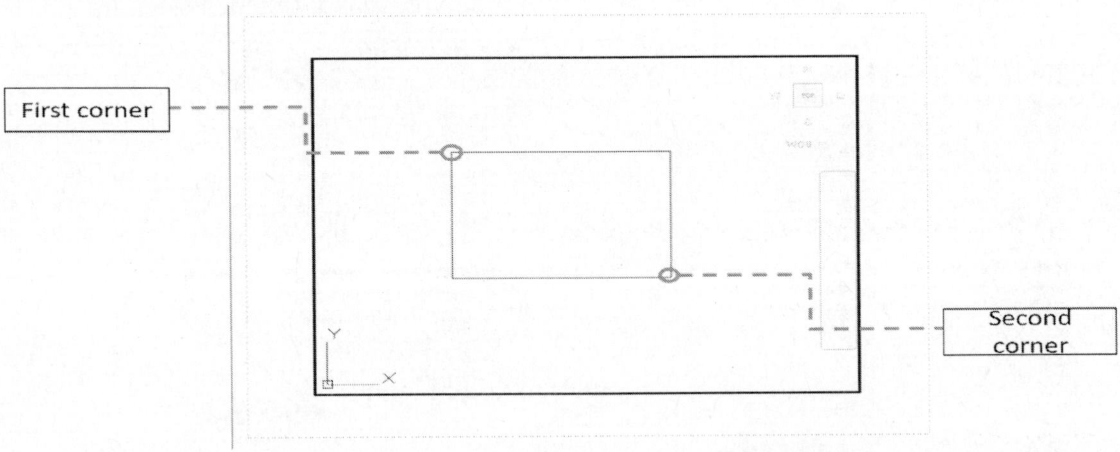

Fig. 5.12: Create first and second corner of the rectangle

Step 6: Once you have created the viewport, you will want to activate the model space inside it. Click the PAPER button on the status bar to do this. Note that the viewport frame will appear thicker when you are in the model space.

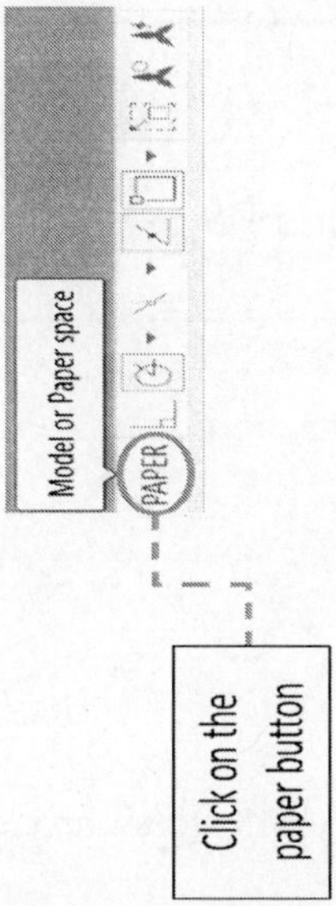

Fig. 5.13: Activate the model space inside the viewport

Step 7: With the model space activated, you can set the scale of your drawing. Click the Viewport Scale button and select 1:2 from the menu. This will zoom out your drawing.

Step 8: After adjusting the scale, position the drawing in the center of the viewport using the Pan tool.

Step 9: Once you are satisfied with the position and scale of your drawing inside the viewport, you can lock it to prevent any unintentional changes. Do this by clicking the Lock/Unlock Viewport button on the status bar.

Step 10: Lastly, to switch back to paper space, click the MODEL button on the status bar.

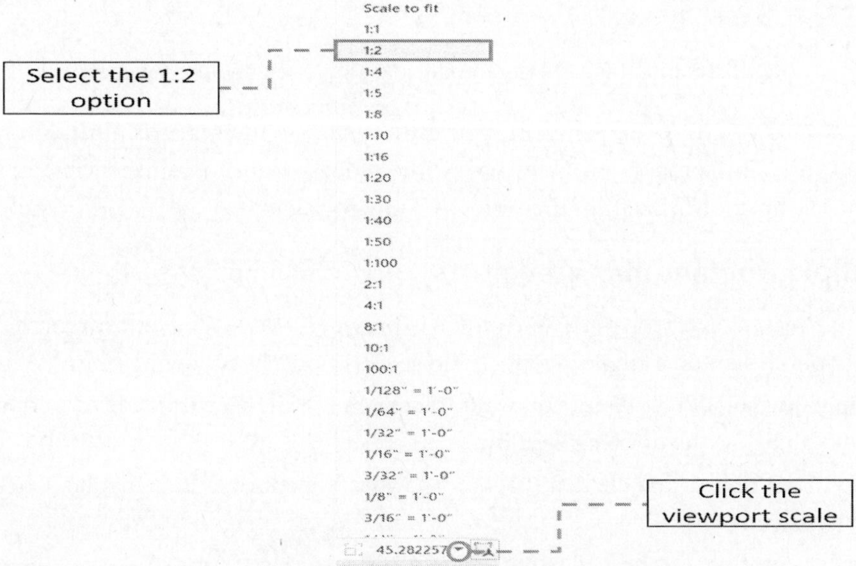

Fig. 5.14: Set the viewport scale

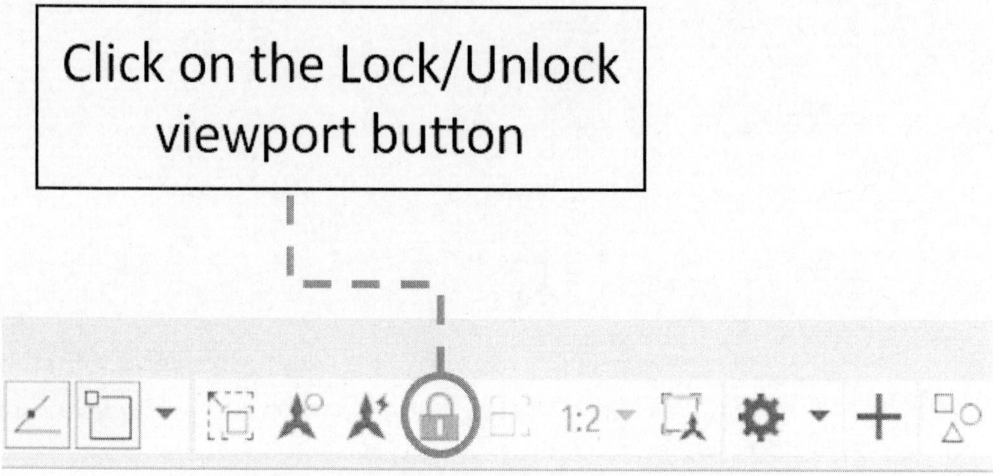

Fig. 5.15: Lock/unlock the drawing inside the viewport

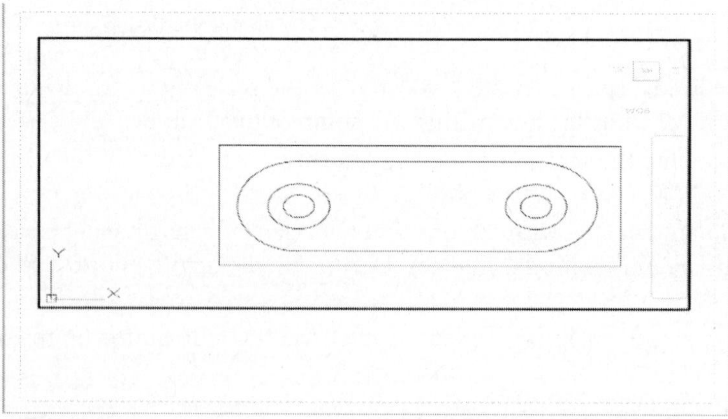

Fig. 5.16: Insert the created model into single rectangular viewport

Remember, once you lock the viewport, you cannot change the scale or position of the drawing within the viewport. This process provides you with a focused and organized view of your design, allowing for more efficient work within the AutoCAD environment.

5.2.3 Multiple Rectangular Viewports

Creating multiple rectangular viewports in AutoCAD provides a versatile platform for viewing various perspectives of your design on a single layout. To do so, carry out the following steps:

Step 1: Open the layout tab where you want to create multiple viewports. If it is not already open, click the ISO A1 tab below the drawing window.

Step 2: Go to the ribbon and click on Layout > Layout Viewports > Rectangular. Click and drag to create your first viewport.

Step 3: After creating the first viewport, select the viewport frame. You can then modify this viewport by using the grip.

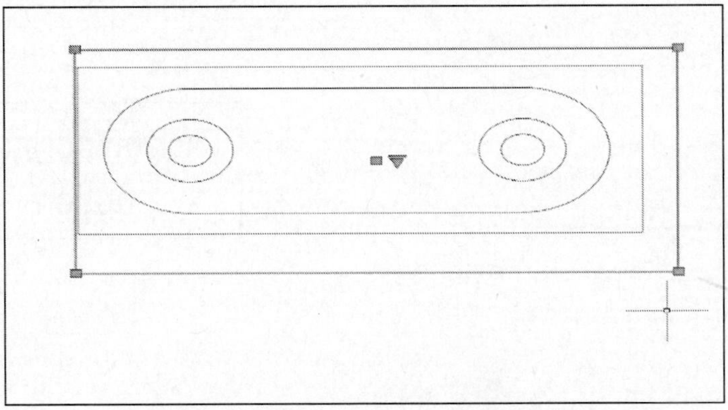

Fig. 5.17: Modify the viewport using grip

Step 4: With the first viewport selected and modified, double-click inside it to switch to the model space.

Step 5: Use the Zoom and Pan tools to adjust the view and position of your drawing inside this viewport. This will ensure the drawing is centralized in the viewport, giving you a clear view.

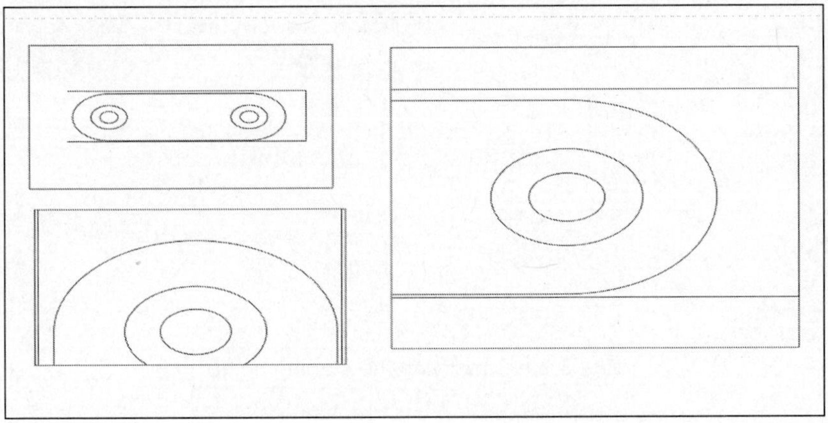

Fig. 5.18: Multiple rectangular viewport

Step 6: Repeat this process to create additional rectangular viewports as per your need. Each viewport can contain a different view or a different part of your drawing, depending on the requirements of your design.

Remember, you can also quickly create multiple viewports in a specific arrangement by using the 'Viewport' > 'Two' or 'Three' options in the 'Layout' panel.

This approach gives you a comprehensive overview of your design in a single layout, enabling an efficient and effective workflow within the AutoCAD environment.

5.2.4 Polygonal Viewports

Creating polygonal viewports in AutoCAD can provide more flexibility for viewing designs that involve irregular shapes. Here is a step-by-step guide on how to create a polygonal viewport:

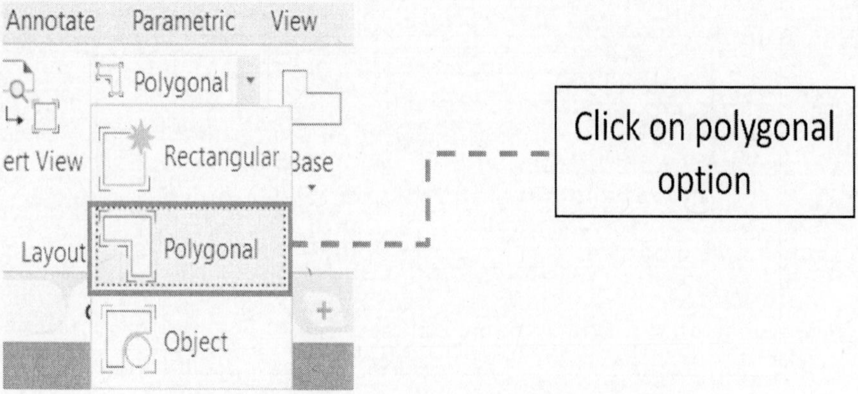

Fig. 5.19: Select polygonal option from the ribbon

Step 1: Open the layout tab where you intend to create your polygonal viewport.

Step 2: Go to the ribbon and click on Layout > Layout Viewports > Polygonal.

Step 3: After selecting the 'Polygonal' option, you will need to specify the corner points for your polygonal viewport. Click to set the first corner point of your polygonal viewport.

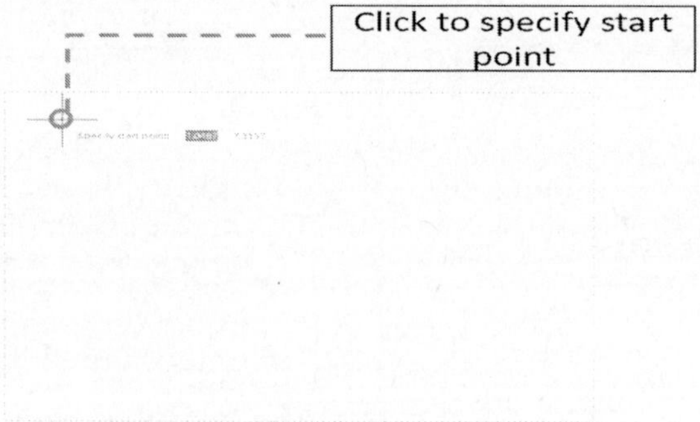

Fig. 5.20: Specifying the starting point

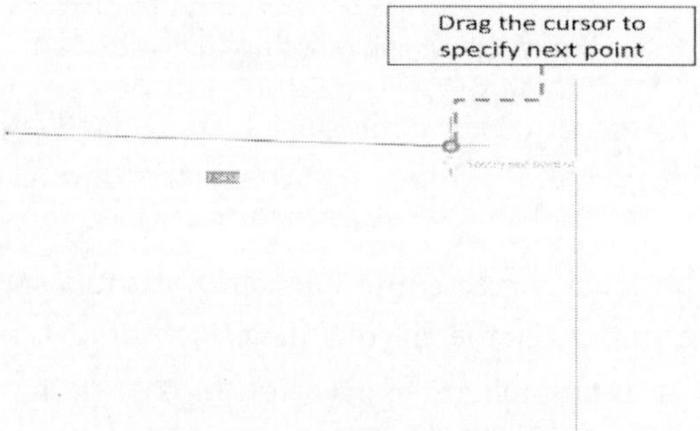

Fig. 5.21: Specifying next point

Step 4: Continue clicking to specify additional corner points, creating the shape of your viewport. Remember, these points can be placed as needed to form the polygon shape that best suits your design needs.

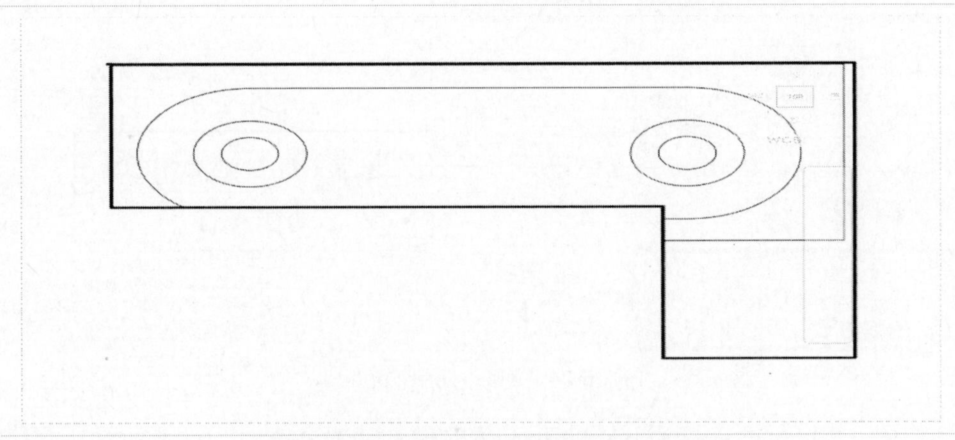

Fig. 5.22: Creation of polygonal viewport

Step 5: Once you have outlined the desired shape with corner points, complete the creation of the viewport by either clicking on the first point or using the 'Close' command.

Step 6: Press 'Enter' to finalize the creation of your polygonal viewport.

By following these steps, you can create viewports of different shapes to best fit the specific requirements of your designs, offering more versatility and utility within the AutoCAD workspace.

5.2.5 Converting Existing Objects

AutoCAD enables the transformation of existing closed objects into viewports, which is especially beneficial when creating viewports with irregular shapes. Here is the step-by-step guide:

Step 1: Open the layout tab where you have created your closed objects. These objects can be of any shape and size, fitting the requirements of your design.

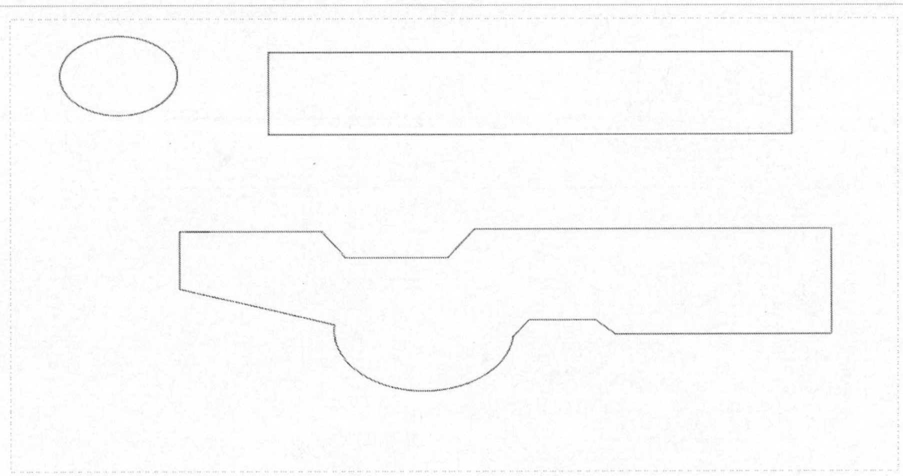

Fig. 5.23: Create the objects

Step 2: Go to the ribbon and click on Layout > Layout Viewports > Object. This will set you up to convert an existing object into a viewport.

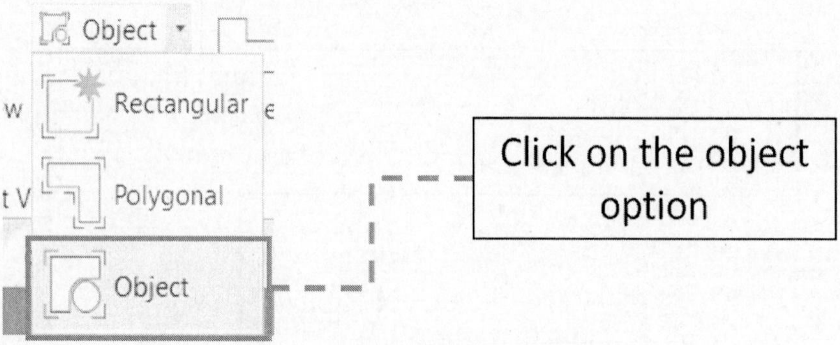

Fig. 5.24: Select object option

Step 3: Now, select one of the closed objects that you wish to convert into a viewport. This can be done by clicking directly on the object.

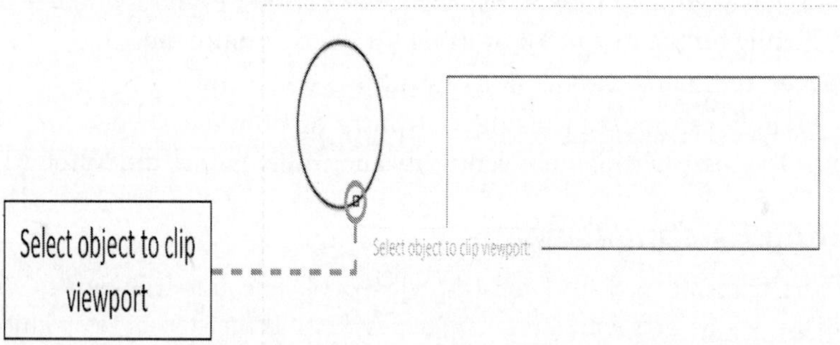

Fig. 5.25: Select an object to clip viewport

Step 4: Upon selection, your object will be instantly transformed into a viewport, retaining its original shape and size.

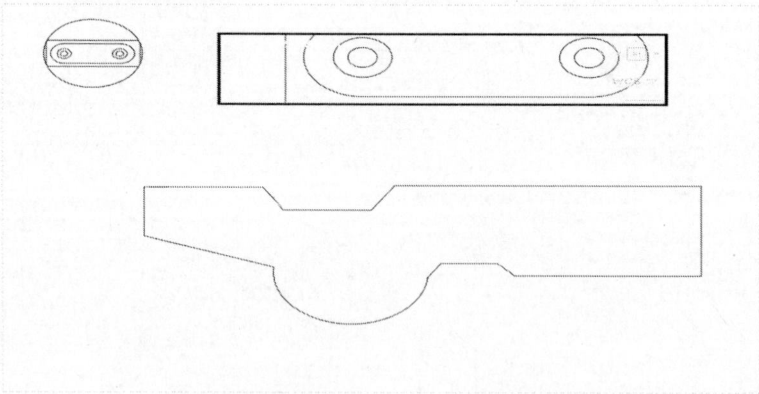

Fig. 5.26: Objects converted into viewport

This process enables you to take full advantage of existing design elements, repurposing them into functional viewports without the need for additional drawing or shaping.

5.2.6 Clipping Existing Viewports

Clipping existing viewports in AutoCAD allows you to adjust the shape and extent of your viewports, making them fit perfectly with your design requirements. Here are the steps to clip a viewport:

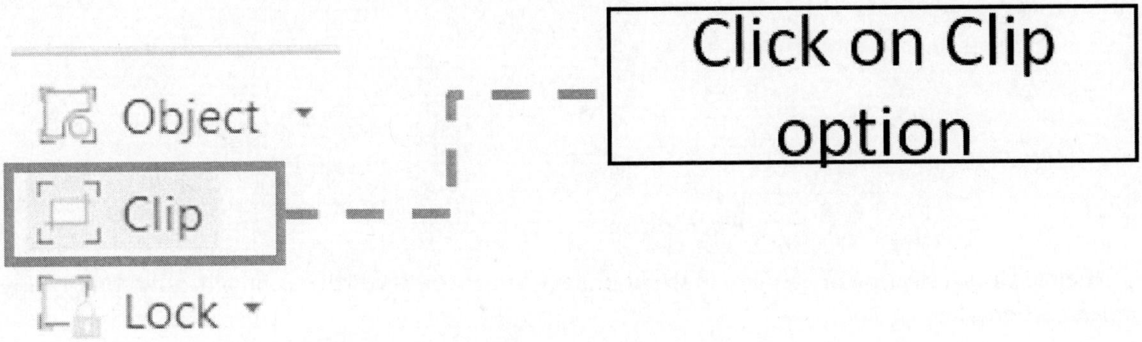

Fig. 5.27: Select the clip option from the ribbon

Step 1: Open your layout tab where your viewport is located. This is the viewport that you are planning to clip.

Step 2: Go to the ribbon and click on Layout > Layout Viewports > Clip. This sets you up to modify the boundary of your viewport.

Step 3: Now, you will need to select the viewport that you want to clip. You can do this by simply clicking on it.

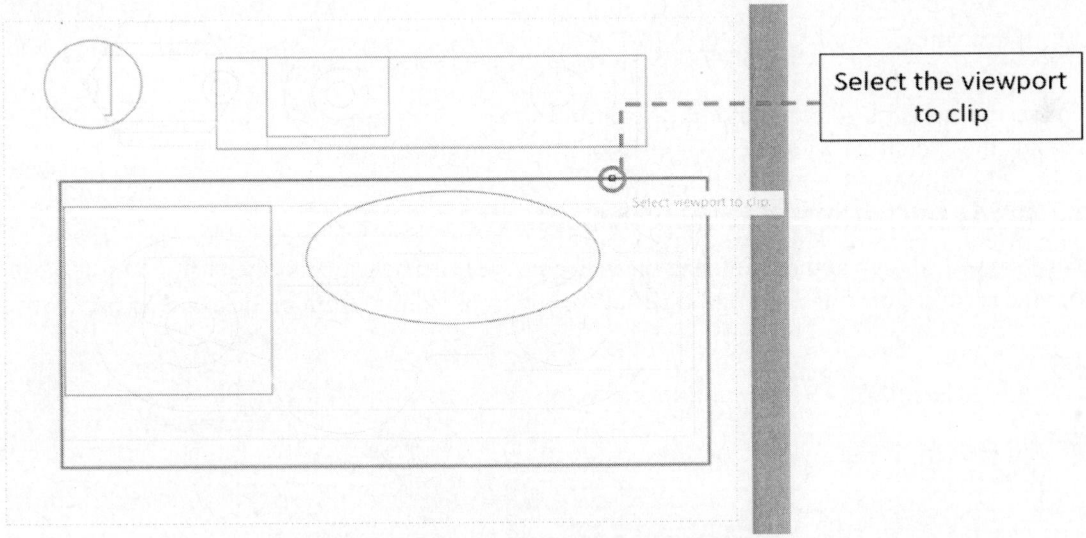

Fig. 5.28: Select the viewport for clipping

Step 4: After selecting the viewport, you will need to select a new boundary. This could be an existing closed object or you could draw a new boundary as needed. Select or draw this boundary by clicking on the points to define it.

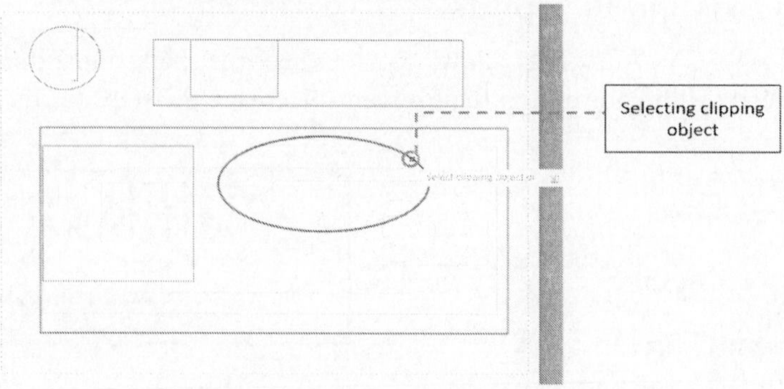

Fig. 5.29: Select clipping object

Step 5: Upon selection or creation of the boundary, your viewport will be clipped, now conforming to the new boundary.

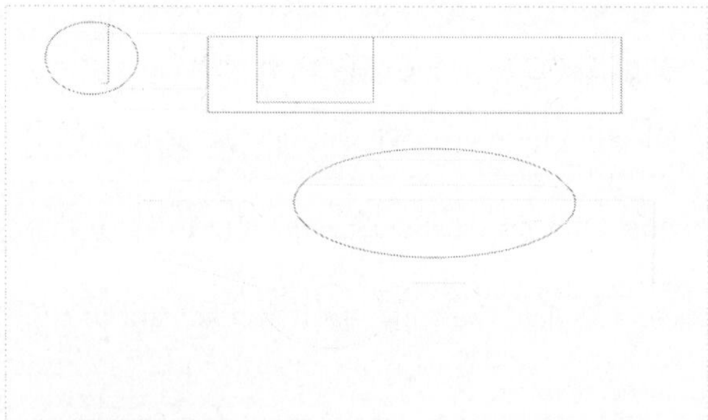

Fig. 5.30: Clipped viewport in the selected object

With this method, you can tailor the shape and size of your viewports to your specific design needs, enhancing the overall clarity and comprehensibility of your drawings.

5.3 SCALING AND MAXIMIZING VIEWPORTS

The scale of your design and maximizing the viewport are crucial elements when using AutoCAD. They ensure the accuracy of your drawings, especially when transitioning from model space to paper space.

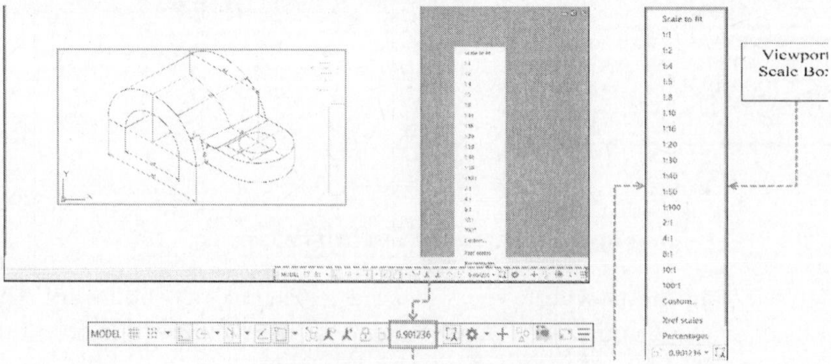

Fig. 5.31: Scaling viewports

Scaling Viewports

Each viewport can have a unique scale that helps to correctly represent different areas of your design. Follow these steps to set and lock a scale for a viewport:

Step 1: Activate the viewport by double-clicking inside it.

Step 2: Locate the scale list in the application status bar at the bottom of your screen. Click on it to open the drop-down list.

Step 3: Choose the scale you desire from the drop-down list. The viewport will automatically zoom in or out to match your chosen scale. In AutoCAD, objects in model space are usually created at full size, so this step is crucial for maintaining the correct proportions when viewed in paper space.

Step 4: To lock the scale and prevent any accidental changes, select the padlock icon near the scale list. This action ensures that the scale remains consistent, even when you are working within the viewport.

Maximizing Viewports

Sometimes, you might need to make detailed changes to a specific part of your drawing without affecting other viewports. In such cases, maximizing a viewport can be beneficial. To do so, carry out the following steps:

Step 1: To maximize a viewport, simply double-click within it. The viewport will then fill your screen, allowing you to focus on the details.

Step 2: Once you have made the necessary changes, you can minimize the viewport and return to the layout view. To do this, double-click anywhere outside the viewport.

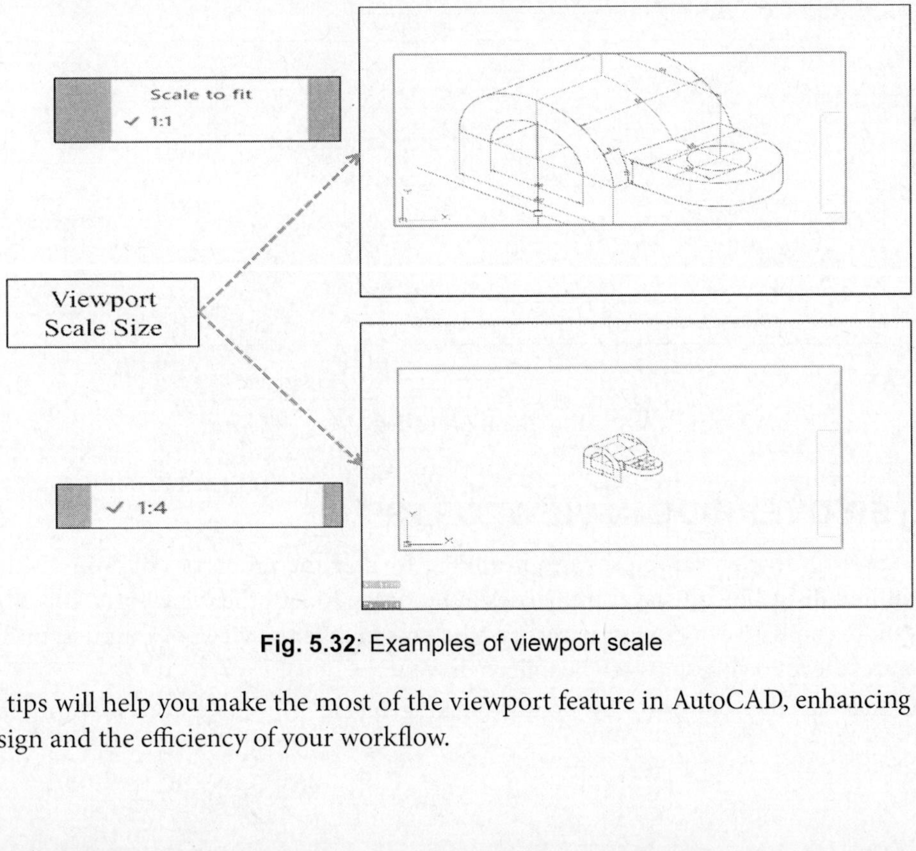

Fig. 5.32: Examples of viewport scale

These tips will help you make the most of the viewport feature in AutoCAD, enhancing the quality of your design and the efficiency of your workflow.

5.4 FREEZING LAYERS IN VIEWPORTS

Controlling the visibility of different design elements across multiple viewports enhances your workflow in AutoCAD. A vital tool for managing this visibility is the ability to freeze layers in specific viewports. When you freeze a layer in one active viewport, it does not impact the same layer in other viewports; the effect is restricted to the selected viewport only. To do so, carry out the following steps:

Step 1: Activate the viewport where you want to freeze the layer by double-clicking inside it.

Step 2: Open the Layer Properties Manager. You can do this by typing 'LAYER' in the command line or 'VPLAYER' in the model space. This manager allows you to control various aspects of the layers in your design.

Step 3: Within the Layer Properties Manager, look for the column labeled "Freeze". This column enables you to freeze layers independently for each viewport.

Step 4: Find the layer you wish to freeze and click in the Freeze column next to it.

Step 5: Once you have selected the layers to freeze, deactivate the viewport by double-clicking outside it.

The layers you have chosen will now be frozen in the specific viewport, but remain visible in all other viewports. This function offers you a flexible way of controlling which aspects of your design are displayed in different viewports, thus improving clarity and efficiency.

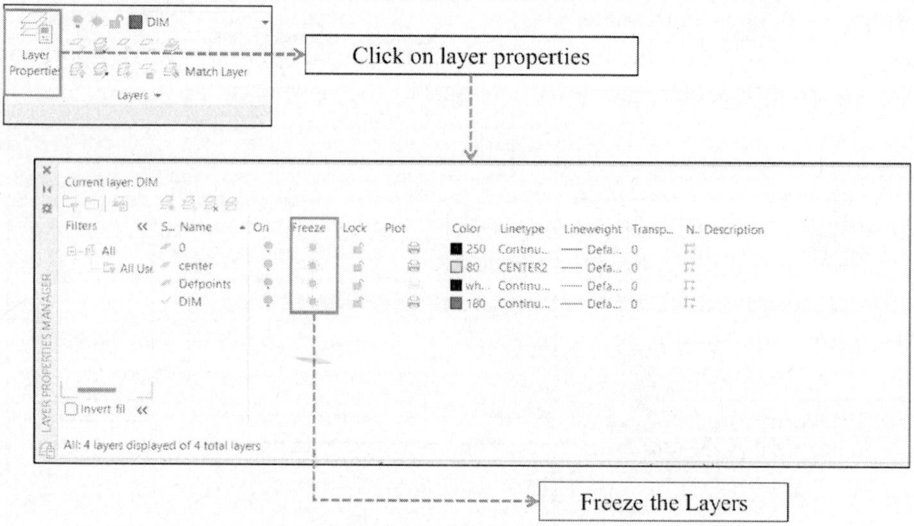

Fig. 5.33: Freeze layer

5.5 LAYER OVERRIDE IN VIEWPORTS

AutoCAD's Layer Override feature provides flexibility to alter the properties of layers within specific viewports without disturbing the layer properties in model space or other viewports. It is a helpful tool when you want to emphasize or downplay certain elements in different views of your design. To use layer override feature, the following steps can be followed:

Step 1: Activate the viewport (by double-clicking inside it), where you wish to change layer properties.

Step 2: Open the Layer Properties Manager, which allows you to modify various attributes of your layers.

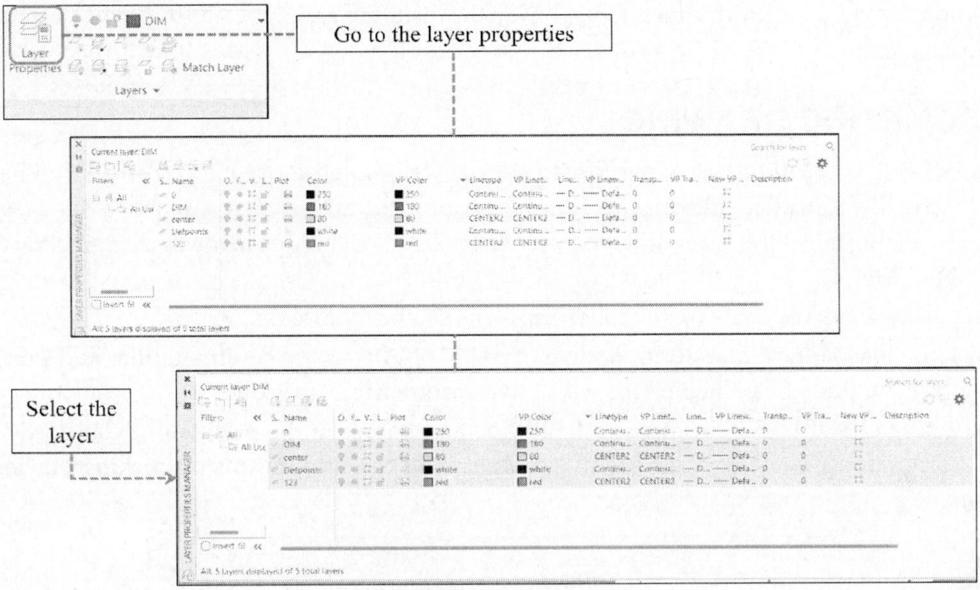

Fig. 5.34: Layer override

Step 3: From the Layer Properties Manager, select the layer you want to alter.

Step 4: You can modify aspects like color, line type etc. Look for the Viewport Override column within the Layer Properties Manager. Click on the desired setting to change the property value. For example, if you want to change the color of selected layers, go to 'VP Color' and choose a new color.

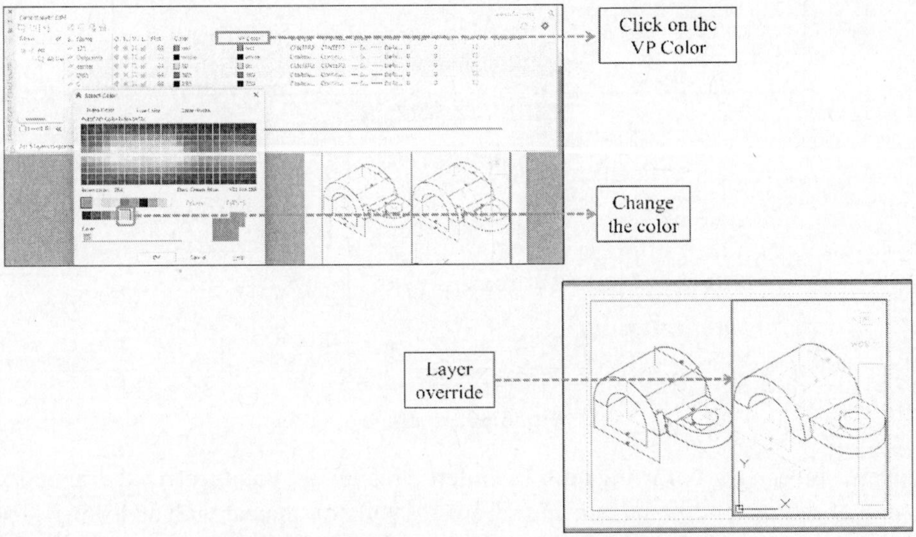

Fig. 5.35: Change the color of selected layers

Step 5: After making your changes, deactivate the viewport by double-clicking outside it.

As a result of these steps, the properties of the chosen layer will now be overridden in the active viewport, giving you the ability to highlight or obscure specific elements in different views of your design. This function provides you with more control over the visibility and appearance of your layers within different viewports.

5.6 ISOMETRIC DRAWING

Isometric drawing is a method that allows you to depict three-dimensional objects on a two-dimensional surface. Essentially, isometric drawings in AutoCAD are 2D drawings that simulate a 3D perspective, aiding in visualizing the object from different orientations. This technique involves changing the drawing plane in AutoCAD.

To set up an isometric drawing the following steps can be followed:

Step 1: Enable Isometric drafting mode by typing 'SNAP' in the command line and press Enter. Then type 'ST' and press Enter again. This will activate isometric snap mode.

Step 2: Type 'ISOPLANE' in the command line and based on the orientation you need for your drawing select among 'Left,' 'Top,' or 'Right' isoplanes. You can also do this by clicking on "Isometric drafting mode".

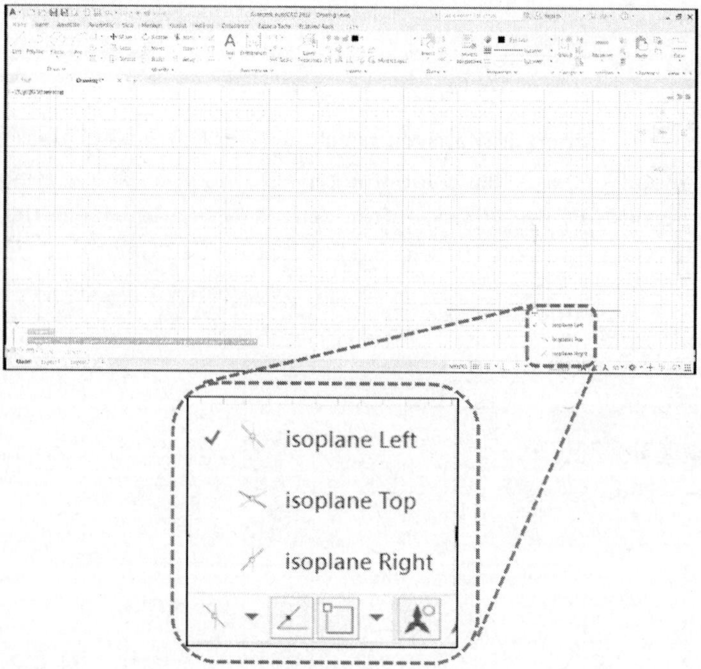

Fig. 5.36: Isoplane

Step 3: Now, you can start drawing on the isometric plane. Use your preferred drawing tools as you would in any other setting, but remember, your drawings will now appear with an isometric perspective.

Remember: It is important to turn off the isometric snap mode when you want to return to regular 2D drafting. You can do this by entering the 'SNAP' > 'ST' command again.

Here is an example of creating an isometric drawing:

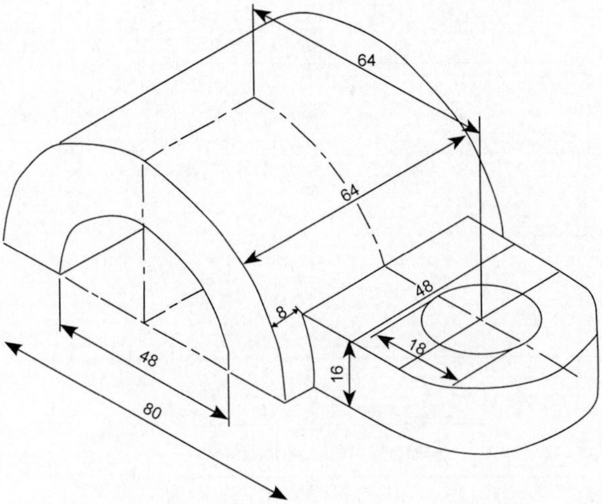

Fig. 5.37: Example of an isometric drawing

The detailed steps to draw the above example is given below:

1. Begin by enabling isometric drafting mode and selecting the isoplane you want to use.

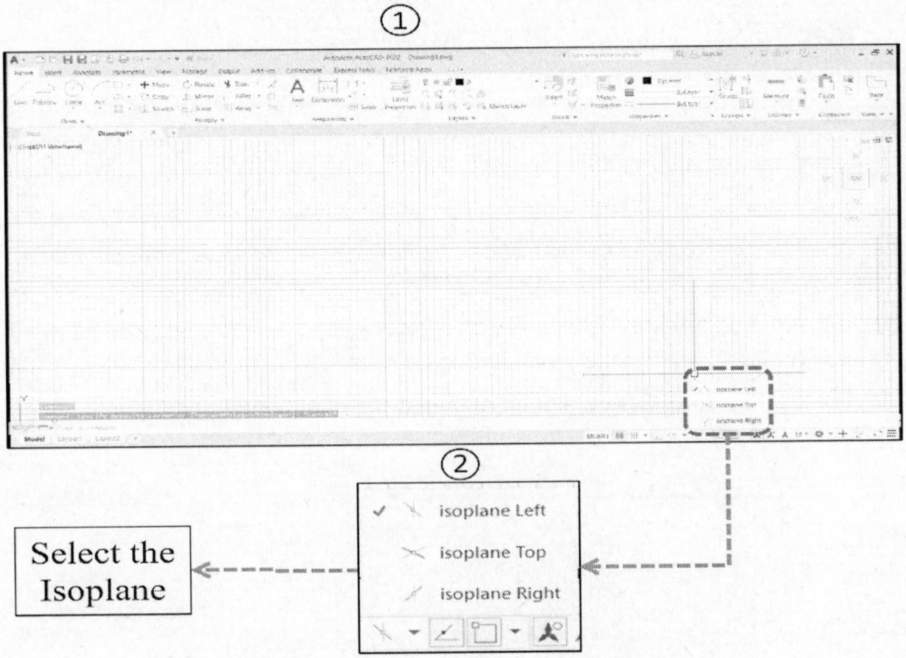

Fig. 5.38: Example of an isometric drawing (Step 1 and 2)

2. Use the line command, isocircle option and other drawing tools to create the geometry of your drawing.
3. Modify your drawing as necessary using the trim and extend commands.
4. Remember to switch the isoplane as needed to draw in different orientations.

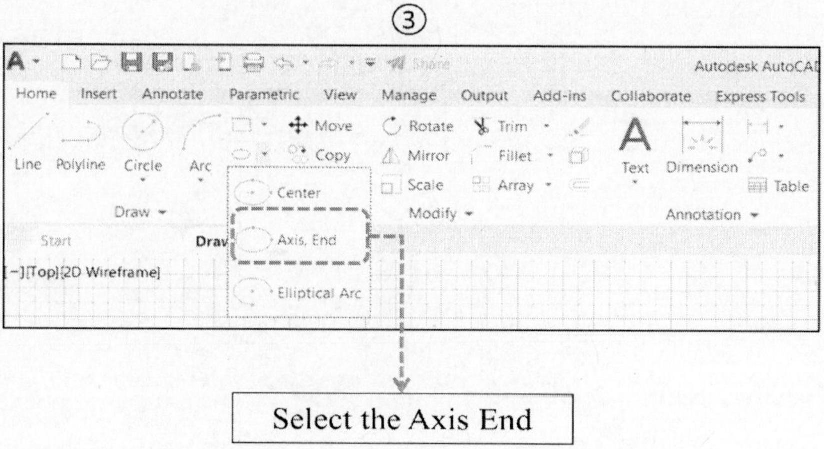

Fig. 5.39: Example of an isometric drawing (Step 3)

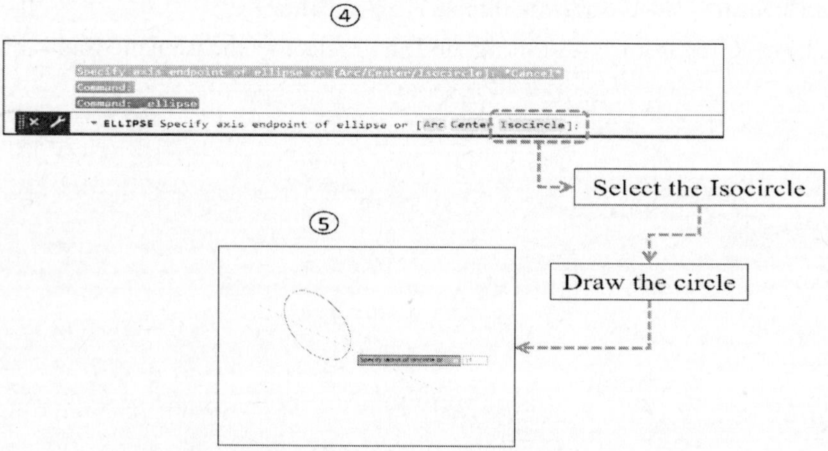

Fig. 5.40: Example of an isometric drawing (Step 4 and 5)

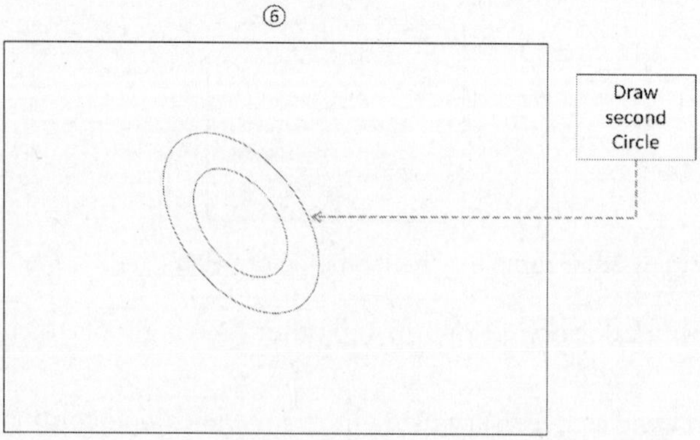

Fig. 5.41: Example of an isometric drawing (Step 6)

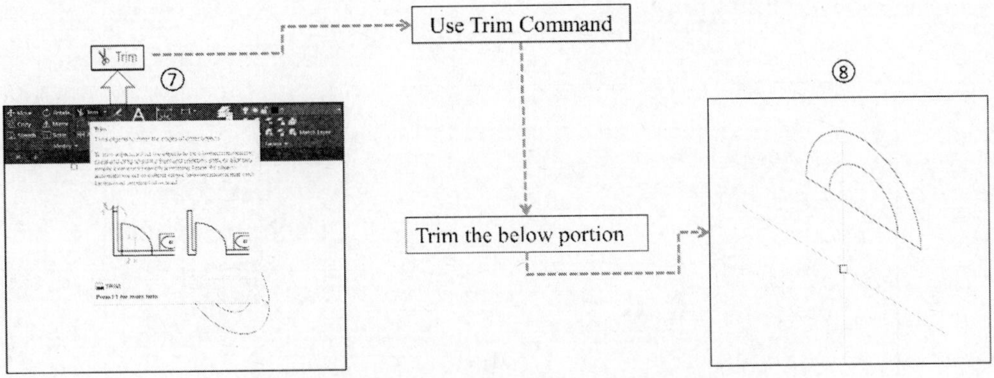

Fig. 5.42: Example of an isometric drawing (Step 7 and 8)

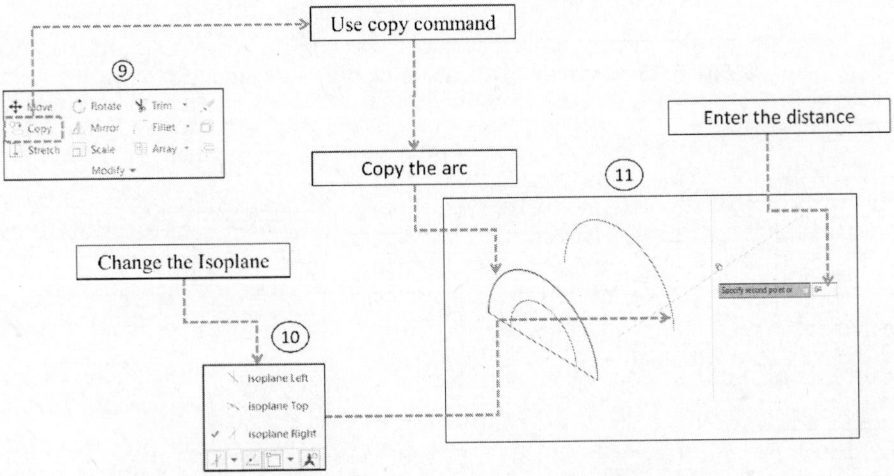

Fig. 5.43: Example of an isometric drawing (Step 9, 10 and 11)

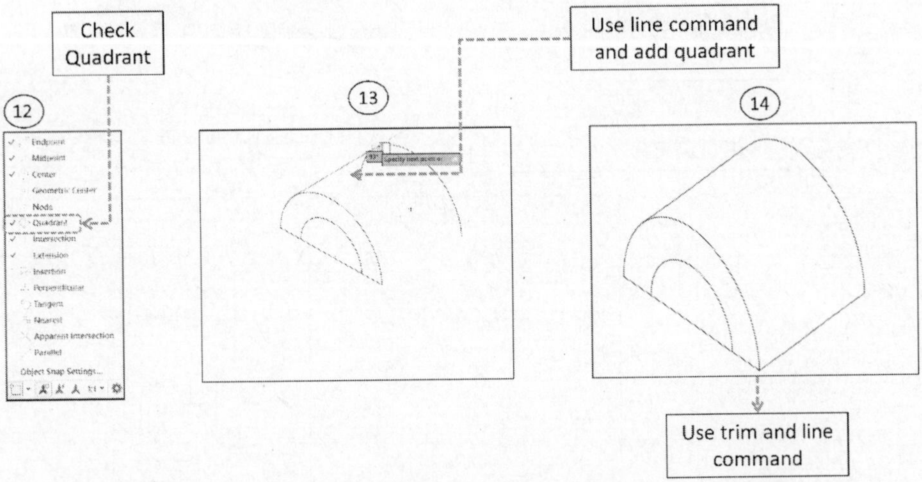

Fig. 5.44: Example of an isometric drawing (Step 12, 13 and 14)

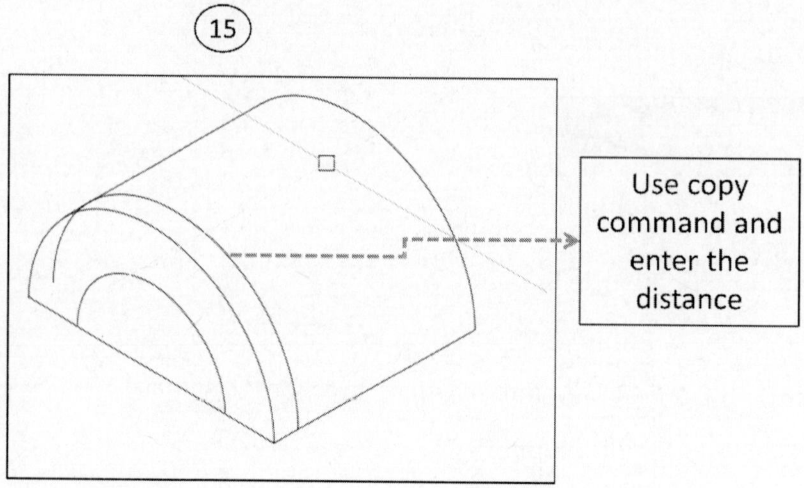

Fig. 5.45: Example of an isometric drawing (Step 15)

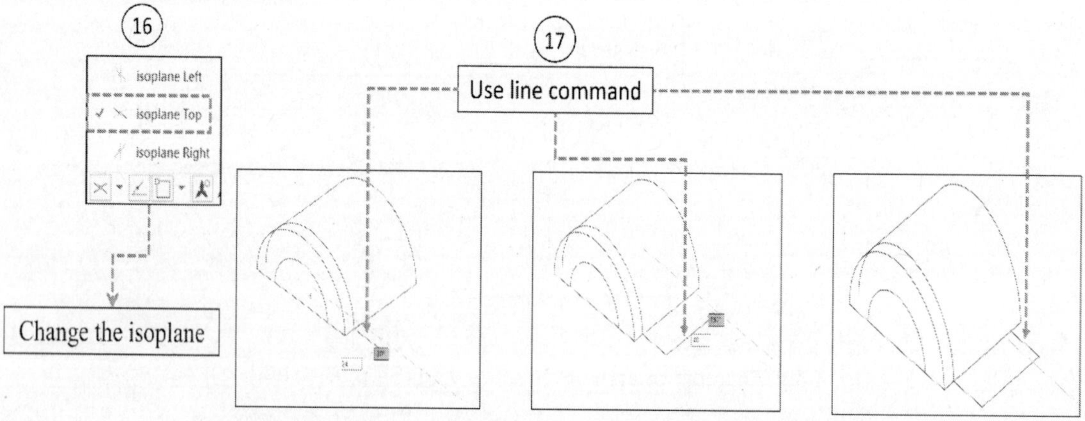

Fig. 5.46: Example of an isometric drawing (Step 16 and 17)

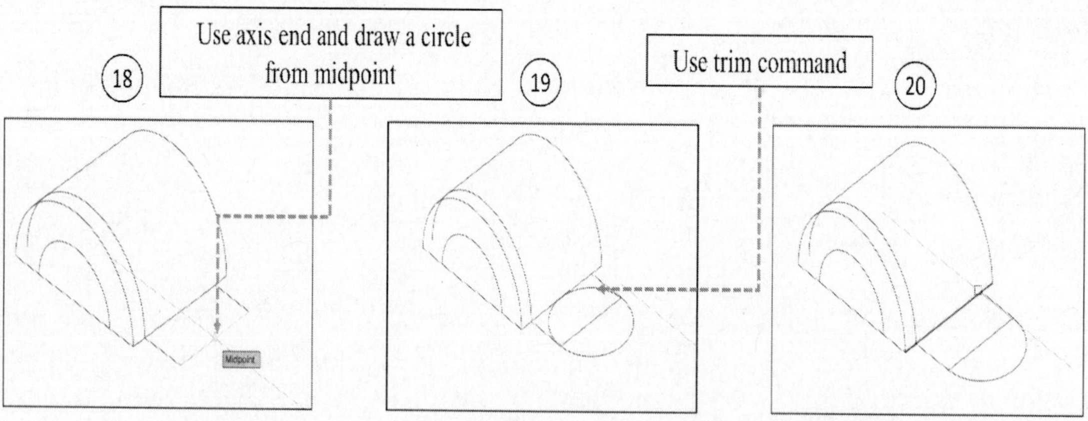

Fig. 5.47: Example of an isometric drawing (Step 18, 19 and 20)

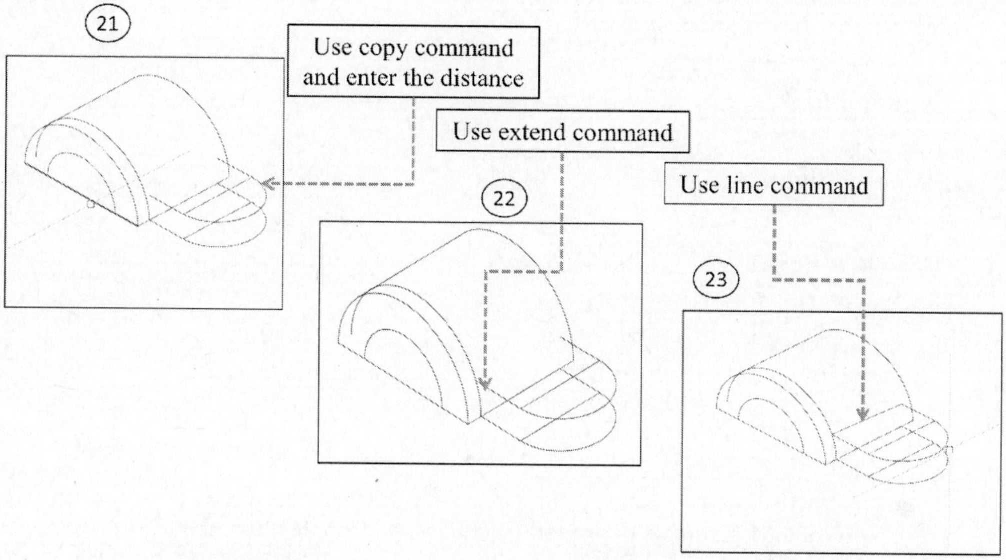

Fig. 5.48: Example of an isometric drawing (Step 21, 22 and 23)

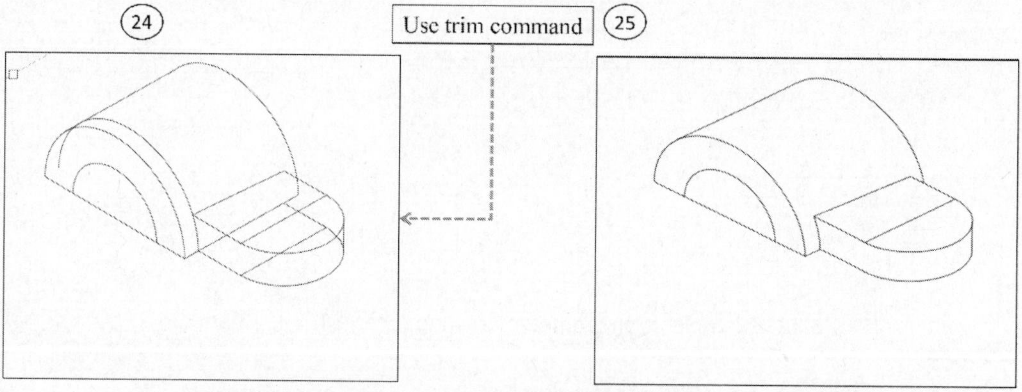

Fig. 5.49: Example of an isometric drawing (Step 24 and 25)

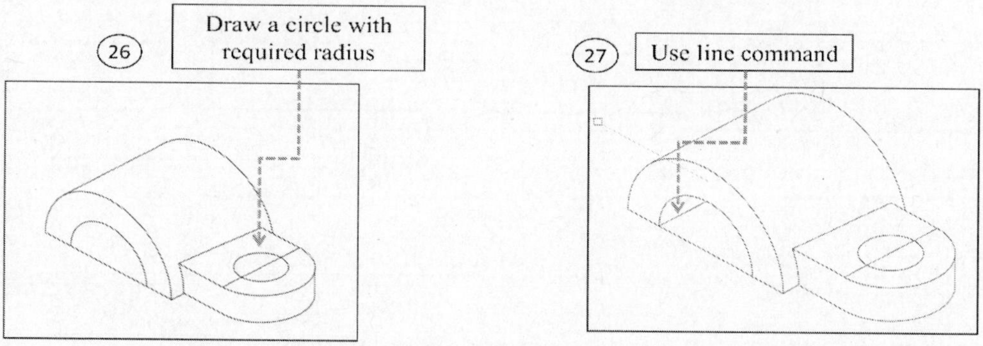

Fig. 5.50: Example of an isometric drawing (Step 26 and 27)

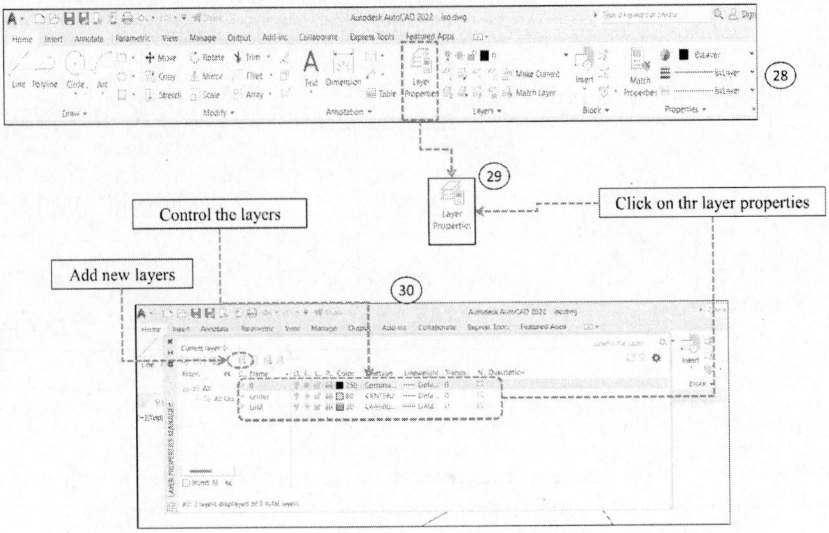

Fig. 5.51: Example of an isometric drawing (Step 28, 29 and 30)

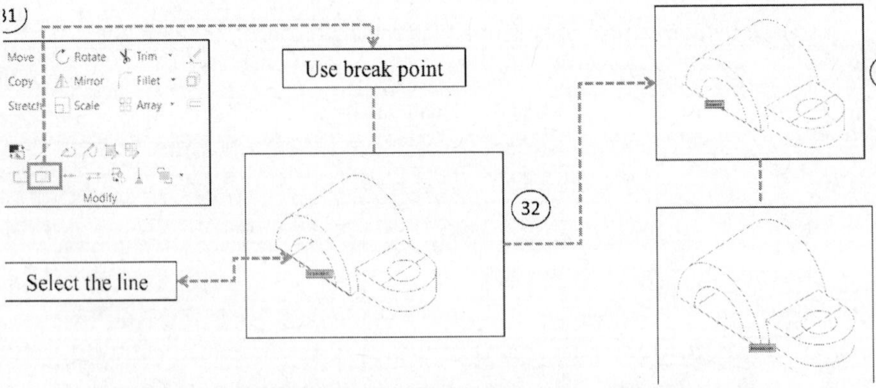

Fig. 5.52: Example of an isometric drawing (Step 31, 32, 33 and 34)

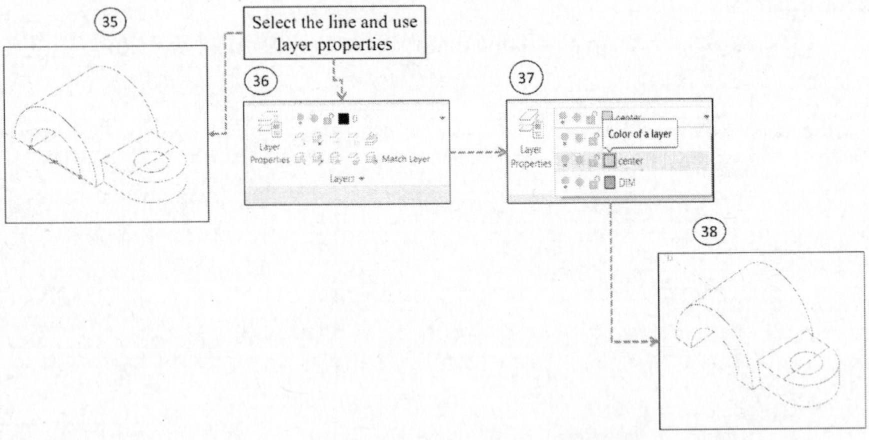

Fig. 5.53: Example of an isometric drawing (Step 35, 36, 37 and 38)

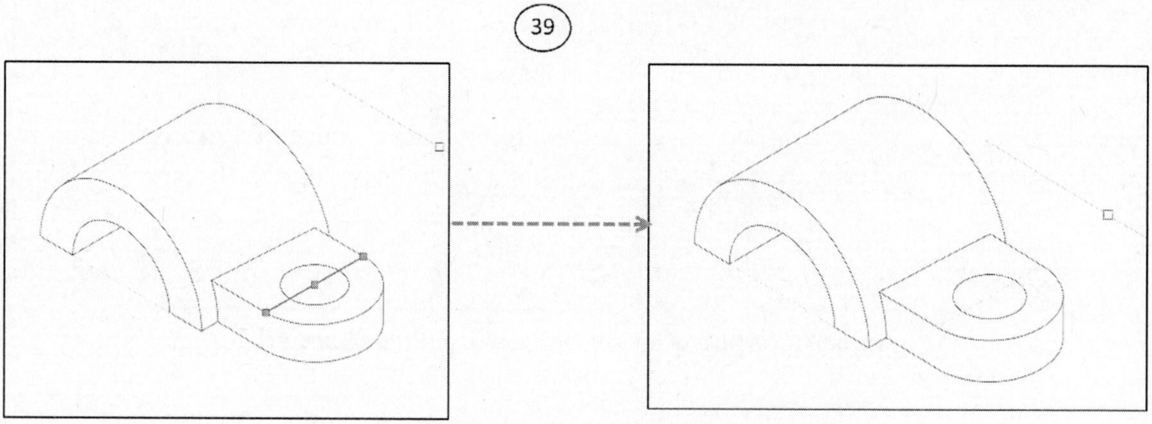

Fig. 5.54: Example of an isometric drawing (Step 39)

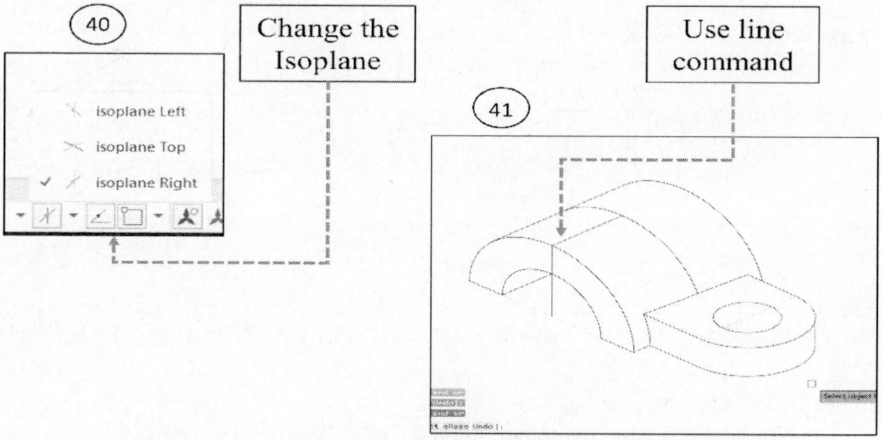

Fig. 5.55: Example of an isometric drawing (Step 40 and 41)

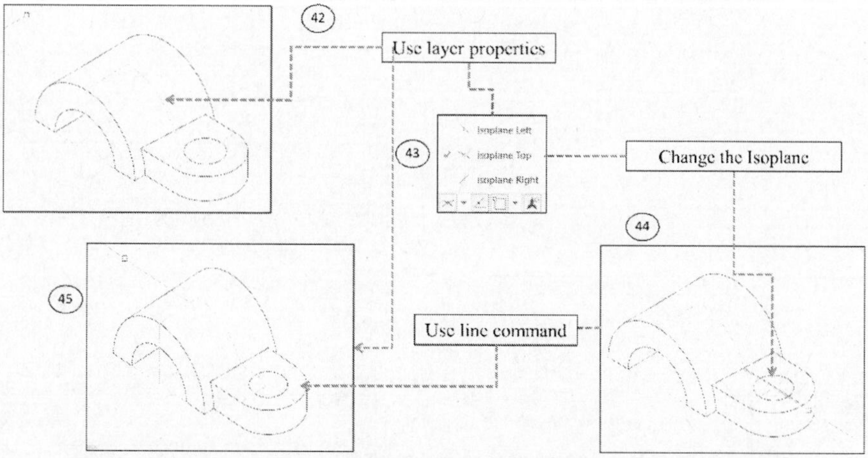

Fig. 5.56: Example of an isometric drawing (Step 42, 43, 44 and 45)

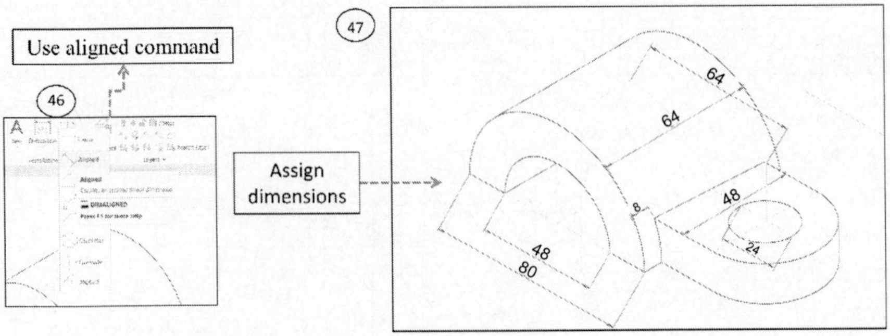

Fig. 5.57: Example of an isometric drawing (Step 46 and 47)

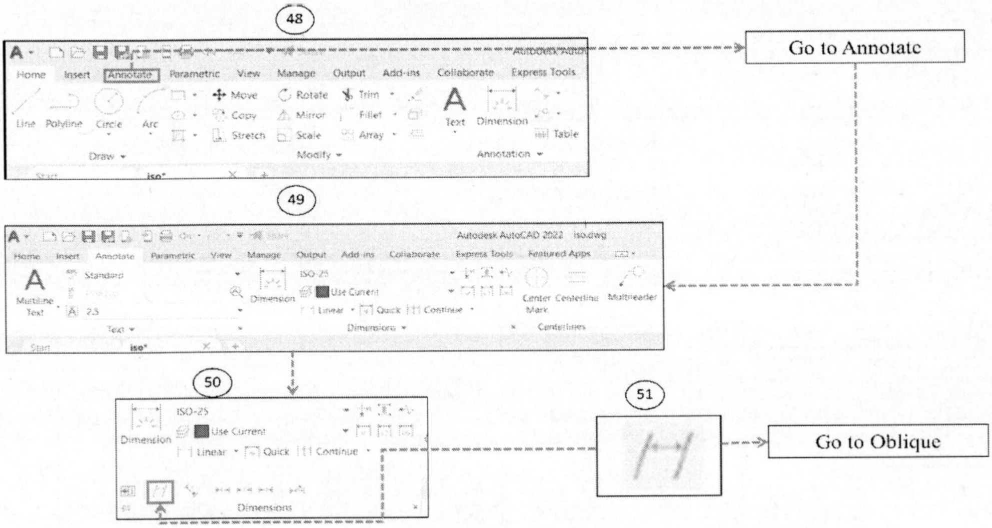

Fig. 5.58: Example of an isometric drawing (Step 48, 49, 50 and 51)

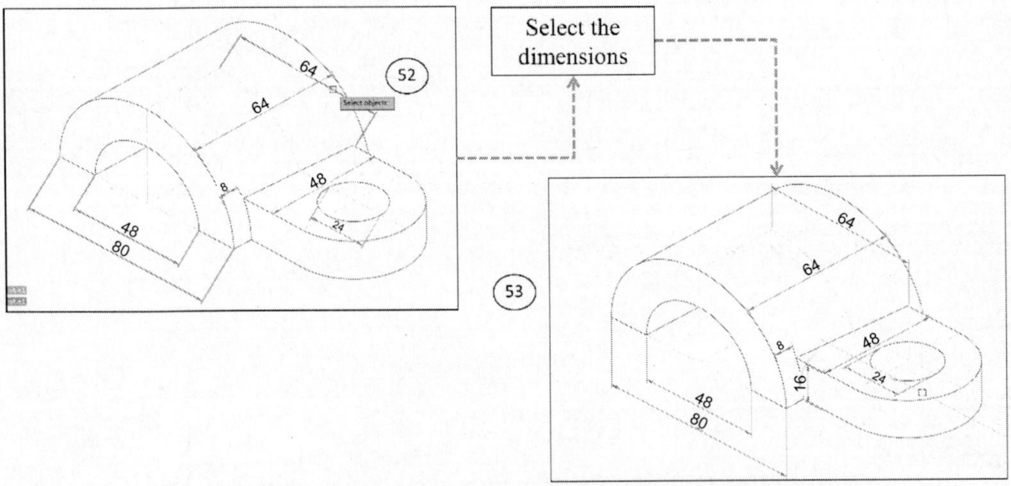

Fig. 5.59: Example of an isometric drawing (Step 52 and 53)

5. Copy elements of your drawing using the copy command and specify the distance to create identical elements at precise distances.
6. Make use of the quadrant snap option to ensure accurate drawing.
7. Layer properties can be used to differentiate parts of your drawing by assigning different colors to different layers.
8. Finally, add dimensions to your drawing using the dimensioning tools available in the Annotate tab. Use the oblique dimension for accurate isometric dimensions.

Through these steps, isometric drawing allows you to create detailed, spatially accurate 2D representations of your 3D designs, enhancing your ability to visualize and present your work.

UNIT SUMMARY

Based on the contents covered in this unit you should remember that:

- Viewports are windows within AutoCAD that display different parts of a design or drawing at varying scales or angles.
- Single Rectangular Viewports are the most common type of viewport and display a single view of a design.
- Multiple Rectangular Viewports allow for displaying multiple views of a design simultaneously, which can be useful for complex projects.
- Polygonal Viewports offer the flexibility of creating non-rectangular viewport shapes to suit specific design needs.
- Existing objects within a drawing can be Converted into Viewports, providing another way to work with your designs.
- You can Clip Existing Viewports to limit the visible portion of a drawing within a viewport, allowing focus on specific details.
- The Scaling and Maximizing of Viewports is a crucial aspect of managing the visibility and proportions of your designs within a layout.
- You can Freeze Layers in Viewports to hide non-essential parts of a drawing in specific viewports without affecting the overall design.
- Layer Overrides in Viewports allow for changes to be made to layer properties in specific viewports without altering the properties in the entire drawing.
- Isometric Drawing in AutoCAD allows for the representation of three-dimensional objects in two dimensions, giving a better visual understanding of designs.
- The creation and modification of Layouts are central to organizing your drawings for plotting or printing and the Page Setup Manager is a crucial tool in this process.

EXERCISES

A. Multiple Choice Questions

5.1 What is the primary function of viewports in AutoCAD?

(a) They allow you to draw in 3D

(b) They allow you to plot multiple drawings

(c) They allow you to view and work on different sections of a design simultaneously.

(d) They allow you to print your work.

5.2 When working with single rectangular viewports, how do you create a viewport?
  (a) By using the "Single" option from the "Viewport" menu.
  (b) By using the "Multiple" option from the "Viewport" menu.
  (c) By converting an existing object.
  (d) By using the "Polygonal" option from the "Viewport" menu.

5.3 How do you create multiple rectangular viewports?
  (a) By repeating the process of creating single rectangular viewports.
  (b) By using the "Single" option from the "Viewport" menu.
  (c) By converting an existing object.
  (d) By using the "Polygonal" option from the "Viewport" menu.

5.4 Which of the following options allows you to create irregular viewport shapes?
  (a) Single rectangular viewport.
  (b) Multiple rectangular viewport.
  (c) Converting an existing object into a viewport.
  (d) Polygonal viewport.

5.5 How do you clip an existing viewport in AutoCAD?
  (a) By selecting the viewport and choosing "Clip" from the "Modify" panel.
  (b) By using the "Single" option from the "Viewport" menu.
  (c) By converting an existing object.
  (d) By using the "Polygonal" option from the "Viewport" menu.

5.6 How can the viewport scale be locked?
  (a) By clicking on the padlock icon near the scale list.
  (b) By selecting the "Single" option from the "Viewport" menu.
  (c) By creating a polygonal viewport.
  (d) By converting an existing object.

5.7 What is the benefit of maximizing a viewport in AutoCAD?
  (a) It makes the viewport invisible.
  (b) It fills your screen with the viewport, allowing you to make detailed changes.
  (c) It allows you to work on different sections of a design simultaneously.
  (d) It locks the viewport scale.

5.8 How do you freeze layers in specific viewports?
  (a) By using the "Viewport" > "Freeze" command.
  (b) By using the "Single" option from the "Viewport" menu.
  (c) By converting an existing object.
  (d) By using the "Polygonal" option from the "Viewport" menu.

5.9 What are layer overrides in AutoCAD used for?
  (a) To change the properties of layers in specific viewports.
  (b To create multiple rectangular viewports.
  (c) To lock the viewport scale.
  (d) To convert an existing object into a viewport.

5.10 In the context of AutoCAD, what is an isometric drawing?
- (a) A drawing that allows you to represent three-dimensional objects in two dimensions.
- (b) A drawing that allows you to plot multiple drawings.
- (c) A drawing that allows you to view and work on different sections of a design simultaneously.
- (d) A drawing that allows you to print your work.

5.11 Which of the following options allows you to set up an isometric drawing?
- (a) By using the "ISOPLANE" command.
- (b) By using the "Viewport" menu.
- (c) By converting an existing object.
- (d) By using the "Polygonal" option from the "Viewport" menu.

5.12 What is the purpose of a layout in AutoCAD?
- (a) It allows you to change the properties of layers.
- (b) It allows you to draw in 3D.
- (c) It represents the conventional drawing sheet for plotting drawings.
- (d) It allows you to lock the viewport scale.

5.13 How can you modify a layout in AutoCAD?
- (a) By using the Page Setup Manager dialog box.
- (b) By using the "Viewport" menu.
- (c) By converting an existing object.
- (d) By using the "Polygonal" option from the "Viewport" menu.

5.14 In the Page Setup dialog box, how do you set the plot style table?
- (a) By choosing "acad.stb".
- (b) By choosing "DWG to PDF.pc3".
- (c) By selecting ISO A1(841.00x594.00MM)
- (d) By setting the "Plot scale" to 1:1.

5.15 How do you set the plot scale in AutoCAD?
- (a) By setting the plot scale to 1:1.
- (b) By choosing "acad.stb".
- (c) By selecting ISO A1(841.00x594.00MM)
- (d) By choosing "DWG to PDF.pc3".

Answers to Multiple-Choice Questions
5.1 (c), 5.2 (a), 5.3 (b), 5.4 (d), 5.5 (a), 5.6 (a), 5.7 (b), 5.8 (a), 5.9 (a), 5.10 (a), 5.11 (a), 5.12 (c), 5.13 (a), 5.14 (a), 5.15 (a)

B. Subjective Questions

5.1 Recall the steps to create a new layout in AutoCAD?

5.2 Describe the concept of "Viewports" in AutoCAD. How do they help in the process of designing?

5.3 If you are given a complex 3D model to work on, how would you use the feature of multiple rectangular viewports to facilitate your design process?

5.4 Discuss the advantages and disadvantages of using single rectangular viewports versus multiple rectangular viewports in AutoCAD.

5.5 Imagine you are working on a complex architectural design and you are using several layers. Evaluate the use of layer freezing and layer override features in this scenario.

5.6 Develop a step-by-step plan on how you would create an isometric drawing of a simple object like a chair or a table in AutoCAD.

5.7 What is the role of the Page Setup Manager dialog box in modifying a layout? How does it enhance your drawing or plotting process?

5.8 In what scenarios would polygonal viewports be more advantageous than rectangular viewports? Provide practical examples.

5.9 Analyze the impact of scaling and maximizing viewports on the accuracy and visibility of your designs in AutoCAD.

5.10 Considering a case where multiple engineers are working on the same project, evaluate how layer override can help in ensuring consistency and avoiding errors in design.

KNOW MORE

In addition to the knowledge shared in this unit about AutoCAD's Viewports and Isometric Drawing features, it is also beneficial to delve into a few complementary concepts and practices that can optimize your overall designing experience.

AutoCAD offers a feature known as "Annotation Scaling." This function enables you to create text, dimensions and other annotations that automatically adjust their size based on the scale of the viewport they are displayed in. Hence, the text appears consistent in size across different viewports, regardless of the scale, improving readability and design professionalism.

The power of viewports extends beyond just viewing different sections of your design; they can also facilitate a unique "collaborative" way of working. AutoCAD's "Viewport Controls" allow for easy navigation between different views, visual styles and coordinate systems right from the viewport itself. This is particularly helpful when working on complex projects where multiple engineers are contributing to different sections of the same design.

Moreover, while isometric drawing allows for the representation of 3D objects in 2D, there is an emerging trend in the CAD world called "Photorealistic Rendering." This technique uses complex algorithms to create a 3D rendered image of your design that closely mimics real-life materials, lighting and shadows. Although this was not discussed in detail in this unit, familiarity with it can greatly enhance your presentation skills, particularly in fields such as architecture and product design.

Lastly, it is crucial to keep abreast of the regular updates and feature enhancements AutoCAD introduces. Autodesk, the parent company, is heavily invested in improving user experience and functionality. Staying updated with these changes will ensure you are always leveraging the full potential of this powerful designing tool.

REFERENCES AND SUGGESTED READINGS

1. Yarwood, Alfred. Introduction to AutoCAD 2012. Routledge, 2011.
2. Fane, Bill. AutoCAD for dummies. John Wiley & Sons, 2016.
3. Cadfolks. AutoCAD 2020 For Beginners, 7th ed.; Kishore, 2019.

Dynamic QR Code for Further Reading

Single Rectangular Viewports

Multiple Rectangular Viewports

Polygonal Viewports

Converting Existing Objects

Clipping Existing Viewports

Scaling Viewports

Freezing Layers in Viewports

Layer Override in Viewports

Isometric Drawing

6 Printing and Plotting in AutoCAD

UNIT SPECIFICS

This unit presents information related to the following topics:

- Plotting Commands
- Plot Styles
- Printing Methods
- Sheet Set Manager

This unit delves into the essentials of plotting, plot styles and printing methods in AutoCAD, starting with the fundamentals of converting digital designs into hard copies or different digital formats. It also explores plot styles, which dictate how various elements appear when printed. The unit introduces various methods of printing and concludes with an overview of the Sheet Set Manager, a tool that streamlines managing and printing multiple drawing sheets. Each topic is elaborated with step-by-step procedures to ensure a comprehensive understanding within AutoCAD.

Apart from this at the end of the unit, a succinct recapitulation of the overall broad concepts is provided in form of a unit summary. Besides, a large number of multiple-choice questions as well as descriptive type questions are included. A list of references and suggested readings are given in the unit so that one can go through them for practice. It is important to note that for getting more information on various topics of interest some QR codes have been provided in different sections which can be scanned for relevant supportive knowledge. Video resources along with QR codes are mentioned for getting more information on various topics of interest which can be surfed or scanned through mobile phones for viewing.

RATIONALE

Plotting is an essential aspect of AutoCAD that enables the transformation of digital designs into hard copies or different digital file formats. It allows designers and engineers to share, present and document their work. Learning how to plot efficiently and accurately is critical for achieving professional results. Plot styles, meanwhile, offer a way to ensure consistent and accurate color, lineweight and other settings across different plots. The Sheet Set Manager aids in the organization and automation of the plotting process. By mastering these elements, students can streamline their workflow and elevate the quality of their output.

PRE-REQUISITES

Before reading this unit, the students are advised to revisit the following:

- Mathematics: Concepts such as scale and ratio (Class XII level).
- Engineering Graphics (ES101)
- AutoCAD commands covered in Unit-1 to Unit-5

UNIT OUTCOMES

After studying this unit students will be able to:

U6-O1: Understanding and utilizing the PLOT command and various plot settings to create accurate and professional plots.

U6-O2: Configuring and applying plot styles to ensure consistent color, lineweight and other properties across different plots.

U6-O3: ifferentiating between the various methods of printing and choosing the appropriate method for different requirements.

U6-O4: Employing the Sheet Set Manager for better organization and efficiency in the plotting process.

U6-O5: Integrating the knowledge and skills acquired in this unit to produce accurate, efficient and high-quality plots in AutoCAD.

Unit-6 Outcomes	EXPECTED MAPPING WITH COURSE OUTCOMES (1 – Weak Correlation; 2 – Medium correlation; 3 – Strong Correlation)				
	CO-1	CO-2	CO-3	CO-4	CO-5
U6-O1	-	1	-	2	1
U6-O2	1	-	-	2	1
U6-O3	1	-	-	2	1
U6-O4	-	-	1	2	1
U6-O5	-	-	-	3	2

6.1 INTRODUCTION

In this Chapter 6 we will explore the key aspects of Printing and Plotting in AutoCAD. These functions are crucial for transforming your digital designs into tangible, physical formats or producing digital outputs for dissemination and review.

6.2 PLOTTING

In AutoCAD, plotting is more than just clicking a button to generate your design on paper. It is a multifaceted process that transforms your digital creations into high-quality, printable outputs. This section delves into the intricacies of configuring plotters, setting up page layouts and creating plot style tables. Mastering these elements ensures that your printed drawings capture the precision and details of your digital design.

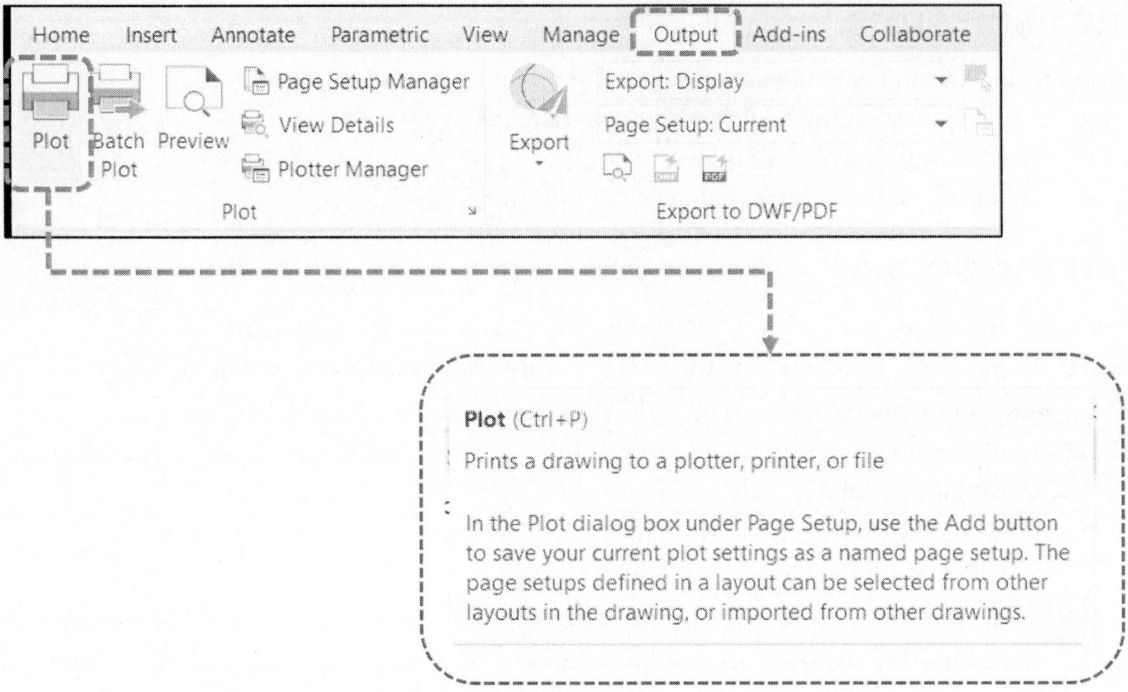

Fig. 6.1: Plot command

6.2.1 Configuring Plotters

Configuring plotters in AutoCAD begins with accessing the plotting interface, which can be done by selecting **Output** > **Click Plot** > **Plot** model or by simply pressing **Ctrl** + **P**.

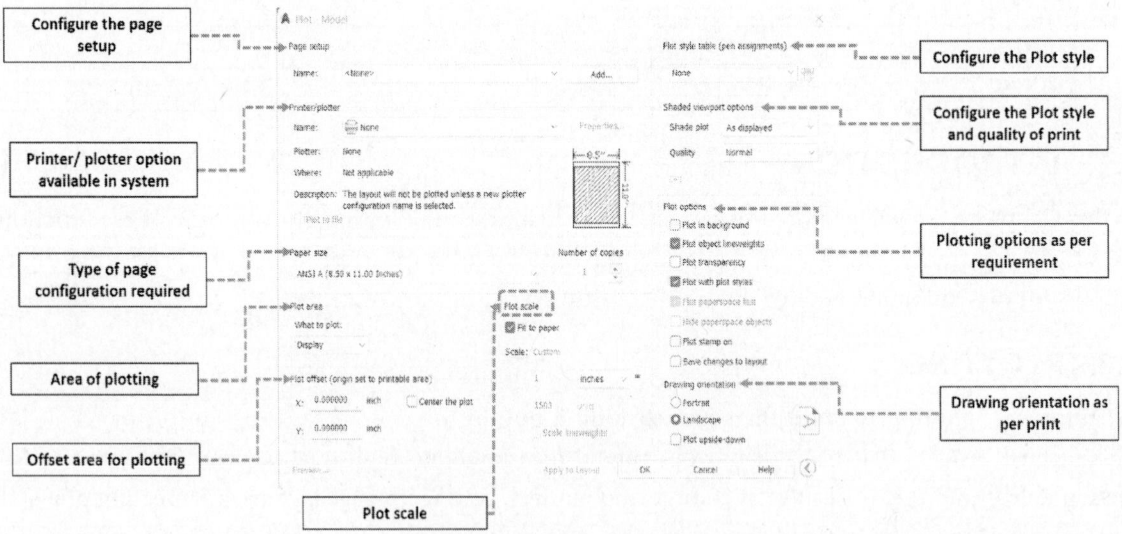

Fig. 6.2: Configuring plotters

One of the fundamental concepts in this phase is understanding the page setup. AutoCAD uses page setups to manage layout settings and plotter properties. These setups can be saved and applied across multiple layouts in the same drawing file, which enhances efficiency and consistency.

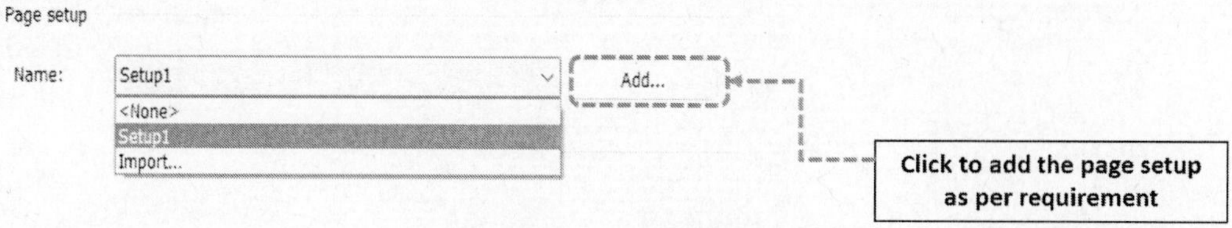

Fig. 6.3: Page setup

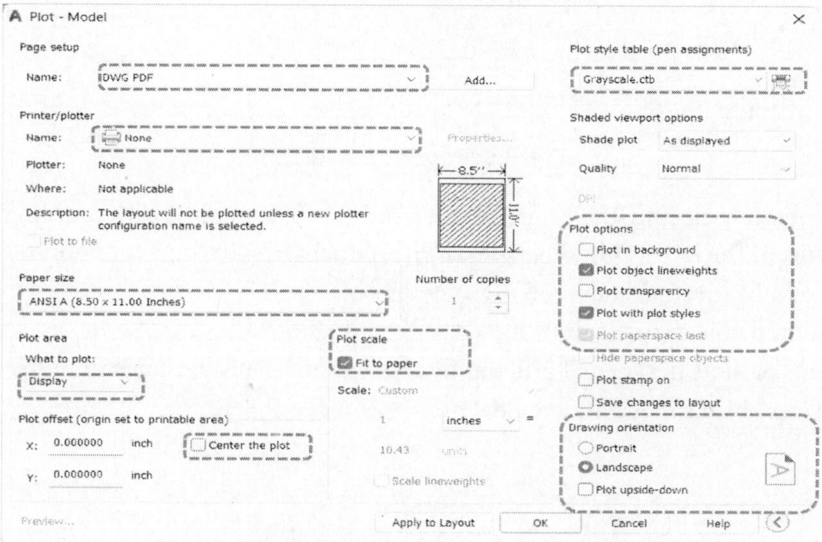

Fig. 6.4: Options for configuring page setup

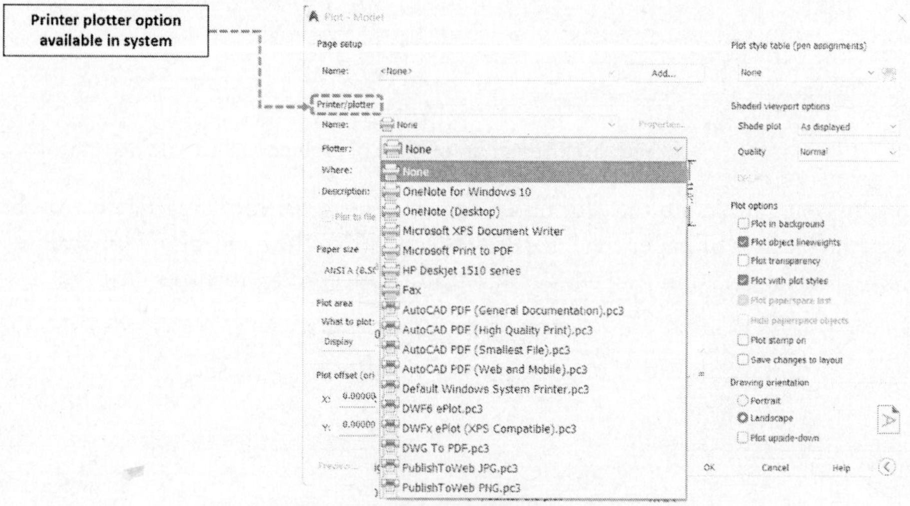

Fig. 6.5: Printer/plotter selection

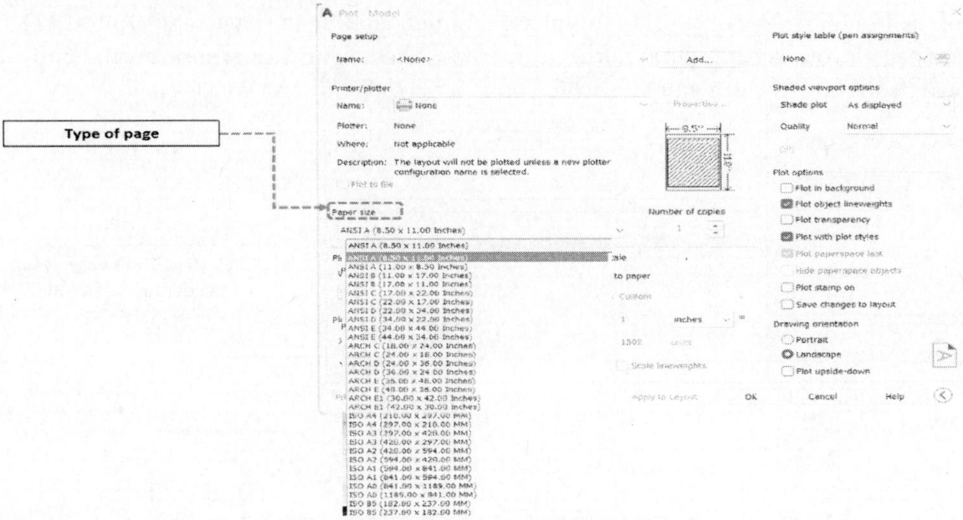

Fig. 6.6: Page size selection

The setup involves selecting the Printer/Plotter and configuring the page according to the size and orientation required. The area of plotting is also a vital parameter, offering four options:

Display: Plots all objects displayed in the drawing area.

Extents: Plots all objects in the drawing.

Limits: Prints or plots the current grid limits. This option is only available from the Model tab.

Window: Plots objects in the area you define.

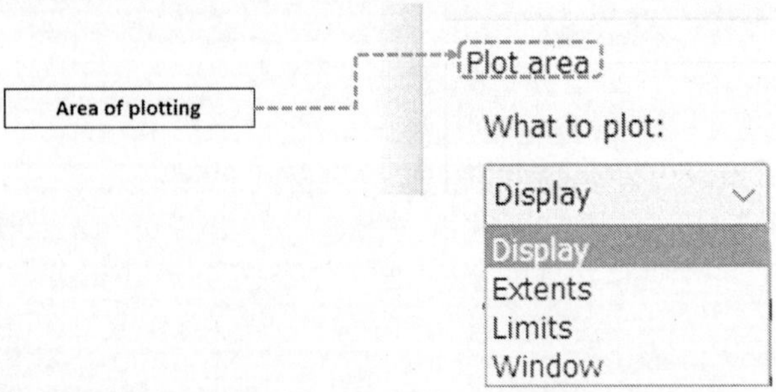

Fig. 6.7: Selection of area of plotting

Additionally, you can define the plot offset, which adjusts the plot area relative to the lower-left corner of the printable area or paper edge. Lastly, understanding the concept of Plot Scale is essential. It scales the entire drawing, including all entities and features, according to a specific ratio.

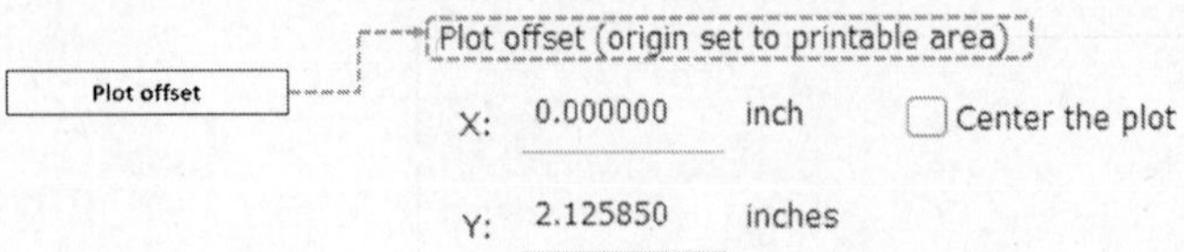

Fig. 6.8: Plot offset

You can offset the geometry on the paper by entering a positive or negative value in the X – and Y – offset boxes. However, this may result in the plot area being clipped.

When you choose this scaling option, it scales the entire drawing, including all the entities and features. As such, you need to be sure of and keep in mind all the scaling factors you might have used while drawing the sketch. If you have used no scale while drawing, this option is handy to do it all at once.

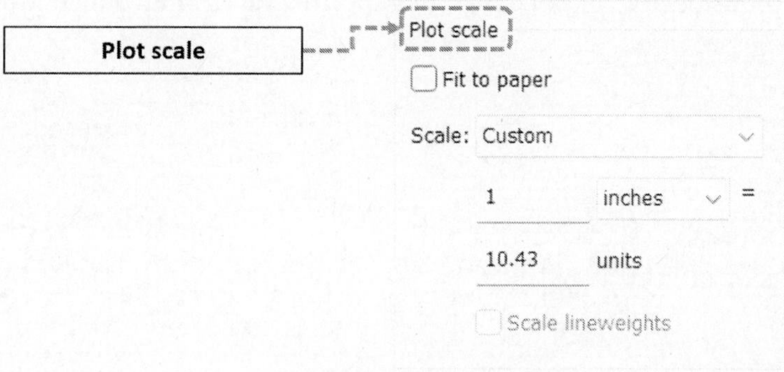

Fig. 6.9: Plot scale

6.2.2 Creating Plot Style Tables

Plot styles in AutoCAD serve as a comprehensive tool to manage how your drawing gets translated into printed or PDF form. They determine the characteristics of your lines, texts and other objects, controlling factors like color, lineweight and dithering. This section provides an in-depth guide to creating plot style tables, including Named Plot Styles (STB) and Color-Dependent Plot Styles (CTB).

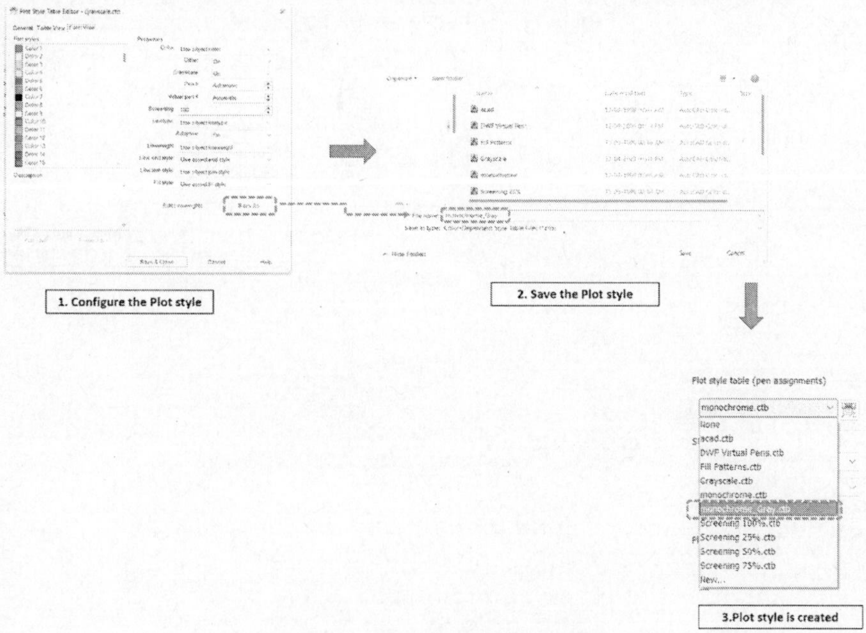

Fig. 6.10: Creating plot style

6.2.2.1. *Understanding Plot Styles*

Before diving into the creation of plot style tables, it is important to understand what they are and why they are useful. Plot styles in AutoCAD come in two types:

1. **Named Plot Style (STB)**: Here, plot styles are attached directly to objects and layers, irrespective of their color. You can customize as many plot styles as you need.
2. **Color-Dependent Plot Style (CTB)**: In this case, plot styles are tied to the 256 standard AutoCAD colors. Every object colored in a specific way is plotted identically according to the color-based plot style.

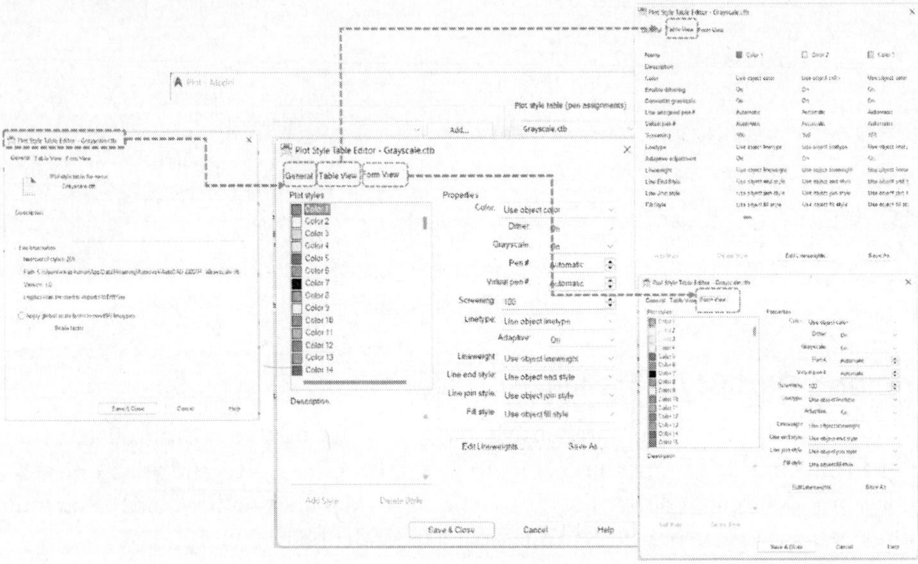

Fig. 6.11: Plot style tables editor

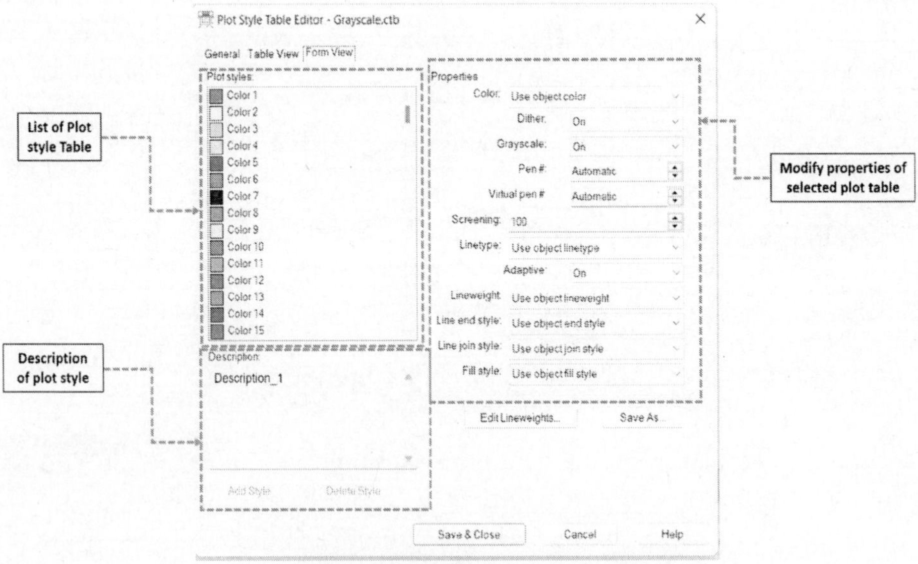

Fig. 6.12: Plot style tables editor options

6.2.2.2. *Process of Creating Plot Style Tables*

Step 1: Select a color to modify: In a Color-Dependent table, selecting a color overrides that color's properties for all objects assigned to it. For example, if you change the plot style color of red objects to black, then all red objects will plot as black. Named plot styles work similarly, but you are modifying a style that could be applied to any layer or object.

Step 2: Select an object color: The color of an object can be modified to suit your requirements. This color selection will serve as the base for the plot style being created or modified.

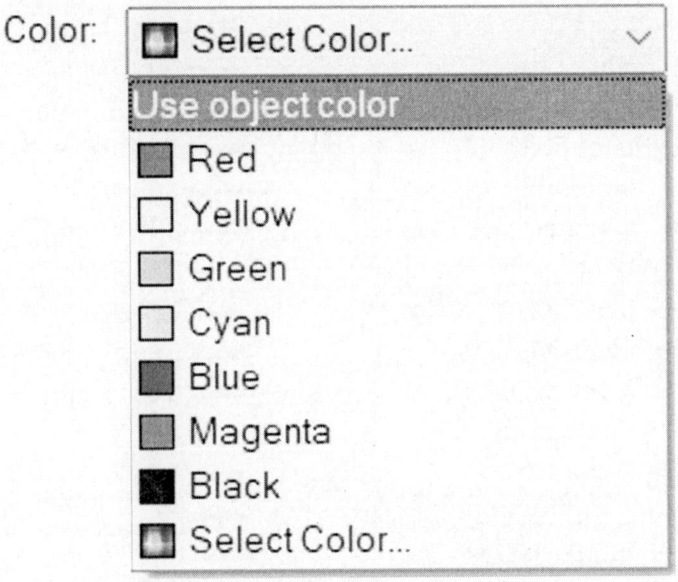

Fig. 6.13: Selecting an object color

Step 3: Dithering: Dithering enables you to make lines appear as a series of dots to simulate shading or to enhance the representation of intricate objects. It is particularly useful for plotters that may not offer full grayscale printing. Note that if the selected plotter or printer does not support dithering, this setting will be ignored.

Step 4: Grayscale: Turning on the grayscale setting will convert all AutoCAD object colors into shades of gray. This is useful when you are working with complex drawings where color might be distracting or unnecessary.

Step 5: Set a screening value: Screening alters the intensity of the colors in the final plot. A value of 100 will preserve the color at its original intensity, while a value of 0 will make it completely white. This is useful for emphasizing or de-emphasizing particular elements of a drawing.

Step 6: Line modifications: The various line modifications are—

- ***Line type:*** You can override the line type applied to the object. The 'adaptive adjustment' setting can complete the line type pattern, ensuring it doesnot appear fragmented.
- ***Line Weight:*** Here, the weight of the line can be altered. This will override the object's existing line weight. You can easily adjust these values in the Plot Style Table Editor.

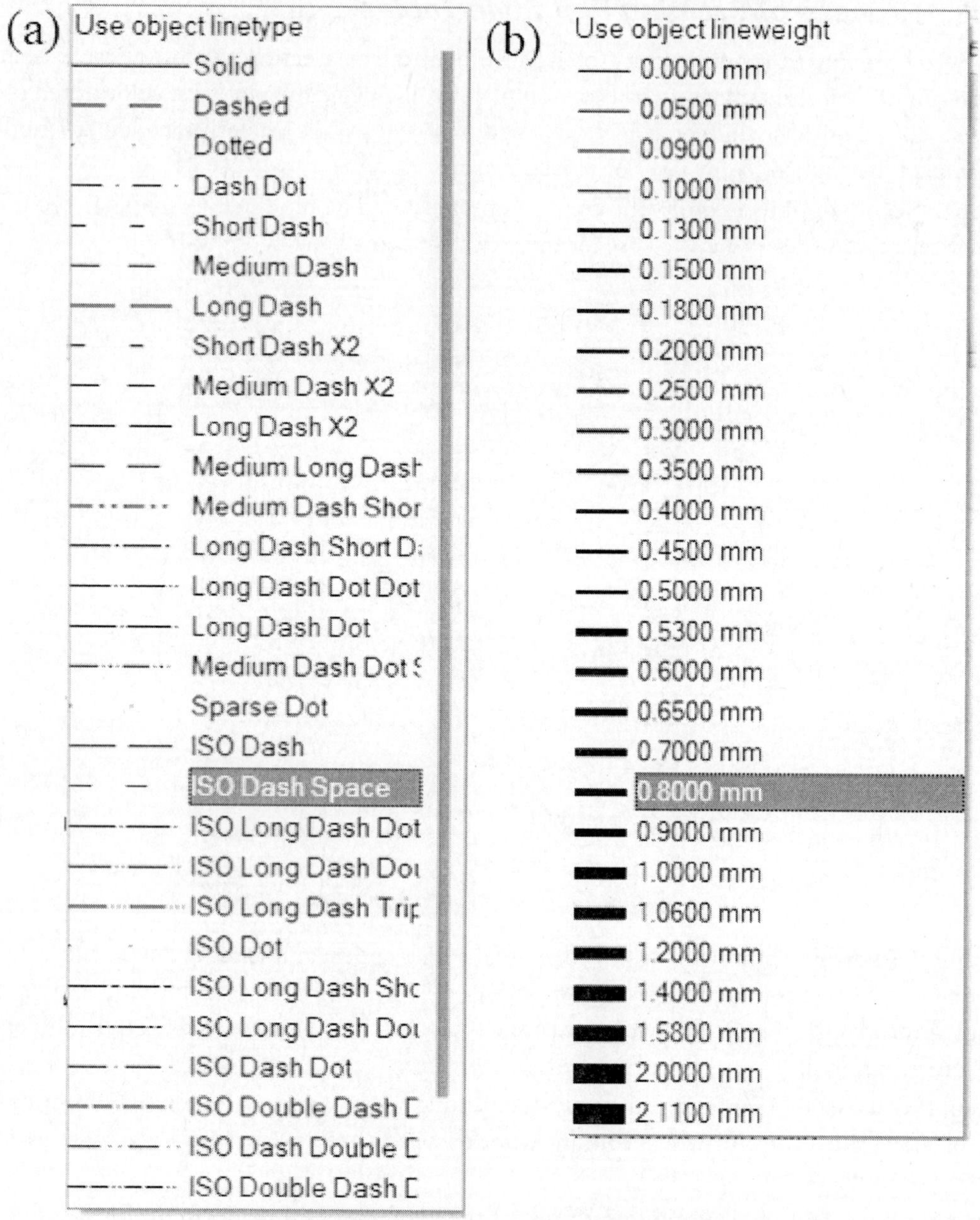

Fig. 6.14: (a) Line type (b) Line weight

- ***Line end style:*** This adjusts the style at the ends of lines, polylines and arcs. Thicker lines make these adjustments more noticeable.
- ***Line join style:*** You can set the style of joins between two lines that share an endpoint. Options include Miter, Bevel, Round and Diamond.
- ***Line fill style:*** This feature allows you to change how a filled object is displayed in the plot. This includes solid hatches, donuts and wide polylines.

Fig. 6.15: (a) Line end style (b) Line join style (c) Line fill style

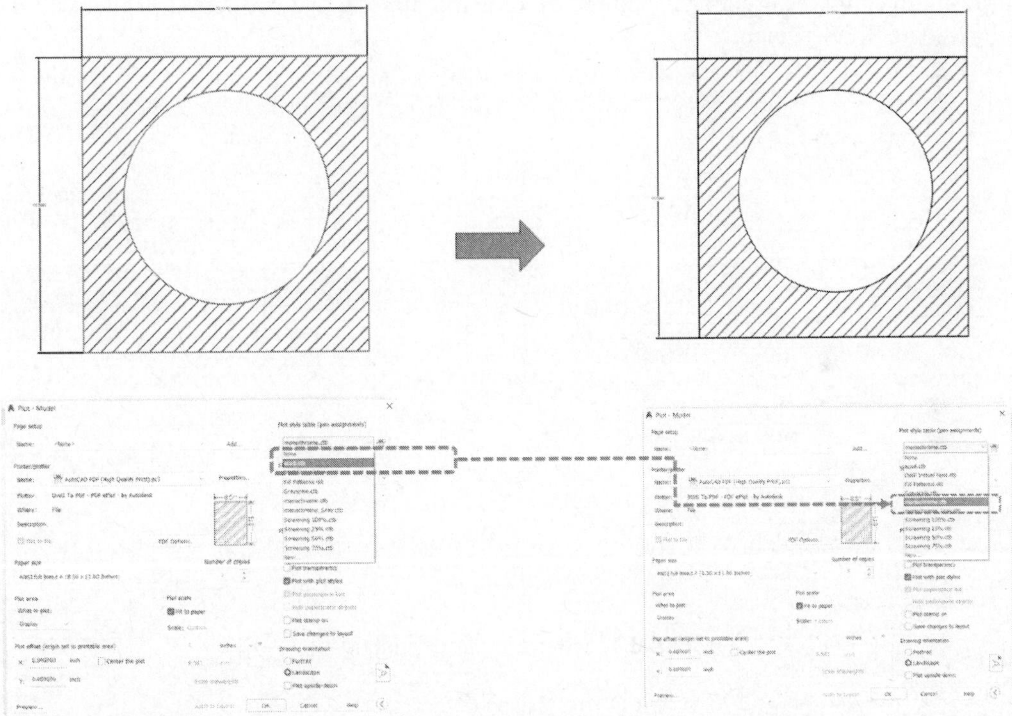

Fig. 6.16: Example of changing the plot style leads to change in graphics (colour) of the object

6.3 PRINTING

Printing in AutoCAD is not just a matter of hitting "Ctrl + P." Several options and file formats exist to ensure that your printed material accurately represents your digital drawings. This section will guide you through the different file formats, methods of direct printing, printing by layout and the Sheet Set Manager's role in AutoCAD printing.

6.3.1 File Formats for Printing

AutoCAD accommodates a variety of non-editable file formats for printing, each with its distinct set of attributes:

- ***DWF6 ePlot:*** Creates a Drawing Web Format (.DWF) file, a vector-based, lightweight format developed by Autodesk. These files can be viewed using a free Autodesk viewer.
- ***DWG to PDF:*** Produces a PDF (Portable Document Format) file, which is widely used across multiple industries. Adobe provides a free viewer for PDF files.
- ***Publish To Web JPG:*** Outputs a JPG image, which is a raster-based format. The quality degrades when zooming in and is not recommended for professional drawing issuance.
- ***Publish To Web PNG:*** Creates a PNG image file, another raster-based format. Unlike JPG, PNG is uncompressed, making it suitable for web usage but not for CMYK printing.

6.3.2 Direct Printing

Direct printing in AutoCAD refers to sending the drawing directly to a connected printer without first converting it to another format.

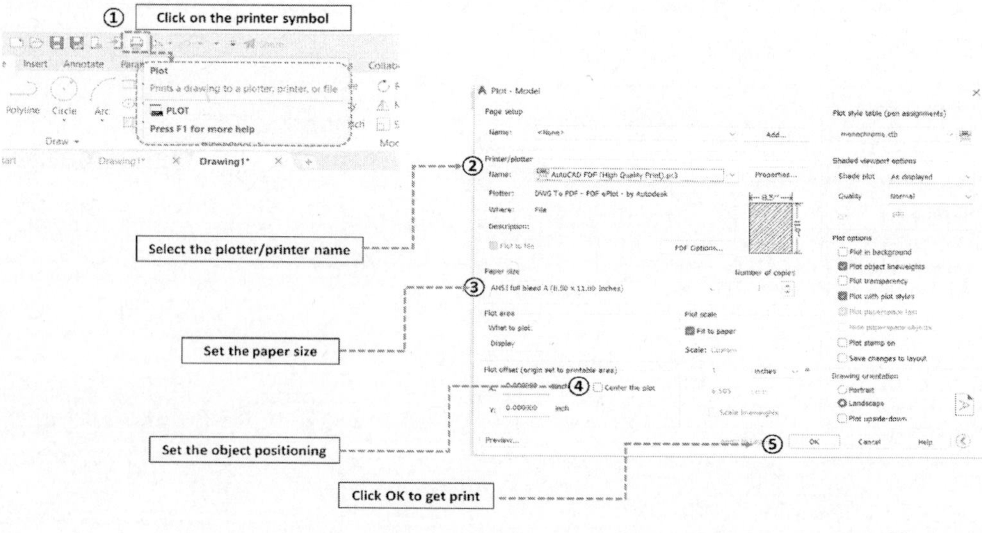

Fig. 6.17: Steps in direct printing

6.3.3 Print by Layout

- ***Open the Drawing:*** Launch the AutoCAD file you intend to print.
- ***Layout Tab:*** Navigate to the "Layout" tab in the AutoCAD ribbon.
- ***Print Area Setup:*** In the Layout tab, click on "Print Area" in the "Page Setup" group.
- ***Define the Area:*** A dialog box will appear offering various methods to define the print area, either by coordinates, dimensions, or object selection.
- ***Confirmation:*** Once the print area is set, click "OK" to close the dialog box.
- ***Preview:*** Use the "Zoom" button to view the defined print area for accuracy.
- ***Plotting:*** Click the "Plot" button and select your desired printer and settings to initiate printing.

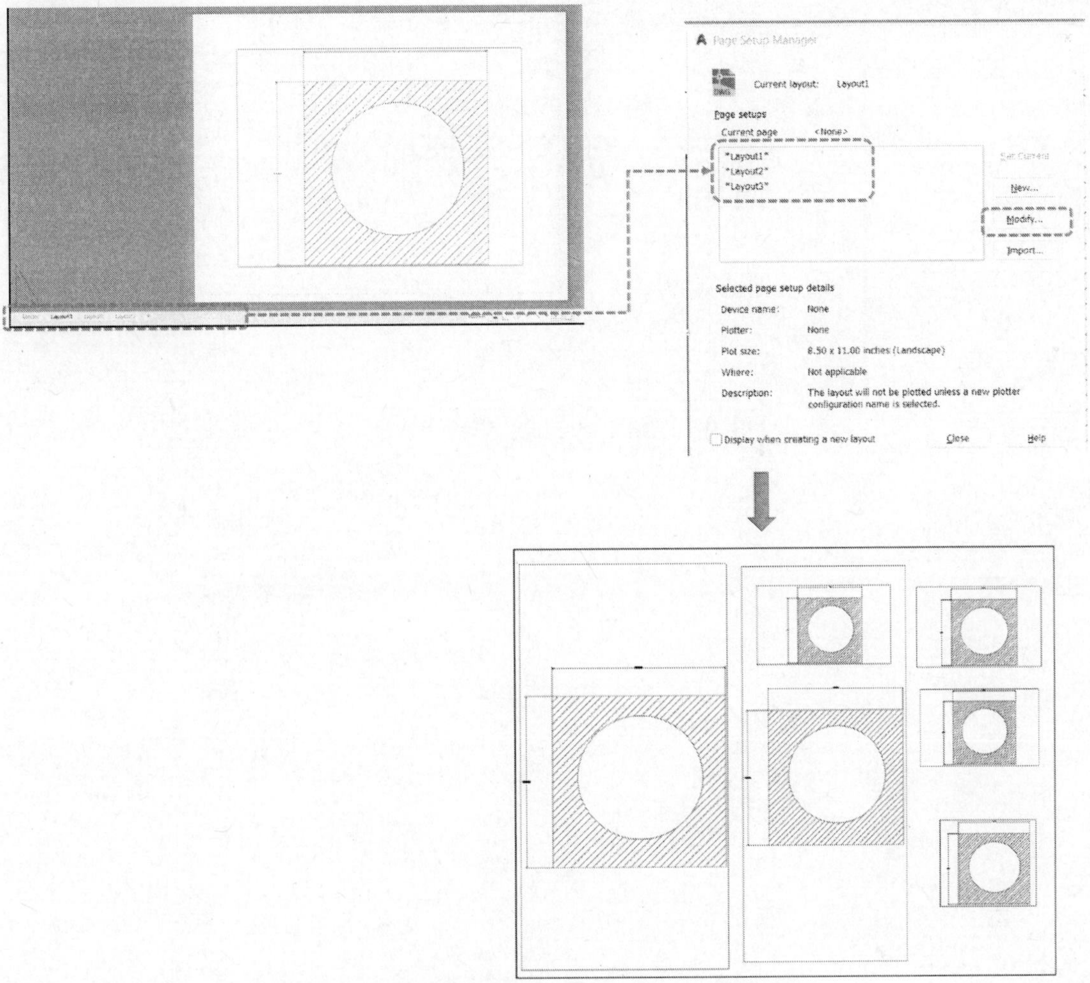

Fig. 6.18: Example of print by layout

6.3.4 Sheet Set Manager

The Sheet Set Manager serves as a comprehensive file management system for digital projects. It allows you to collate all project drawings in a unified location.

- ***New Sheet Set:*** Go to "New > Sheet Set."
- ***Predefined Sets:*** Autodesk offers three example sheet sets for Architectural, Civil and Mechanical designs.
- ***Name and Location:*** Assign a name to the new sheet set and designate its storage location.
- ***Select Layouts:*** Browse through your drawing layouts that need to be printed and select them.
- ***Preview and Finish:*** Review your sheet set and click "Finish."
- ***Publishing:*** To print, right-click on the new set and navigate to "Publish > Publish to PDF."
- ***Final Step:*** The drawings in the sheet set will be published to PDF according to the layouts assigned, ready for printing.

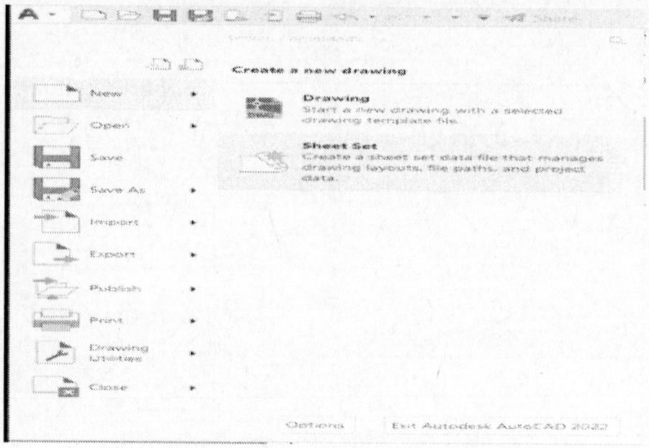

Fig. 6.19: Selecting new sheet set

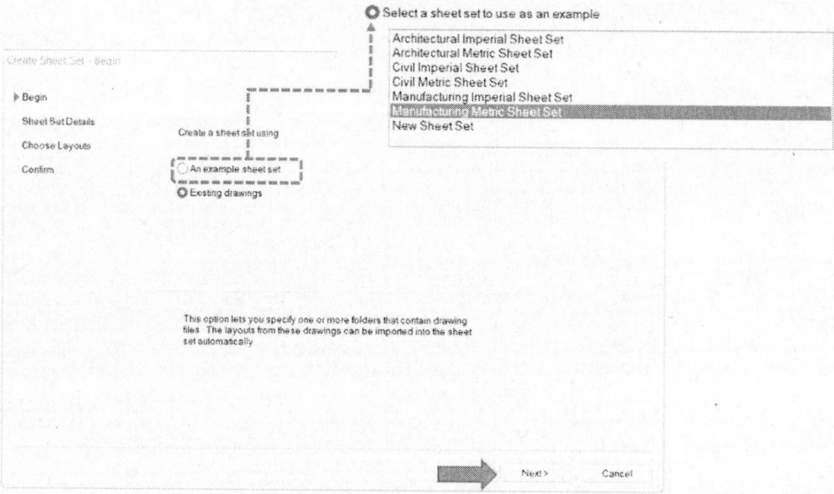

Fig. 6.20: Selecting predefined sets

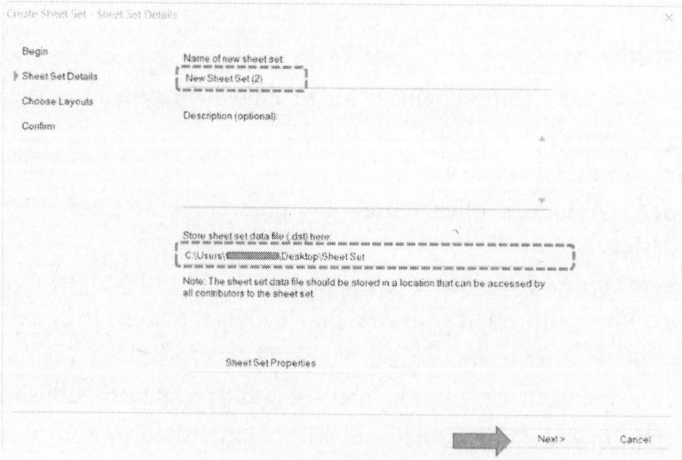

Fig. 6.21: Assigning name and storage location

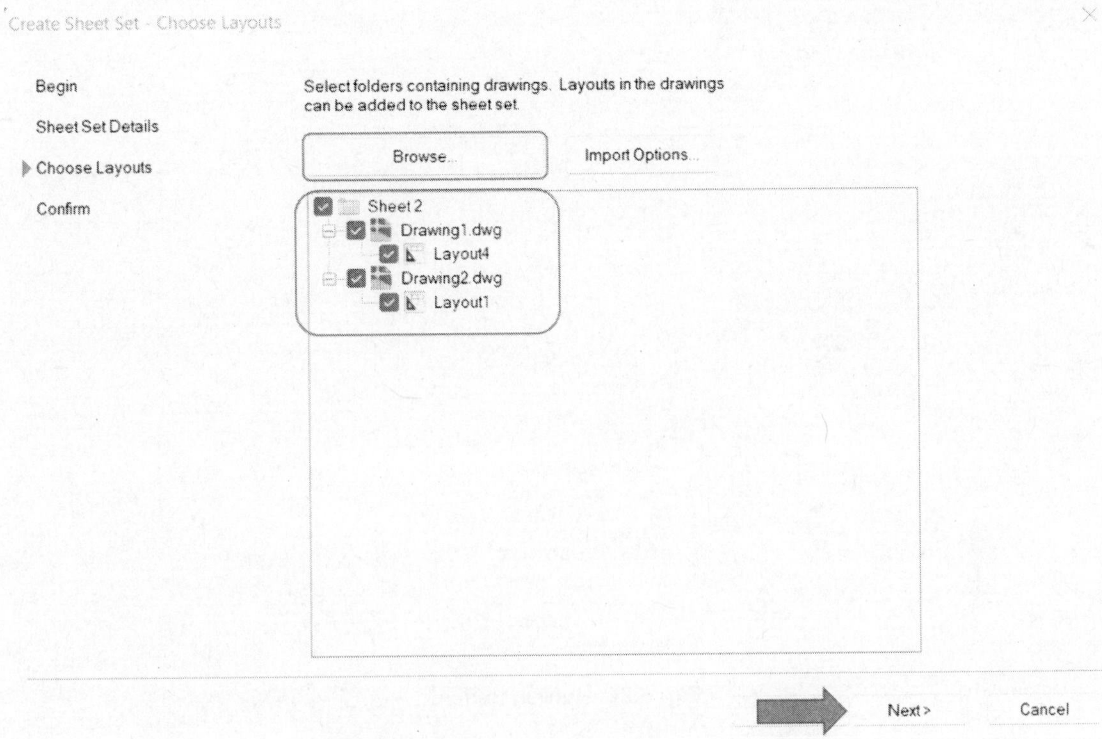

Fig. 6.22: Browsing drawing layouts for printing

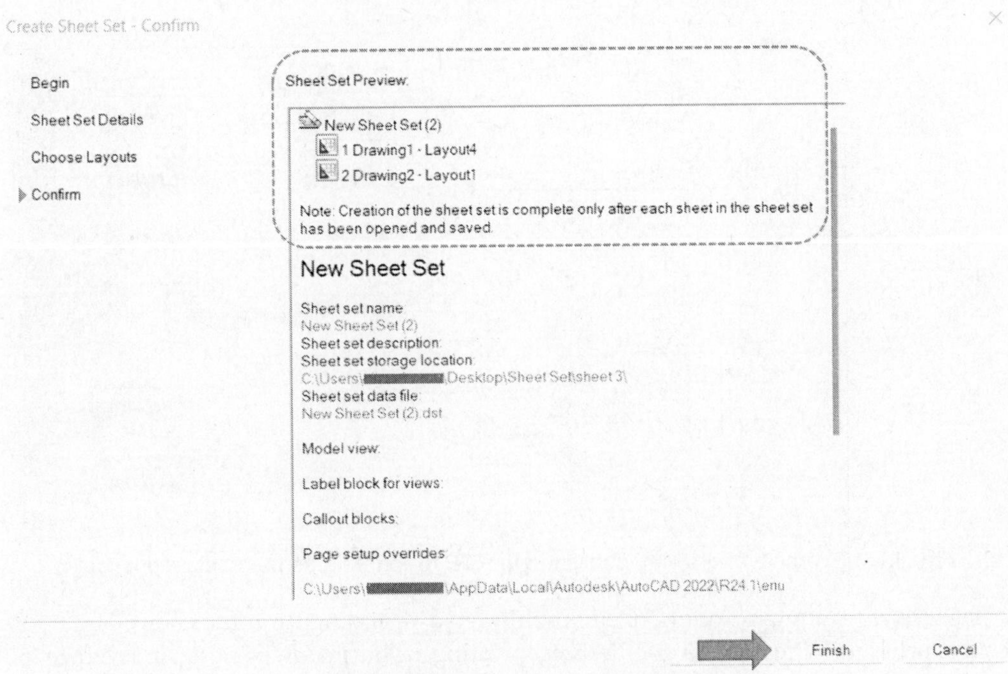

Fig. 6.23: Previewing and finalizing

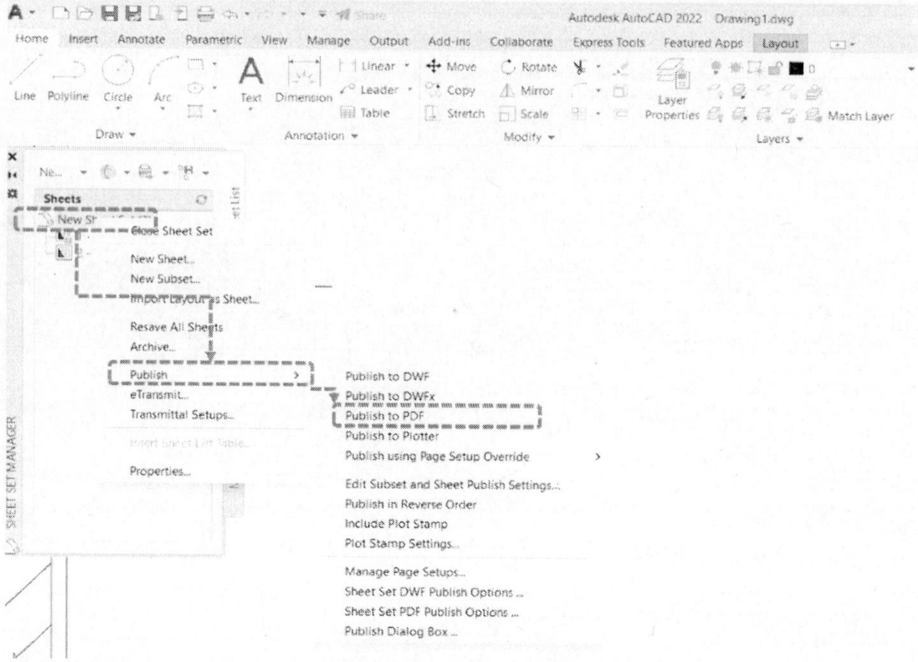

Fig. 6.24: Publish to PDF

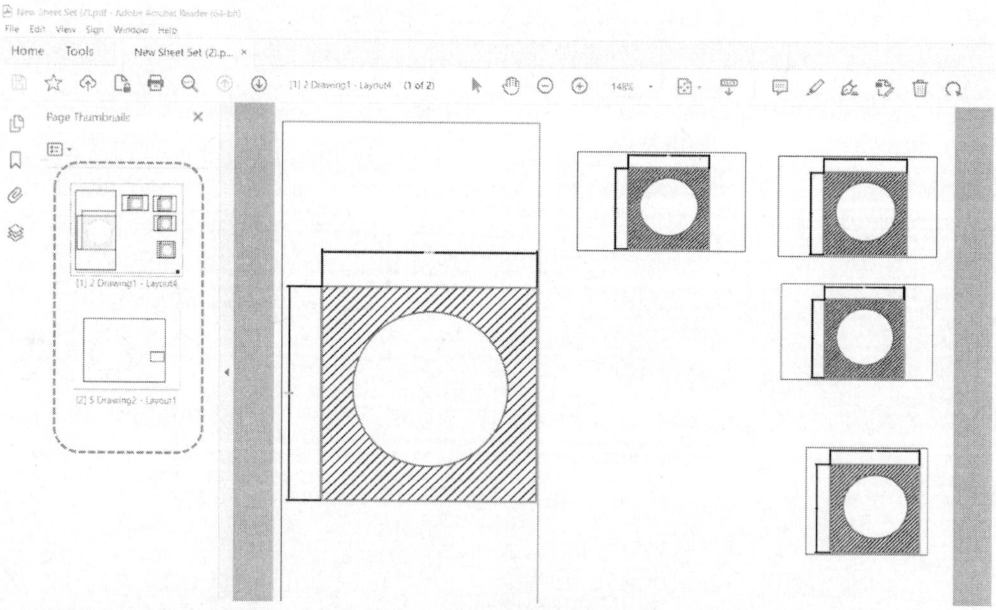

Fig. 6.25: Example of drawing published to PDF for printing as per assigned layout

The nuances of printing in AutoCAD are multifaceted, spanning from the choice of file format to the method of printing. Whether you opt for direct printing, layout-based printing, or leverage the Sheet Set Manager for larger projects, understanding these aspects will equip you with the skills needed for effective and accurate printouts.

UNIT SUMMARY

Based on the contents covered in this unit you should remember that:

- Plotting in AutoCAD involves a multi-step process to convert digital designs into high-quality, printable outputs. It encompasses configuring plotters, setting up page layouts and defining plot style tables.
- Page setups in AutoCAD manage layout settings and plotter properties, allowing for consistency and efficiency across multiple layouts in a drawing file.
- The area for plotting can be defined using options like Display, Extents, Limits and Window. Plot offsets and scaling are important parameters for precise printing.
- Plot Style Tables, including Named Plot Styles (STB) and Color-Dependent Plot Styles (CTB), offer comprehensive control over line characteristics, colors and other object properties in the final output.
- Creating or modifying plot style tables involves steps like selecting object colors, setting dithering and grayscale options and adjusting line properties such as line weight and type.
- Printing in AutoCAD isn't just pressing "Ctrl + P." Different file formats like DWF, PDF, JPG and PNG are available for printing, each with its own characteristics.
- Direct printing in AutoCAD allows you to send the drawing directly to a connected printer, while Print by Layout involves defining the print area in the Layout tab and proceeding with various print settings.
- The Sheet Set Manager is a robust tool for managing project drawings. It allows you to create new sheet sets, organize layouts and publish entire sets to PDF for printing.

EXERCISES

A. Multiple Choice Questions

6.1 What is the primary way to access the plotting configuration settings in AutoCAD?

(a) From the "File" menu

(b) By right-clicking on the Layout tab

(c) Typing 'plot' in the command line

(d) Clicking on the Plot icon in the Output toolbar

6.2 Which of the following can be defined in a Page Setup?

(a) Object colors (b) Area for plotting

(c) Layer visibility (d) Plot offsets and scaling

6.3 What do Named Plot Styles (STB) control?

(a) Layout settings (b) Object colors

(c) Line weight and type (d) Plotter configurations

6.4 What file format cannot be used for printing in AutoCAD?

(a) PDF (b) PNG (c) DWF (d) MP4

6.5 What is the Sheet Set Manager primarily used for?
(a) Changing line properties (b) Managing project drawings
(c) Converting file formats (d) Editing hatch settings

6.6 Which area for plotting cannot be defined in AutoCAD?
(a) Display (b) Extents (c) Limits (d) Circle

6.7 How can you perform direct printing in AutoCAD?
(a) Clicking on "Print" in the File menu
(b) Typing 'directprint' in the command line
(c) Sending the drawing directly to a connected printer
(d) Using the Sheet Set Manager to send to printer

6.8 Which of the following is not a type of Plot Style Table?
(a) STB (b) CTB (c) PTB (d) WTB

6.9 What does Print by Layout involve?
(a) Printing directly to the printer (b) Defining the print area in the Layout tab
(c) Sending the drawing to a PDF (d) Configuring plotters

6.10 What are Color-Dependent Plot Styles (CTB) used for?
(a) Controlling plotter settings (b) Managing layout settings
(c) Adjusting color-based line characteristics (d) Editing text in a drawing

6.11 What does scaling adjust in the plotting process?
(a) File size (b) Plotting speed
(c) Object size in printout (d) Print resolution

6.12 What is the primary function of plot offsets?
(a) To adjust line weight
(b) To change the starting point of the plot on the paper
(c) To change the plot style
(d) To change the paper size

6.13 What does dithering in a plot style table control?
(a) Line weight (b) Grayscale options (c) Color consistency (d) Line type

6.14 How can you publish an entire sheet set to PDF?
(a) From the "File" menu
(b) By right-clicking on the Sheet Set Manager
(c) Typing 'publish' in the command line
(d) Saving the sheet set as PDF

6.15 Which among the following is not a function of the Page Setup?
(a) Managing layout settings (b) Defining plotter properties
(c) Controlling object visibility (d) Editing hatch patterns

Answers to Multiple-Choice Questions
6.1 (d), 6.2 (d), 6.3 (c), 6.4 (d), 6.5 (b), 6.6 (d), 6.7 (c), 6.8 (c), 6.9 (b), 6.10 (c), 6.11 (c), 6.12 (b), 6.13 (b), 6.14 (b), 6.15 (d)

B. Subjective Questions

6.1 What is Page Setup in AutoCAD and how does it influence the plotting process?

6.2 Explain how to define areas for plotting in AutoCAD.

6.3 Describe the difference between Named Plot Styles (STB) and Color-Dependent Plot Styles (CTB).

6.4 How do you adjust plot scaling in AutoCAD?

6.5 Discuss the concept of Plot Offsets.

6.6 What is the purpose of dithering in a plot style table?

6.7 What does direct printing mean in the context of AutoCAD

6.8 Explain the function of the "Publish" feature in AutoCAD.

6.9 What are the advantages of using the Sheet Set Manager for plotting?

6.10 Describe how you can publish an entire sheet set to PDF.

KNOW MORE

- *High-Quality Plotting*: Beyond the standard plotting features, AutoCAD offers a High-Quality Plot mode which employs advanced algorithms for rendering sharp, crisp lines and fills. This can be especially beneficial when your project requires photorealistic printouts or when fine details need to be emphasized.
- *Batch Plotting*: One less-known feature is Batch Plotting, which allows you to print multiple drawings or layouts in one go. This can be incredibly efficient when you have to produce a series of plots for project documentation or presentations.
- *Plot Stamps*: AutoCAD allows users to add plot stamps to their drawings automatically. A plot stamp can include vital information like username, date and filename, which can be instrumental in project management and auditing.
- *3D Plotting*: Although the chapter primarily focused on 2D plotting, AutoCAD is also capable of 3D plotting. The software offers options for viewing and plotting 3D models in various styles, including hidden line removal, shading and wireframe modes.
- *Geolocation Plotting*: AutoCAD enables georeferencing your design onto a map, allowing for real-world context in your plots. This feature is useful in fields like urban planning, landscape design and civil engineering where spatial context is key.
- *Print Styles and Presets*: Beyond the basic and advanced plot styles, you can also create and save your own plot style presets. This is incredibly helpful for maintaining consistency across multiple projects or when sharing work among a team.
- *Plotting to Different File Formats*: In addition to printing on paper, AutoCAD also allows you to plot your drawings to various file formats, like PDF, DWF and even image files like PNG and JPEG. This is particularly useful for digital presentations or archiving.
- *PDF Layers*: When plotting to a PDF, AutoCAD can retain the layer information, enabling anyone with a PDF reader to turn layers on and off within the PDF. This makes the document far more interactive and informative.

REFERENCES AND SUGGESTED READINGS

1. Hamad, Munir. AutoCAD 2019 Beginning and Intermediate. Mercury Learning and Information, 2018.
2. Cadfolks. AutoCAD 2020 For Beginners, 7th ed.; Kishore, 2019.
3. Shoukry, Yasser and Jaiprakash Pandey. Practical Autodesk AutoCAD 2021 and AutoCAD LT 2021: A no-nonsense, beginner's guide to drafting and 3D modeling with Autodesk AutoCAD. Packt Publishing Ltd, 2020.

Dynamic QR Code for Further Reading

Printing and Plotting Interface

Understanding File Formats

Create Plot Style Tables

Setting Up Print Areas

Page Setup for Printing

Save or Print as PDF

Sheet Set Manager

Publish Drawings with Sheet Set Manager

Batch Plotting

7 Machine Drawing Practice Using AutoCAD

UNIT SPECIFICS

This unit presents information related to the following topics:

- temporary fasteners like sleeve & cotter joint, spigot & cotter joint, knuckle joint;
- joining of shafts with couplings like protected flanged coupling and universal coupling;
- bearing components like Plummer block and footstep bearing;
- critical components like eccentric, stuffing box and connecting rod;
- complex assemblies like machine vice and screw jack;

This unit is meant to serve as a well-dimensioned and detailed list of various views of certain important machine parts such that it enables the students to carry out effective practice on CAD software. The part drawings are carefully dimensioned along with the bill of materials so that students find it convenient to reproduce them using CAD software. A concise yet informative discussion is included for all the machine parts presented in this unit so that the students develop a basic understanding of the functioning and importance of each machine part.

Apart from this at the end of the unit, a succinct recapitulation of the overall broad concepts is provided in form of a unit summary. Apart from a large number of multiple-choice questions, descriptive type questions with Bloom's taxonomy action verbs are also included. A list of references and suggested readings are given in the unit so that the students can go through them for practice. It is important to note that for getting more information on various topics of interest some QR codes have been provided in different sections which can be scanned for relevant supportive knowledge.

RATIONALE

The drafter, checker and other professionals tasked with reviewing the drawing before release are accountable for making sure it complies with standards, is thorough, clear, correct and ensures proper functional operation. As a way to encourage the creation of high-quality drawings, it is advisable to use standard procedure and industry standard CAD software, as appropriate, when drafting and examining drawings.

The correctness and cost of parts and assemblies are influenced by the drafter, checker and other reviewers' knowledge of the design specifications, the manufacturing procedure involved and drafting processes. To make sure that the designer and responsible engineer have a complete grasp of the function and application of the design, layouts should be carefully examined and, if necessary, debated. Discussions with responsible individuals about design or manufacturing improvements are always recommended. Finished drawings should reflect the objective findings of all responsible reviewers. The task of the drafter and reviewer becomes significantly easier if standard drawing practices and CAD software are used. This unit introduces the students to some critical parts and equipment used in various industrial applications. Detailed drawings and 3D views are provided for various parts and components so that students can have adequate practice and hone their drafting skills.

PRE-REQUISITES

Before reading this unit, the student is advised to revisit the following:

- Mathematics: Coordinate and Plane geometry (Class XII)
- Engineering Graphics (ES101): Unit II and III
- AutoCAD commands mentioned in the last six units of this book

UNIT OUTCOMES

The list of outcomes of this unit is as follows:

U7-O1: Interpret and apply drawing notations used in detailed drawings

U7-O2: Apply modify commands to draft complex parts and components

U7-O3: Create various 2D views like top, side, front etc. of complex 3D parts and components

U7-O4: Create assembly drawings from part drawings

U7-O5: Apply dimensions to 2D part views using AutoCAD software

Unit-7 Outcomes	EXPECTED MAPPING WITH COURSE OUTCOMES (1 – Weak Correlation; 2 – Medium correlation; 3 – Strong Correlation)				
	CO-1	CO-2	CO-3	CO-4	CO-5
U7-O1	3	1	1	1	1
U7-O2	-	1	3	1	-
U7-O3	-	1	2	-	1
U7-O4	-	-	-	3	-
U7-O5	2	1	1	2	3

7.1 COTTER JOINT WITH SLEEVE

A sleeve and cotter joint is a temporary fastening method employed to connect two coaxial shafts that have identical rod-like geometry. The joint utilizes fasteners called 'cotters' that are tapered wedge inserts that can be pushed through slots made in a sleeve and rods assembly – with the rods placed in the sleeve from both sides. Typically, the cotter is only tapered on a single side – with the taper ranging from 1 in 48 to 1 in 24 and rarely at 1 in 8.

It is the simplest type of a cotter-based joint, with others being socket and spigot and gib and cotter types. The sleeve and cotter joint holds a decent amount of strength in both tensile and compressive modes of loading and constrains any angular distortion created by minor torques on the rods – and therefore helps prevent any angular misalignment of the shafts. However, this joint type is not suited for shafts that are under constant rotation and other joints and couplings must be utilized for those cases.

The sleeve and cotter joint is quite versatile and can be used on rectangular bars, hollow pipes, or a variety of cross sections – provided that the sleeve is designed to have the same inner profile as the external profile of the rods inserted in it. Popular applications include piston rod – crosshead, piston rod – pump rod tails and connecting rod strap end connections.

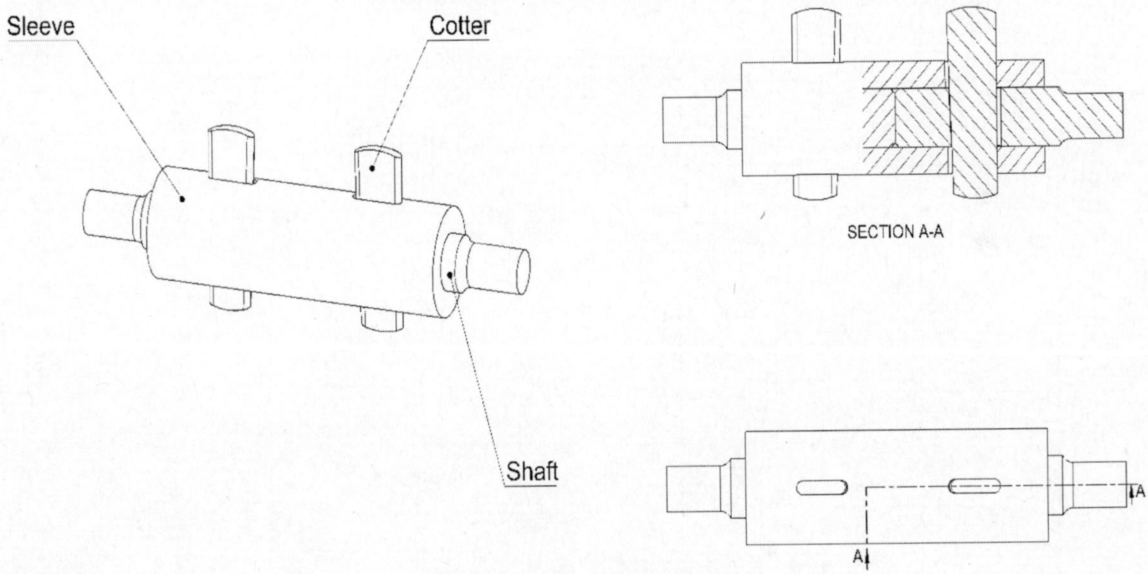

Fig. 7.1: Sleeve and cotter joint

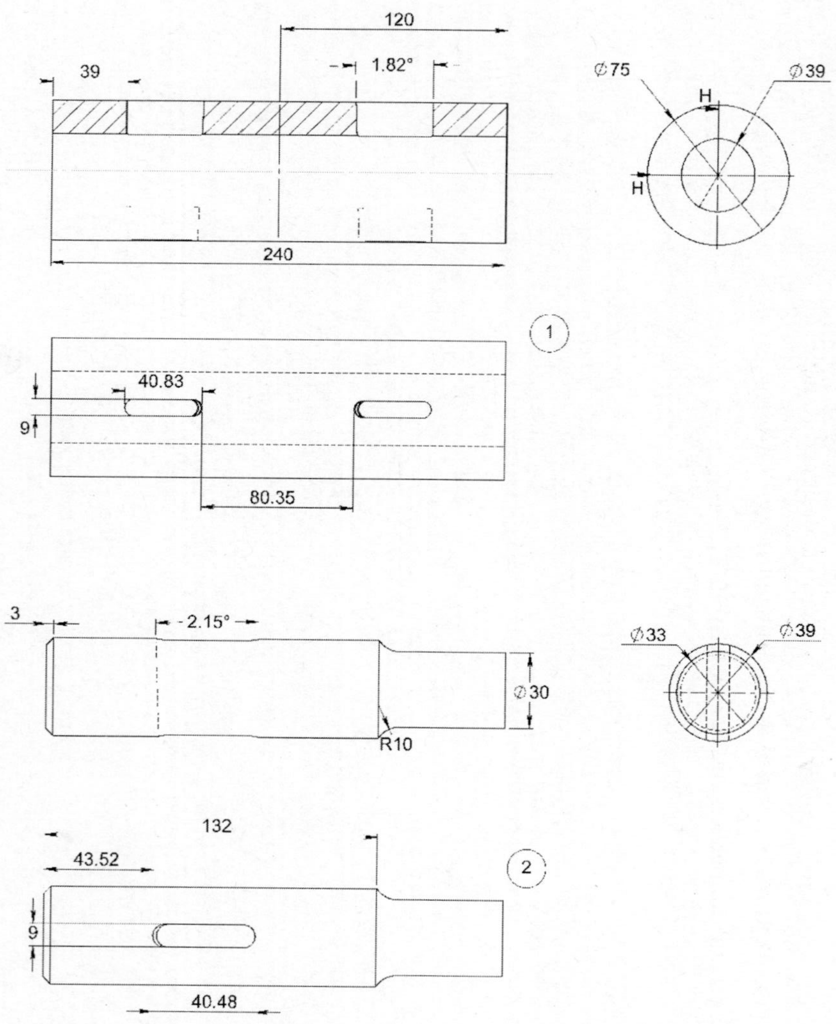

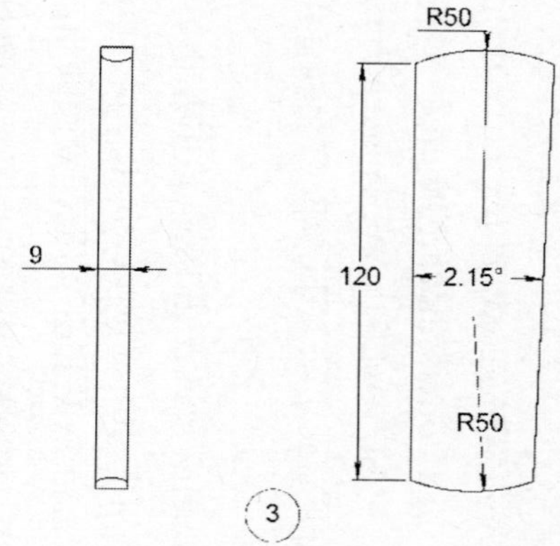

All dimensions are in millimeters.

ITEM NO.	PART NUMBER	QTY.
1	Sleeve	1
2	Shaft	2
3	Cotter	2

Fig. 7.2: Detailed views of sleeve and cotter joint

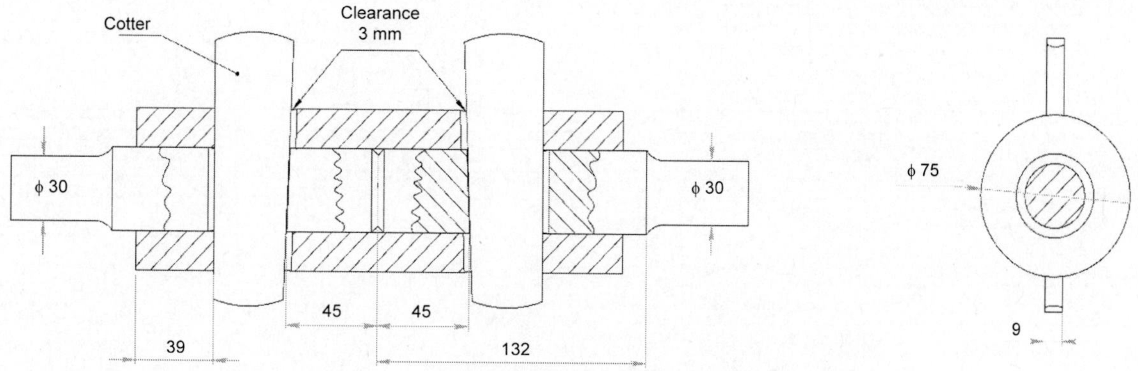

Fig. 7.3: Assembly drawing of sleeve and cotter joint

7.2 COTTER JOINT WITH SOCKET AND SPIGOT

A spigot and cotter joint is a type of temporary cotter-based fastening method, with the first one being the sleeve and cotter joint that was discussed earlier in the first section. The spigot and cotter joint also has similar applications to the sleeve joint and is used to connect two coaxial shafts in a linear arrangement. However, the additional benefit of using this type over the former is the inclusion of dissimilar-sized shaft connections. In this joint, the configuration involves the creation of a socket receptacle as the female connector on the end of a shaft and a spigot rod male connector on the opposing shaft for it to insert and fit into the socket. After coupling the cylindrical spigot into the socket, they are fastened together using cotter inserts that slide into slots that have been made into both the socket and the spigot.

The cotters used here are similar to the ones used in the sleeve and cotter joint and hold the same geometry with the 1 in 48 to 1 in 24 tapers on a side. With the cotters inserted into the rectangular slots, the tight connection ensures that both shafts remain aligned and constrained in the desired plane. Like the previous method, this joint should not be used to connect rotating shafts, as the cotters cannot withstand high centrifugal forces and could pop out of place at high speeds. Such joints have popular applications in foundation bolts for heavy machinery and in connecting bicycle pedals to the sprocket wheel.

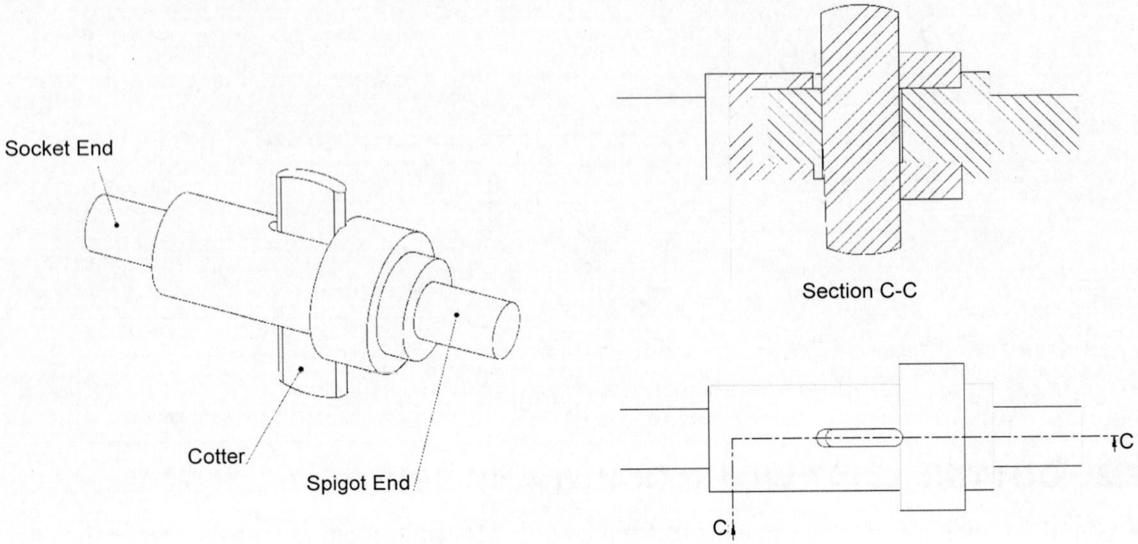

Fig. 7.4: Cotter joint with socket and spigot

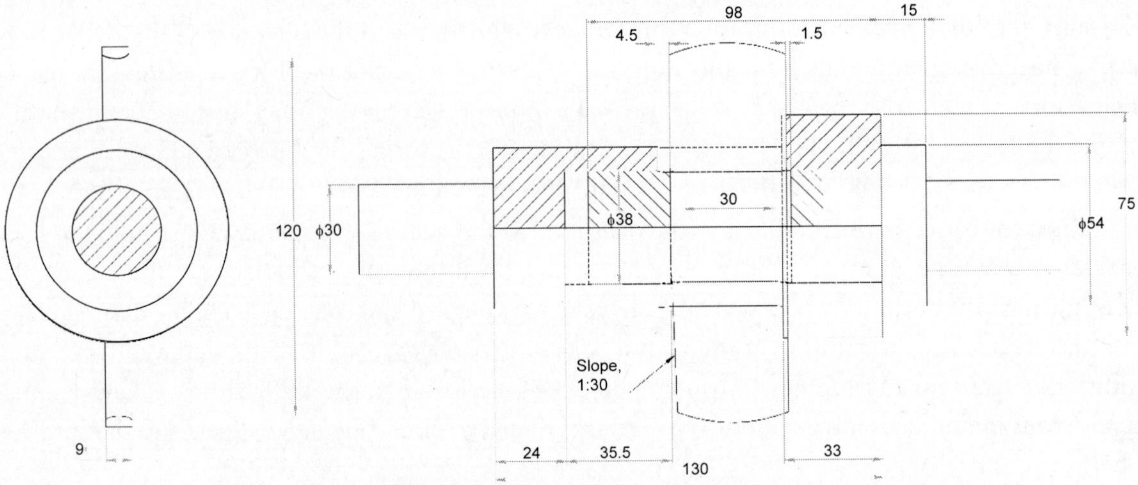

Fig. 7.5: Assembly drawing of cotter joint with socket and spigot

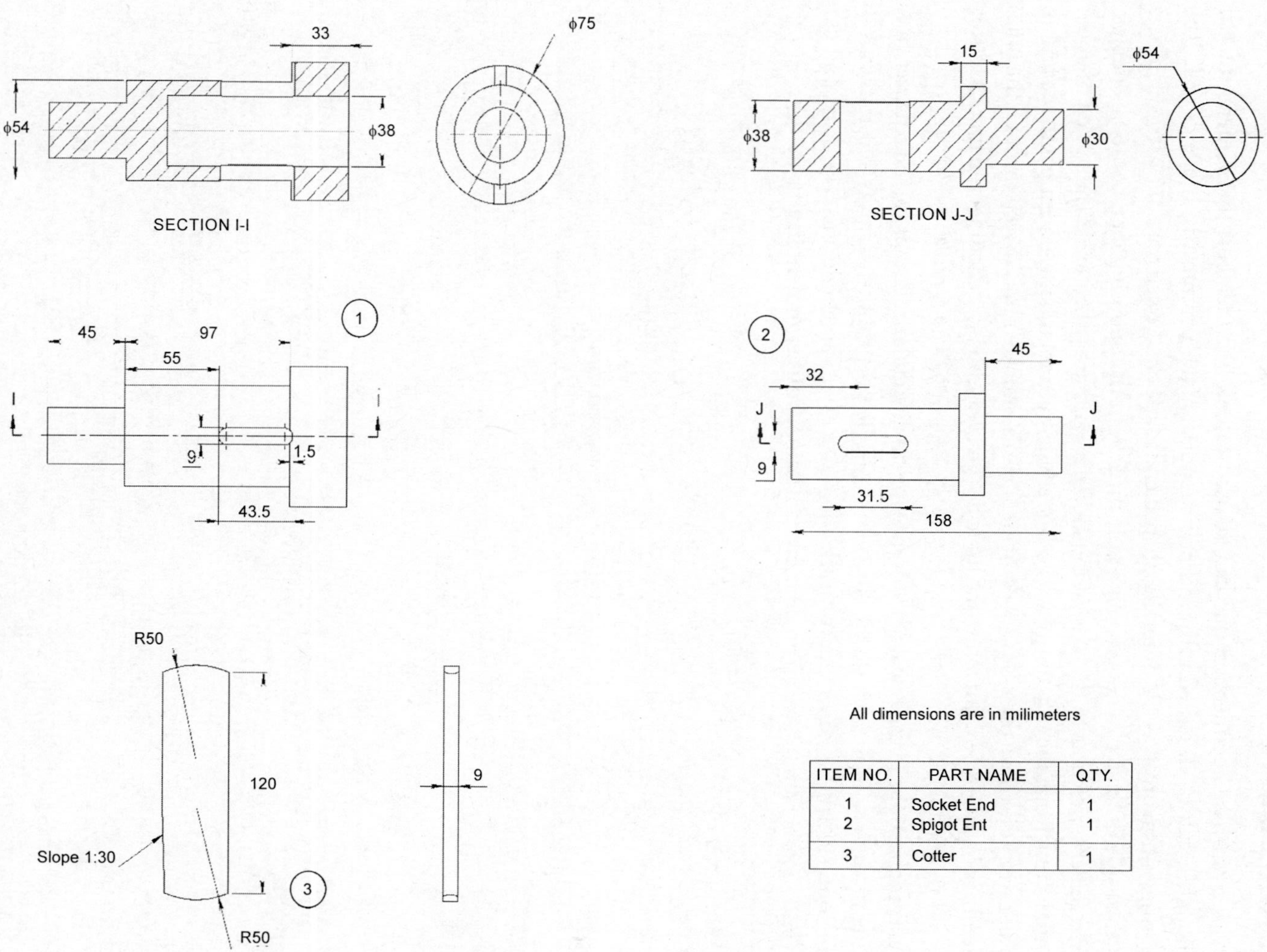

All dimensions are in milimeters

ITEM NO.	PART NAME	QTY.
1 2	Socket End Spigot Ent	1 1
3	Cotter	1

Fig. 7.6: Detailed views of cotter joint with socket and spigot

7.3 KNUCKLE JOINT

A knuckle joint is a pin-based connection that can be used to link two members that undergo tensile loadings – although it can be designed to have some support for the compressive mode as well. The knuckle joint configuration involves a U or forked-shaped end made or screwed onto one of the members with a hole on each side of the fork and a through hole made into the opposing member to be coupled. The holes are made perpendicular to the axes of the members and are called eyes. The forked member has two eyes and the other member has only one eye. The single-eyed member is lined up into the fork in a manner that all three holes have their centres coincident and a knuckle pin is driven through the holes to link both members together via a pin joint. The knuckle pin is essentially a shoulder pin that has a step at one end and a through hole at the other to support a locking pin or a split pin as a fastener.

The joint has applications in cases where allowance for small angular rotations can help prevent any harmful bending moments. These include the chain links of a bicycle, roof truss joints, tension links in bridges, suspensions, steering systems etc. A knuckle joint is easy to assemble and take apart and can connect shafts and members that have an angular or translational misalignment in a plane.

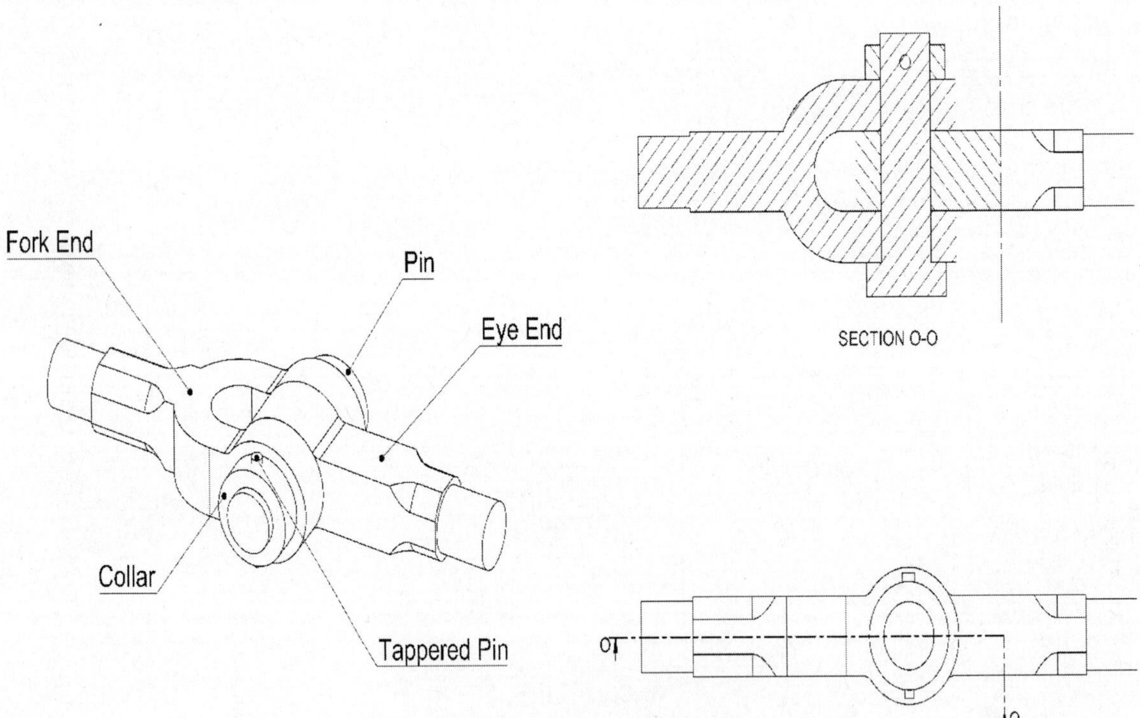

Fig. 7.7: Knuckle joint

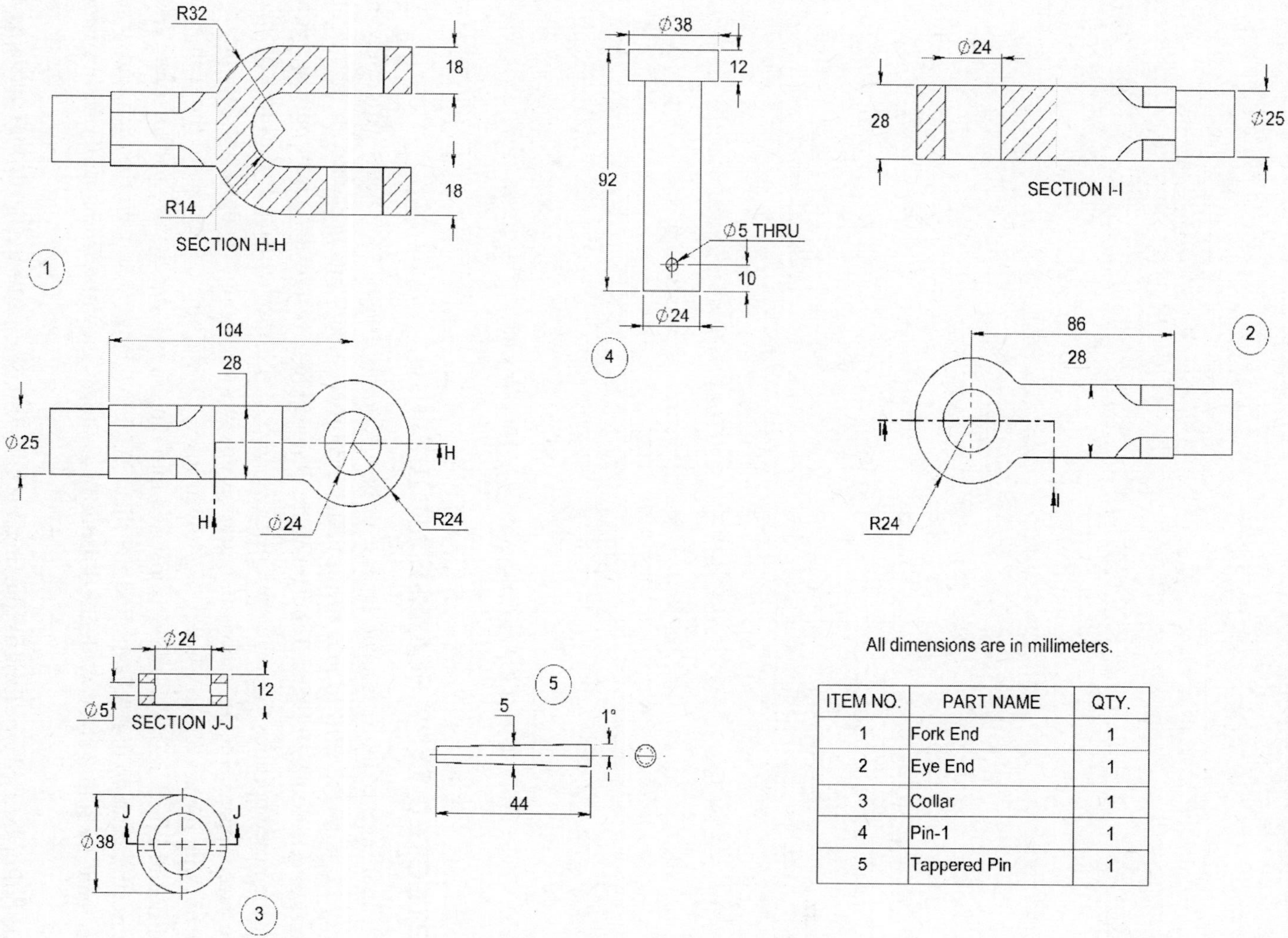

All dimensions are in millimeters.

ITEM NO.	PART NAME	QTY.
1	Fork End	1
2	Eye End	1
3	Collar	1
4	Pin-1	1
5	Tappered Pin	1

Fig. 7.8: Detailed views of knuckle joint

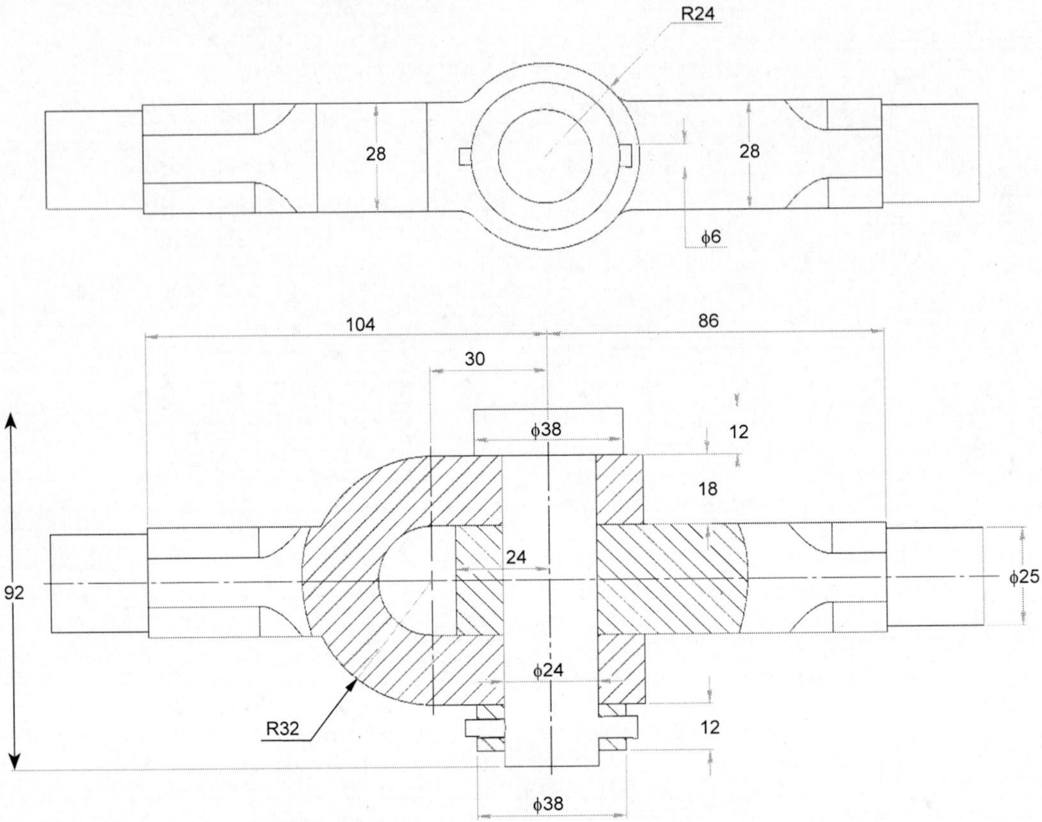

Fig. 7.9: Assembly drawing of knuckle joint

7.4 PROTECTED TYPE FLANGED COUPLING

A protected type flanged coupling is a special case of flanged joints designed for shafts with safety in mind. A flanged coupling is used to join two rotating or non-rotating shafts end-on with the aid of separate flanges attached to both shafts. The flanges are designed to have a male-female coupling, with one flange having raised portions that mate with negative recesses in the other flange to ensure a proper fit without surface slippage. The assembly configuration involves setting up a flanged disk face on the end of the shafts with the aid of a key insert. Keyway grooves are machine cut into both the shaft and the flanged disk and a thick rectangular key is inserted into the slot to lodge the two parts in place and prohibit any relative rotation of the flange and the shaft. With the flanges ready on the shaft ends, the connection is made by bolting the two identical flanged faces together, with the bolts inserted in a ring configuration.

For the specific protected type flanged coupling, the bolts and nuts are seated neatly into the flanged disk by machined counter bores so that the surface remains flush and no fastener protrudes out. This makes the design safer for the workers operating on, or in the proximity of the shafts, as the sharp edges of the hex bolts and nuts can injure and cut.

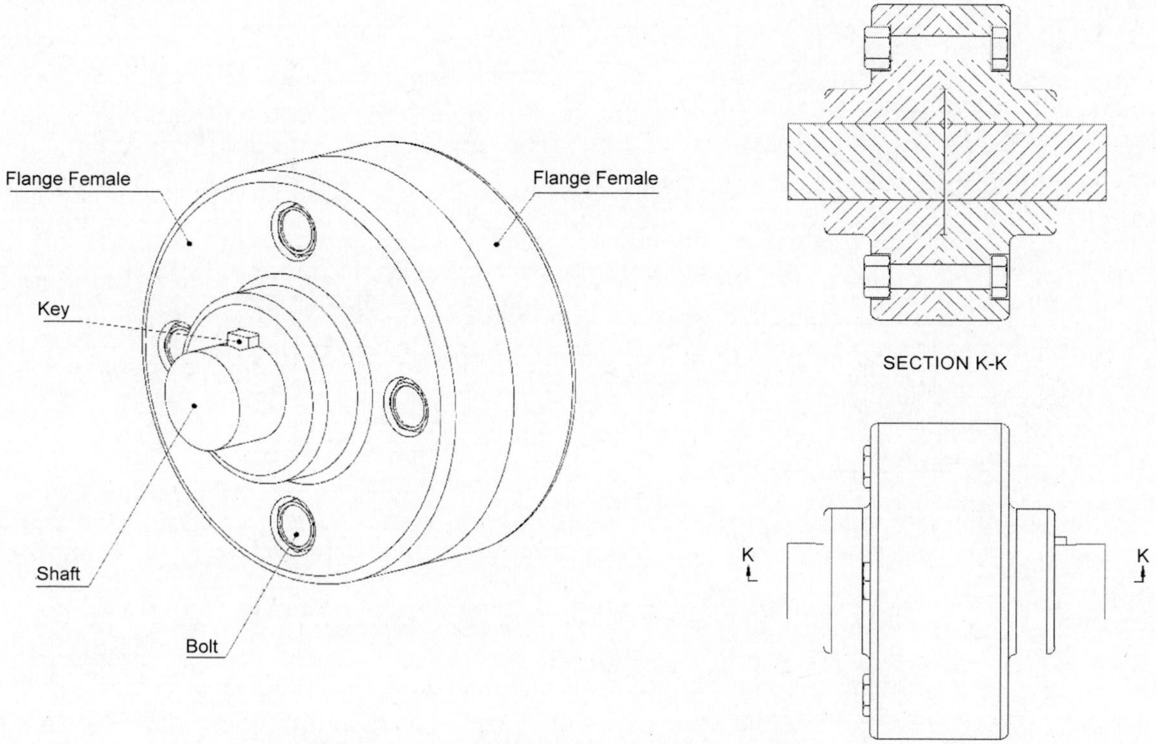

Fig. 7.10: Protected type flanged coupling

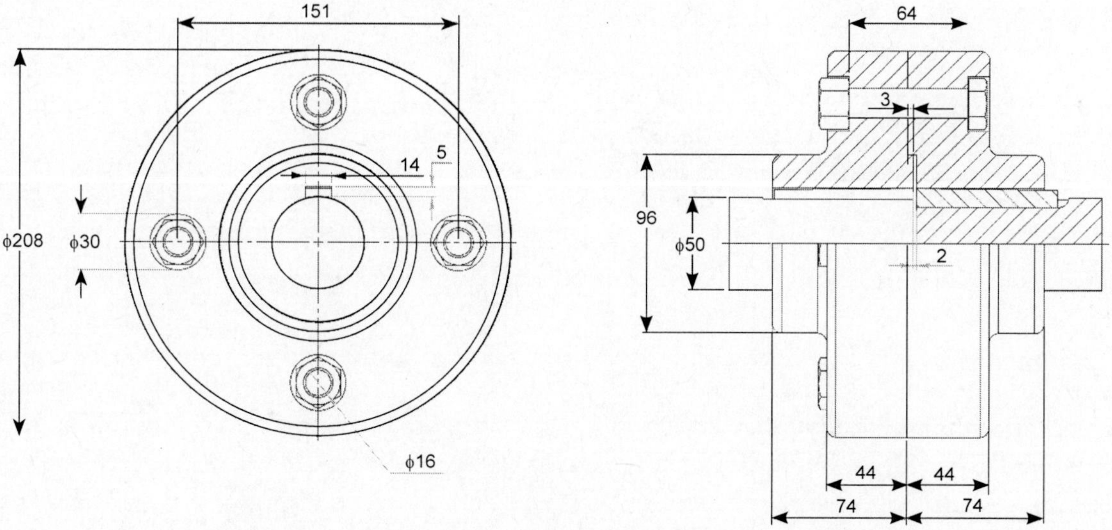

Fig. 7.11: Assembly drawing of protected type flanged coupling

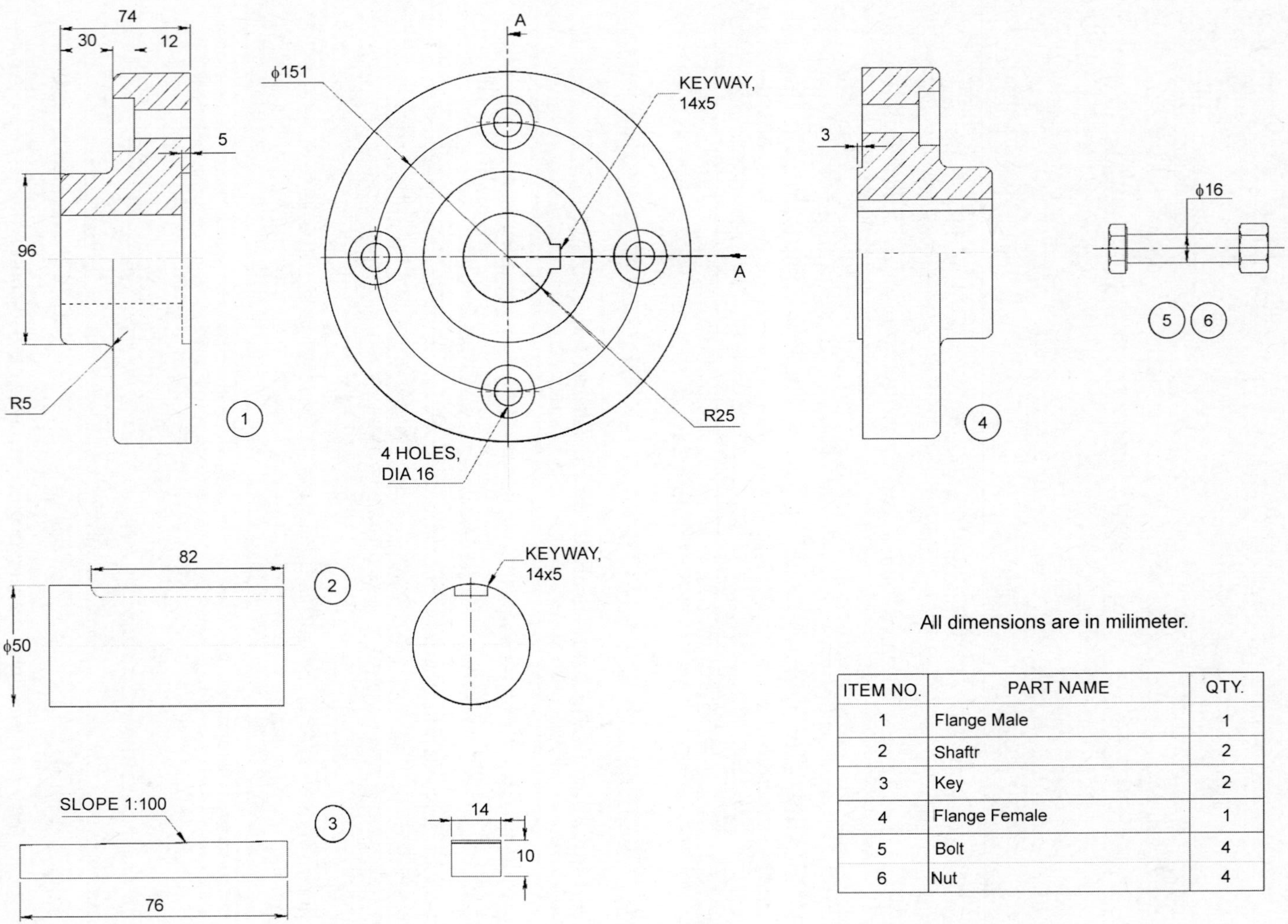

All dimensions are in milimeter.

ITEM NO.	PART NAME	QTY.
1	Flange Male	1
2	Shaftr	2
3	Key	2
4	Flange Female	1
5	Bolt	4
6	Nut	4

Fig. 7.12: Detailed views of protected type flanged coupling

7.5 UNIVERSAL COUPLING

Universal coupling is a combinational double pin joint applied to connect two rotating shafts that are at an angle to each other with their axes having a common point of intersection. It is also referred to as the Cardan joint, Hooke's joint, or the U joint. The joint is capable of transferring high torques and power at an angle but cannot cope with misaligned shafts. Additionally, the velocity that it transmits from the input shaft is not equal to the output unless the shafts are at zero degrees and are coaxial, therefore producing a fluctuating velocity output. Configuration wise it is composed of two U-shaped yokes attached to the ends of the to-be-coupled shafts, oriented perpendicularly to each other using a spider cross between them that provides journals for the two pin joints.

The velocity fluctuation can be managed by employing two universal joints with an appropriate phase relation at both ends of the shaft so that the fluctuations would cancel out. The use of two velocity joints attached in series can also help manage misalignment – something that a single joint cannot treat. These joints are frequently used to transmit power in rear-wheel-drive-based vehicles since the drive shaft and the axles are at an angle – and that too changes all the time due to the action of the suspension.

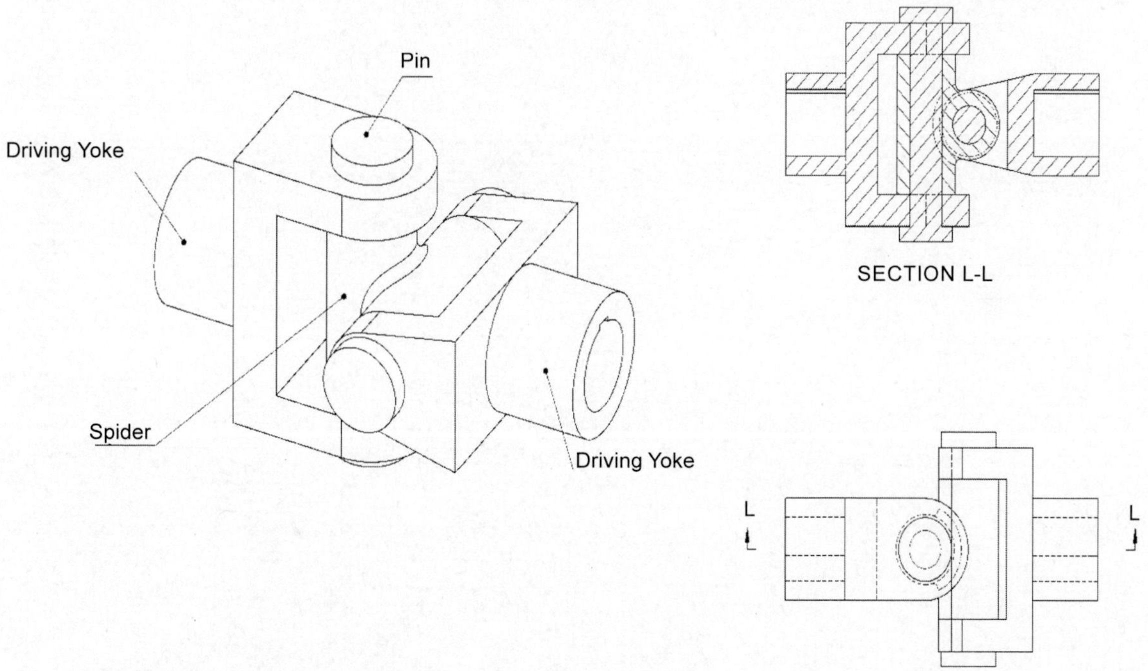

Fig. 7.13: Universal coupling

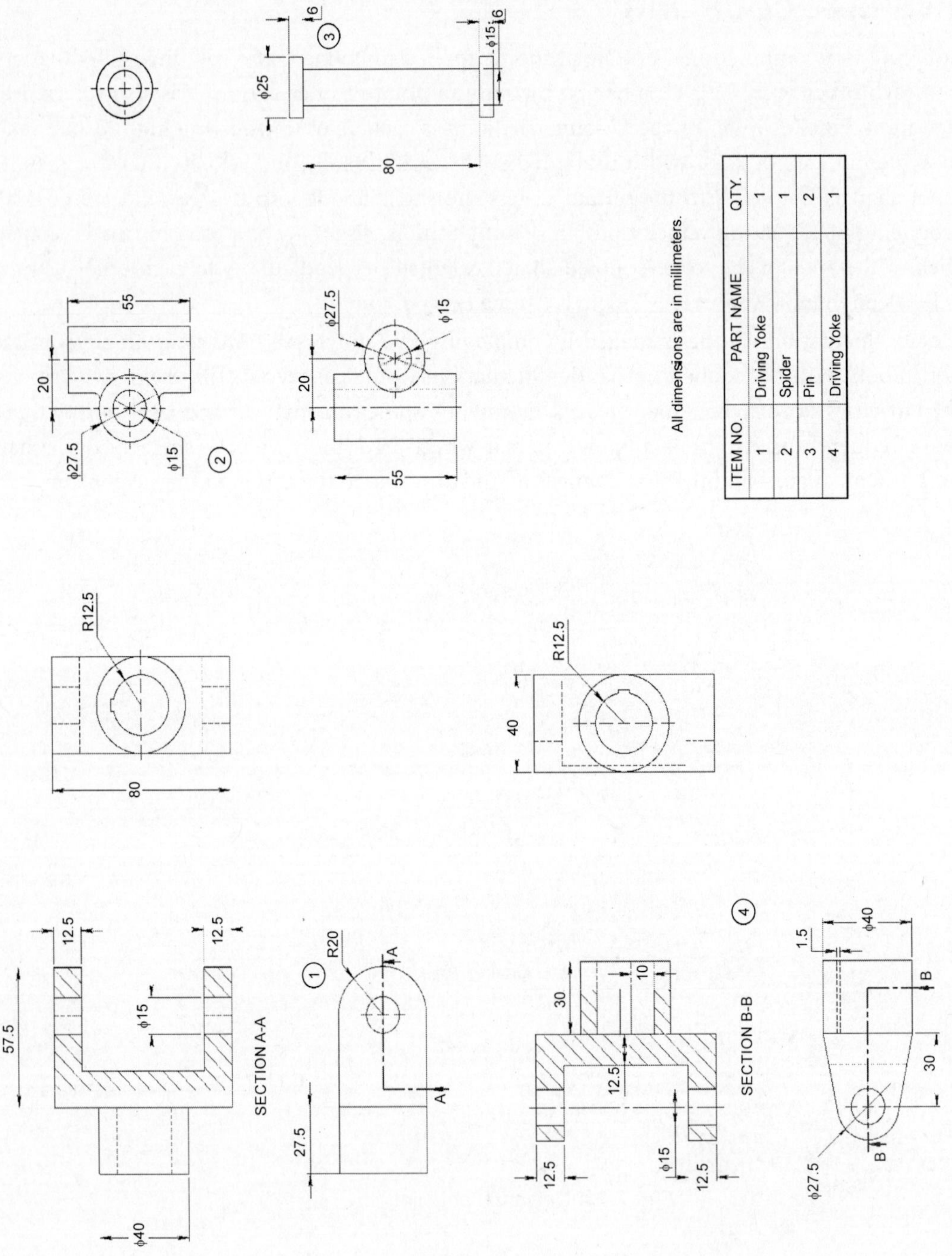

All dimensions are in millimeters.

ITEM NO.	PART NAME	QTY.
1	Driving Yoke	1
2	Spider	1
3	Pin	2
4	Driving Yoke	1

Fig. 7.14: Detailed views of universal coupling

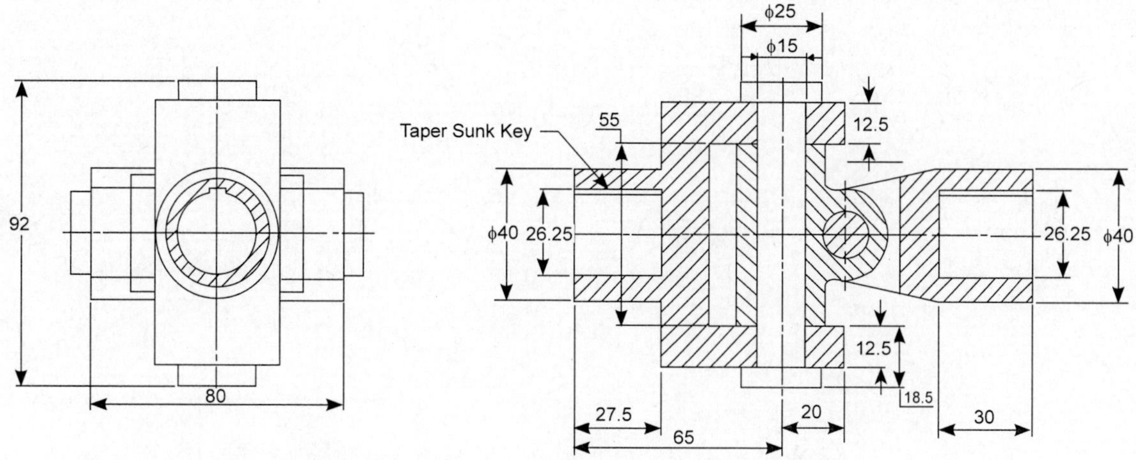

Fig. 7.15: Assembly drawing of universal coupling

7.6 PLUMMER BLOCK

A Plummer block is a type of pillow block bearing housing that has been designed to handle heavier loads and higher power transmission. It is a machined block unit that can support a shaft parallel to a plane using a bearing and is mounted onto surfaces using two bolts that are perpendicular to the base plane. The Plummer block is usually made of cast iron and often has a split design to easily accommodate pre-assembled shaft fittings. The bearing to be housed inside the block unit is usually bought and press-fit separately, depending on the type of rolling element required based on desired parameters such as load rating, life, rpm, operational hours, the size of the shaft, etc. Split bearing types are also available for ease of assembling and taking the system apart.

The Plummer block bearing needs frequent lubrication to remain frictionless and therefore the unit has several holes and channels to periodically feed-in oil or grease. To protect against dust and grime, Plummer blocks can be equipped with cases and seals made exclusively for them to ensure a frictionless run and a longer life span. Applications of a Plummer block include systems for paper mills, steel rolling mills, conveyor belts, belt drives for pumps and motors and industrial machinery setups with reduction gears, etc.

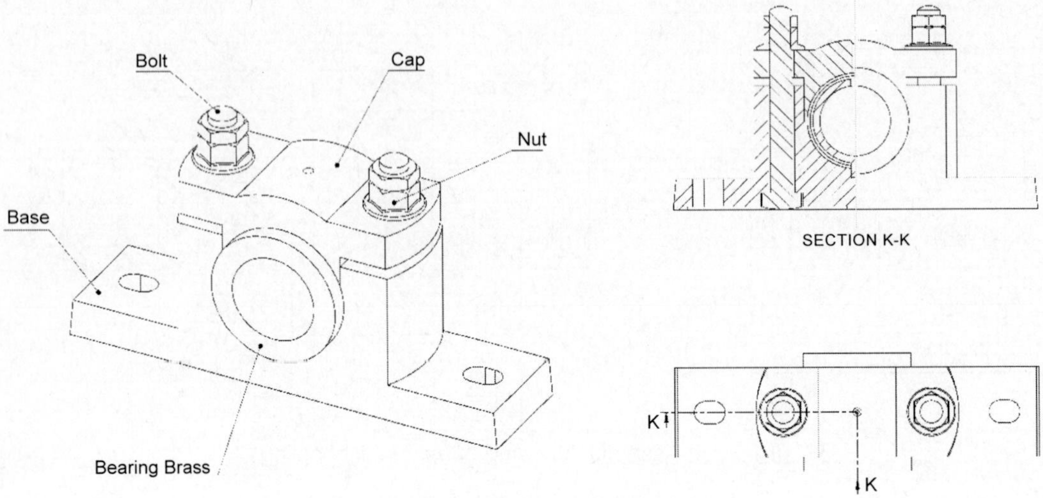

Fig. 7.16: Plummer block

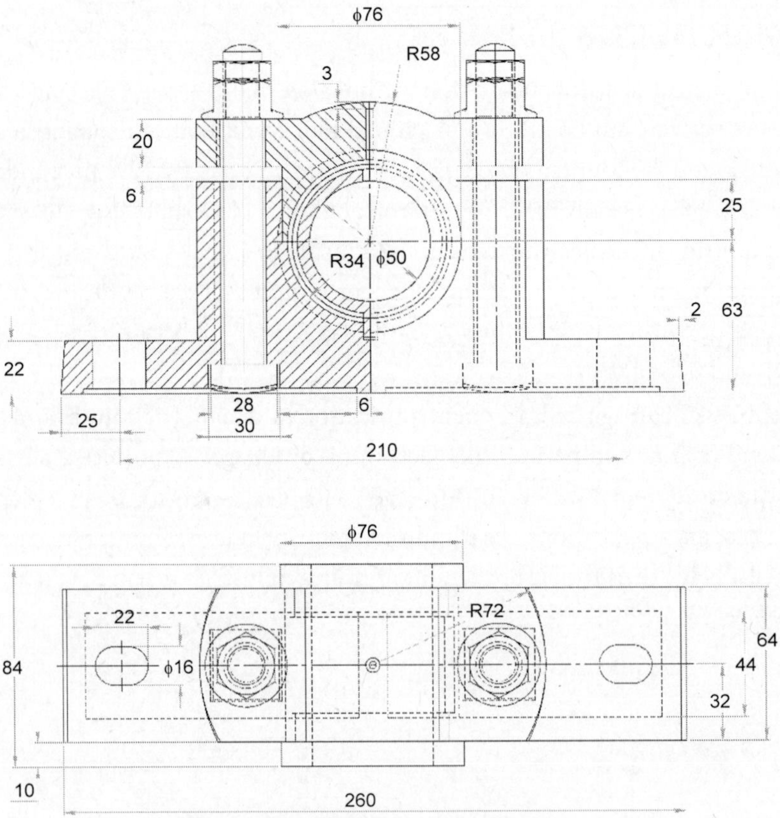

Fig. 7.17: Assembly drawing of plummer block

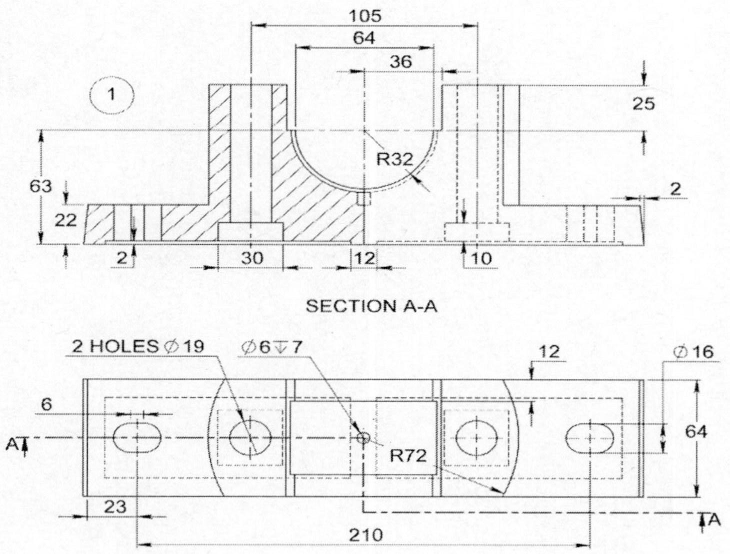

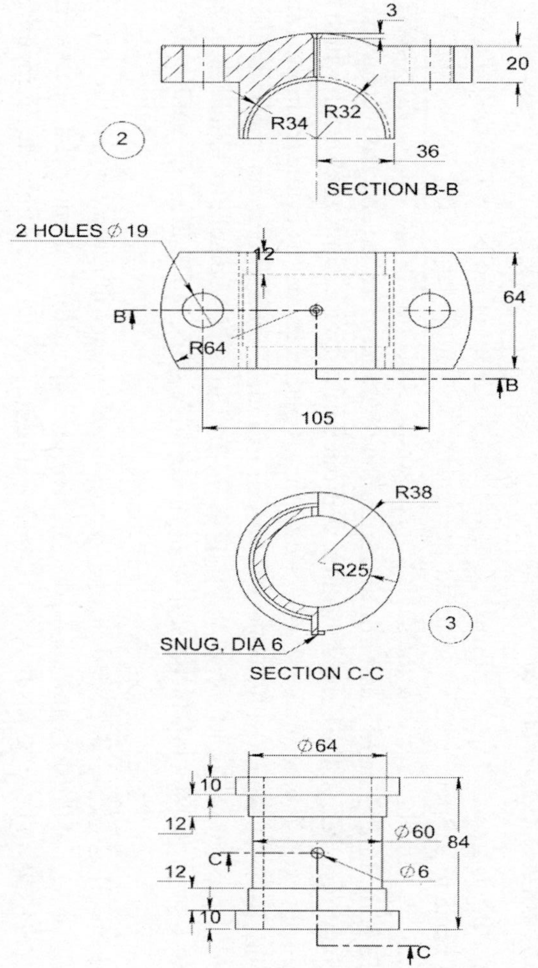

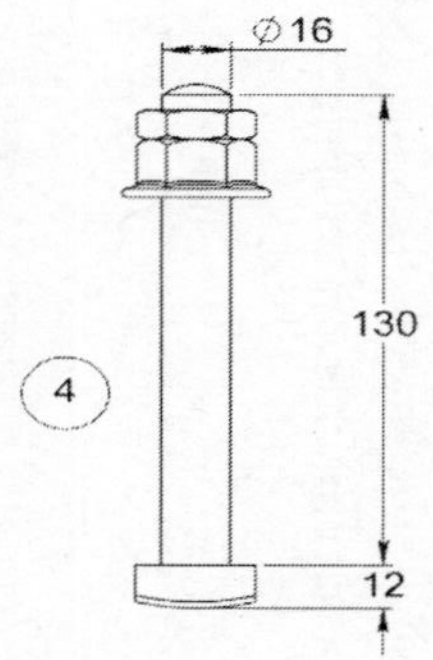

All dimensions are in mm.

ITEM NO.	PART NAME	MATL.	QTY.
1	Base	CI	1
2	Cap	CI	1
3	Bearing Brass	Bronze	1
4	Bolt and Nut	MS	2

Fig. 7.18: Detailed views of plummer block

7.7 FOOTSTEP BEARING

A footstep bearing is a special type of thrust bearing that can be mounted below the end of a vertical overhanging shaft to support it. It is designed particularly to handle large axial loads exerted onto the shaft – which would have caused the shaft to bend and wobble otherwise. The footstep bearing is composed of a body usually made of cast iron and is a hollow cylinder extruded on a flat rectangular base plate with mounting holes. At the base of the hollowed-out part, there is a pinhole through which a pin is inserted to mate with another hole in a buffer disk that is placed inside the cylinder's body. This pin helps constrain the disk from rotating, while the disk acts as an intermediate between the body base and the shaft's end to prevent wear and tear of the body as the shaft rotates. The hollow cylinder is designed with a vertical keyway-like slot in which a stub of a cylindrical bushing can be fitted so that it can be held in place.

Both the bushing and the disk are designed to act as intermediate liners to prevent the main body from the rubbing wear of the shaft and are made from a softer material such as gunmetal. These parts are relatively inexpensive and can be easily replaced once out of service. Once assembled and mounted on support below the shaft, the shaft is then inserted into the bushing and is free to rotate. It has frequent applications in industrial machinery setups.

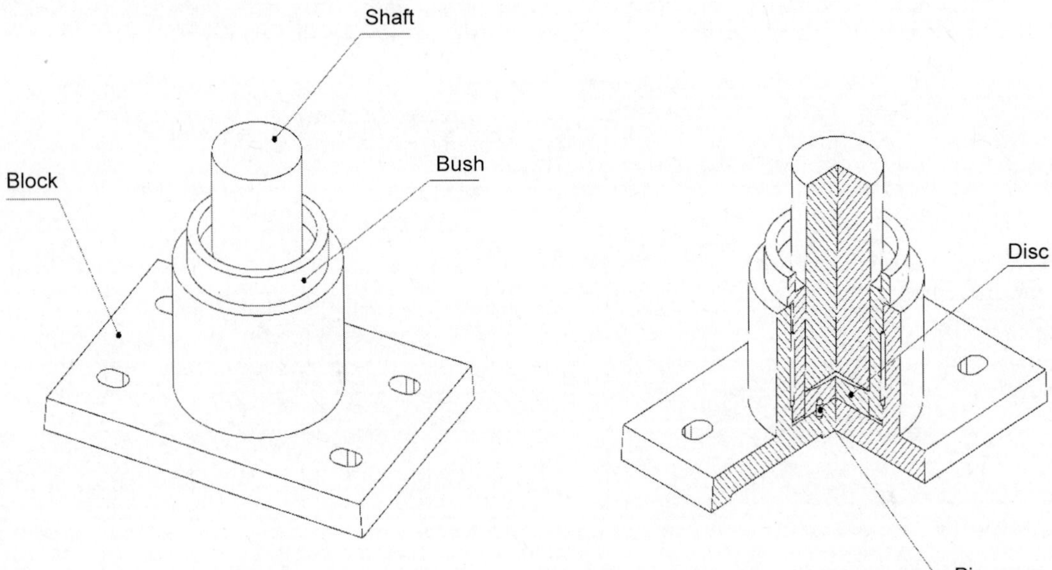

Fig. 7.19: Footstep bearing

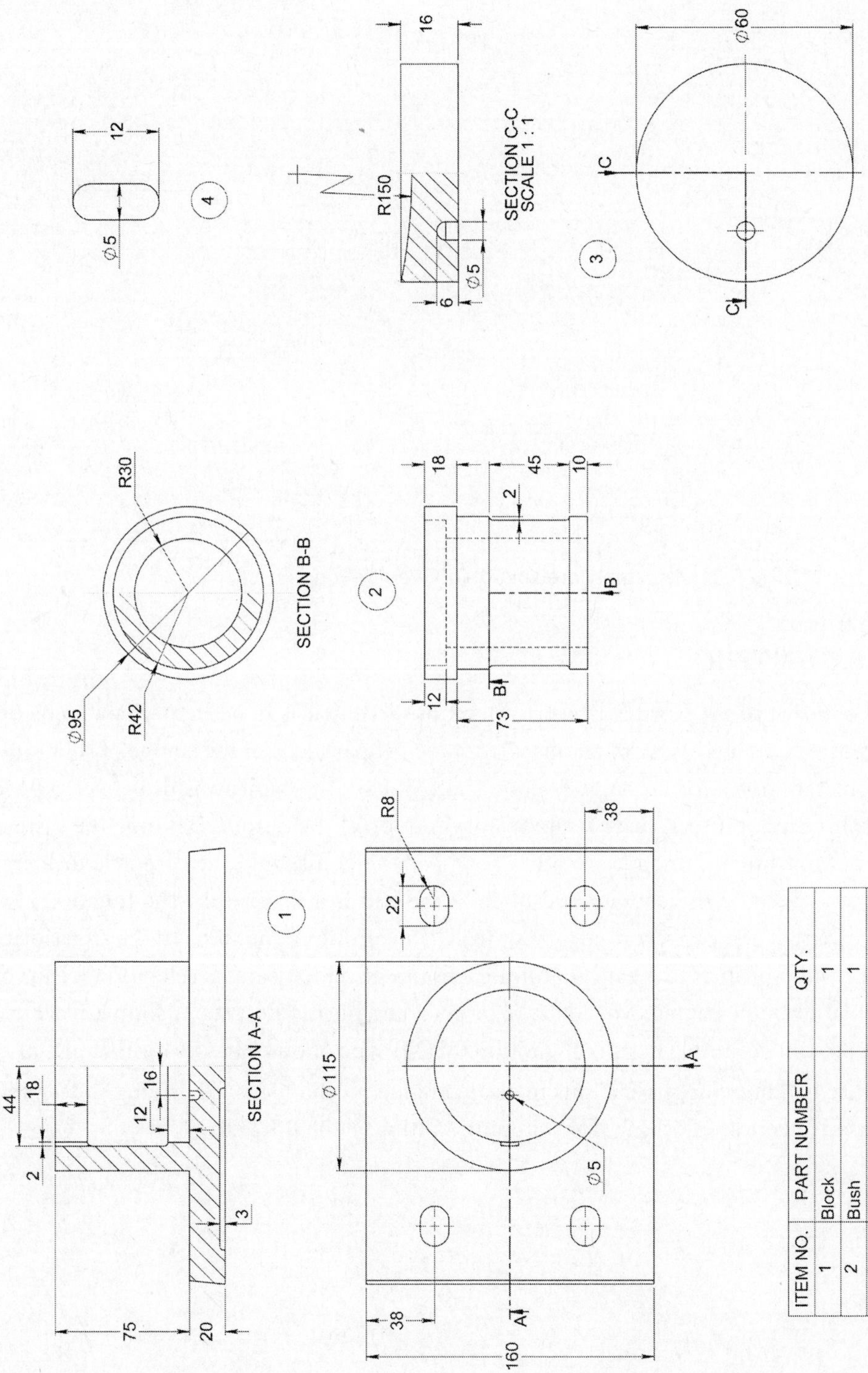

ITEM NO.	PART NUMBER	QTY.
1	Block	1
2	Bush	1
3	Plate	1
4	Pin	1

All Dimensions are in mm.

Fig. 7.20: Detailed views of footstep bearing

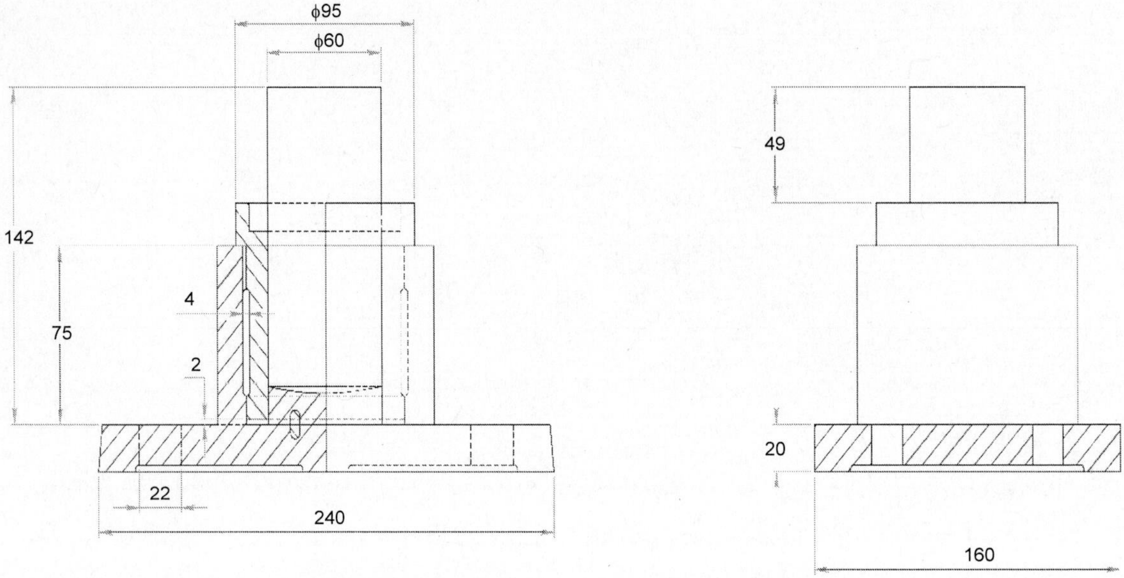

Fig. 7.21: Assembly drawing of footstep bearing

7.8 SIMPLE ECCENTRIC

Eccentric is the literal opposite of the word concentric. A simple eccentric is hence a mechanism based on a disk that has its centre translated away to an offset from the primary axis of the source of rotations. The source can be a pivot, a motor, or an engine's crankshaft. Such a mechanism is designed to pure rotary motion of the source input into a linear reciprocatory motion at the output. Defining the system in terms of mechanics of machinery, the pure rotation source is the crank link, the eccentric disk and intermediate eccentric rod become the coupler and often at the end is a slider link. The function also holds similarities to a cam that can convert rotary input into a linear output, but it lacks the superiority of a cam that can have custom profiles to create an infinite amount of velocity and acceleration trends of the output – with popular examples being cycloidal, polynomial and harmonics, etc. A simple eccentric can only provide an approximate simple harmonic motion (SHM) type of output as its limitation.

The simple eccentric had popular applications in steam engine mechanisms for driving sliders and pumps and is widely used in certain machines for stamping, cutting, stencilling etc.

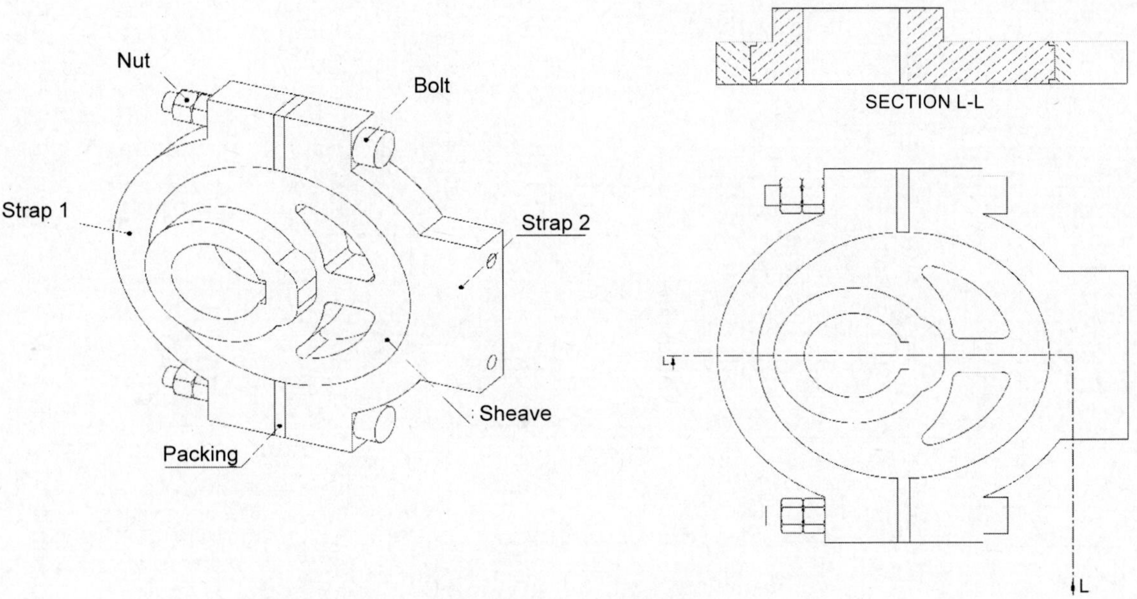

Fig. 7.22: Simple eccentric

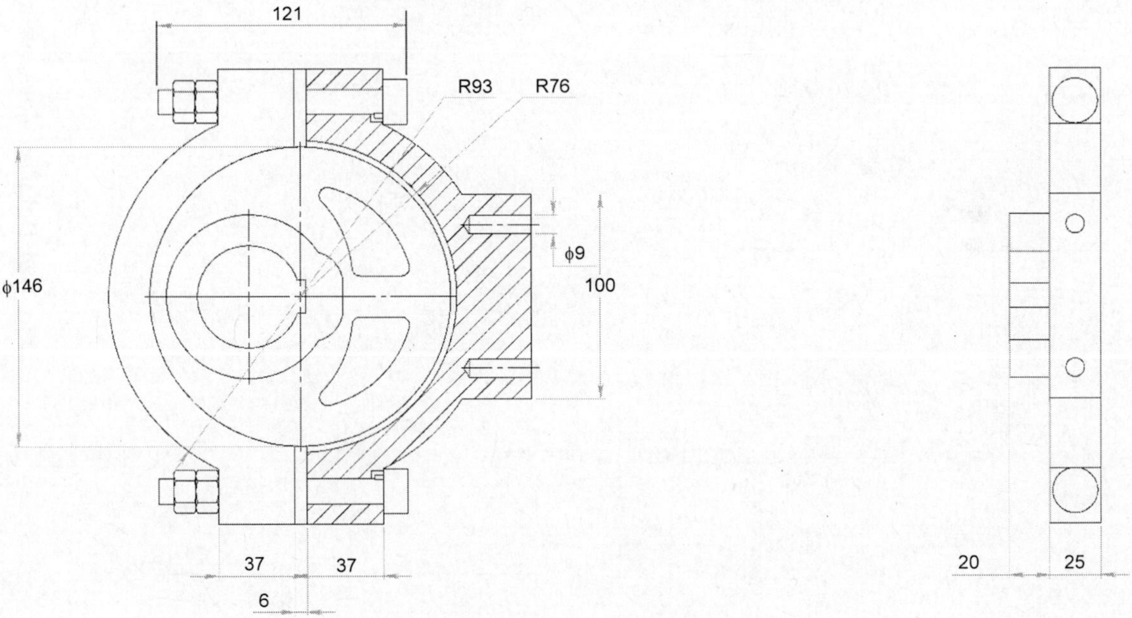

Fig. 7.23: Assembly drawing of simple eccentric

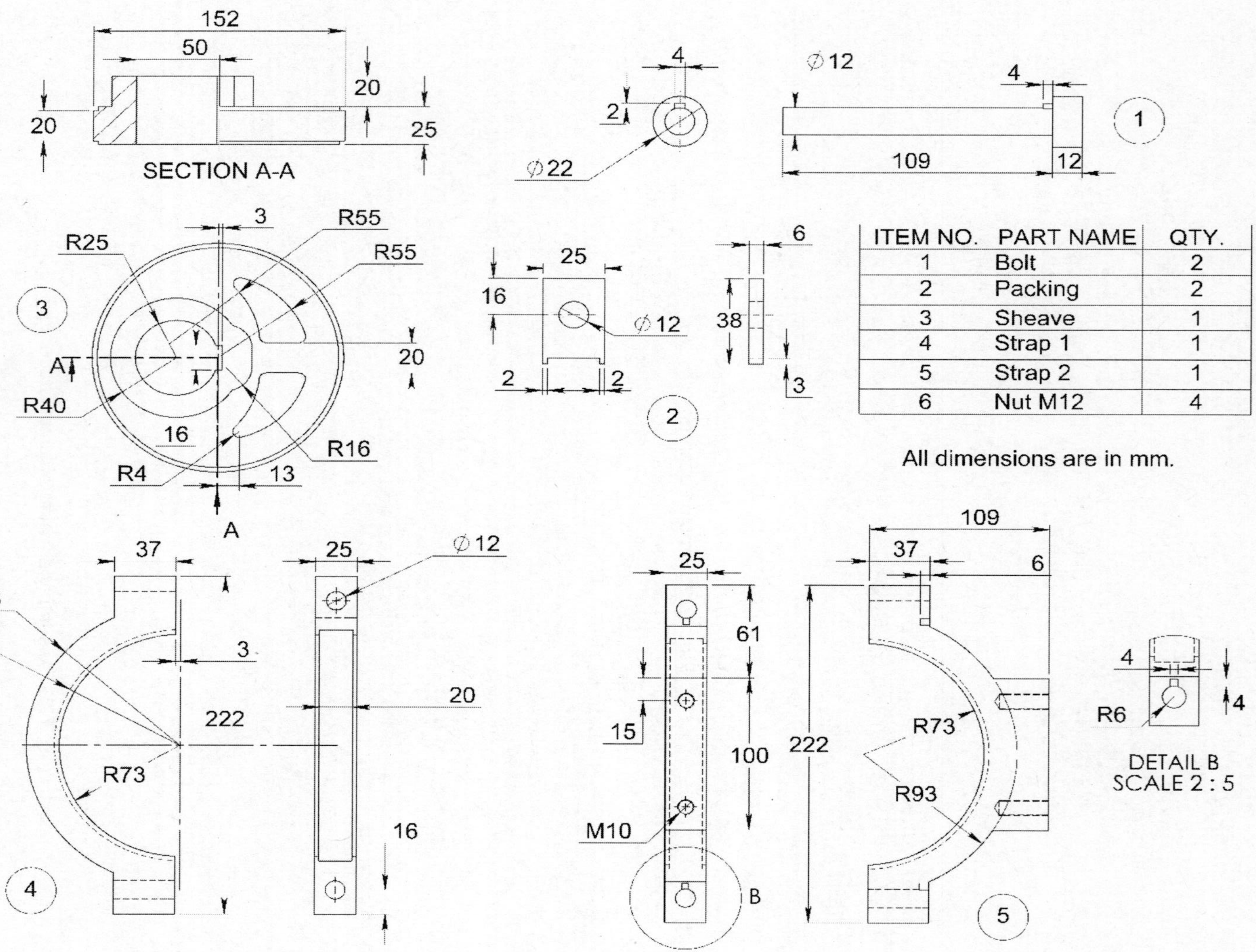

ITEM NO.	PART NAME	QTY.
1	Bolt	2
2	Packing	2
3	Sheave	1
4	Strap 1	1
5	Strap 2	1
6	Nut M12	4

All dimensions are in mm.

Fig. 7.24: Detailed views of simple eccentric

7.9 MACHINE VICE

A machine vice is a job-holding fixture that is used to clamp and hold work pieces during machining operations. The primary difference between a machine vice and a bench vice is the subject location the vice is mounted on – with the machine vice being attached to the machine tool's support bench and the bench vice being mounted on a regular workbench for hand tool operations. Machine vices are designed to machining operations easier and safer and these include, drilling, milling, tapping, sawing, etc. Being attached to the machine itself, they can be accurately positioned and can have better support and integration of jigs – and in some cases eliminate the need for jigs.

Like any other vice, they have a fixed jaw, a movable jaw, a screw to translate the movable jaw for clamping, a handle to rotate the main screw and a cast iron body to hold all components together. The vice is usually hand operated but can be motorized owing to the developments in technology. A vice can be designed to safely handle metal, wood and plastics, with appropriate soft guards attached to the jaws if delicate workpieces are involved and a good surface finish is a high priority. It has popular uses in workshops by craftsmen and at an industrial level where mass operations are conducted in a semi/ assisted automation environment.

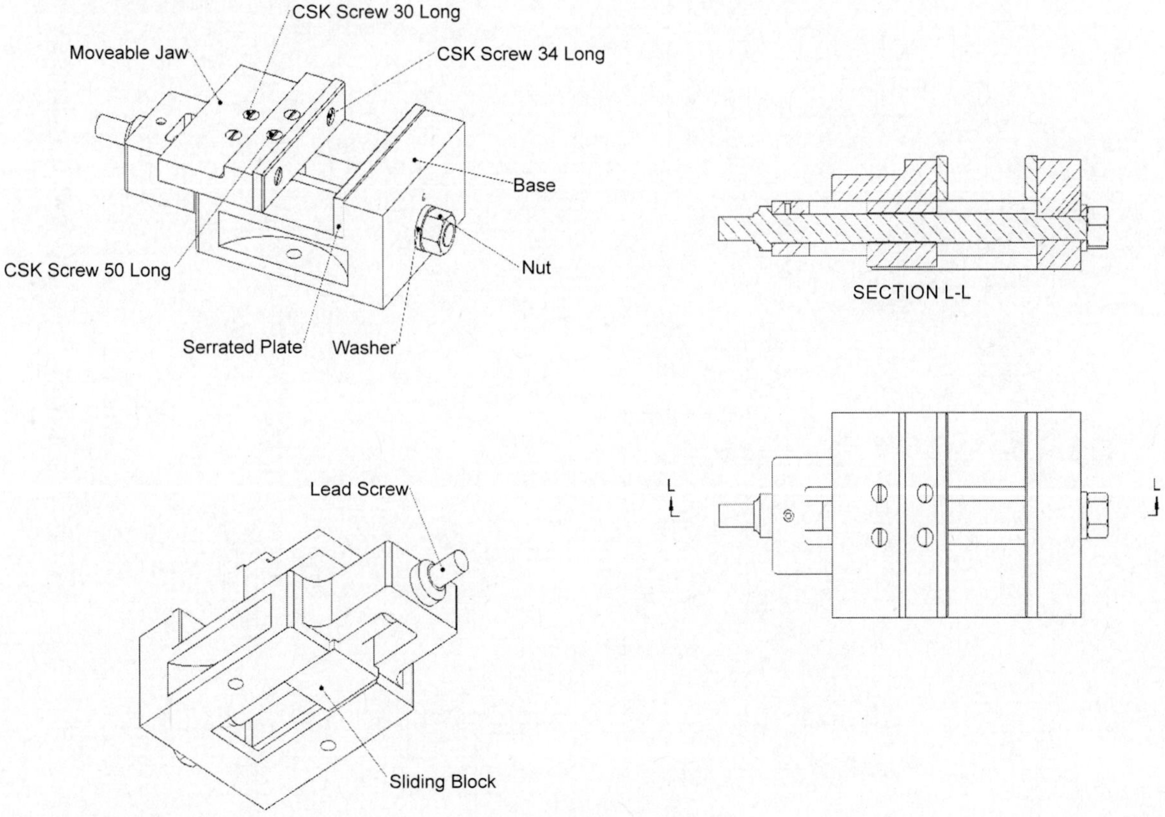

Fig. 7.25: Machine vice

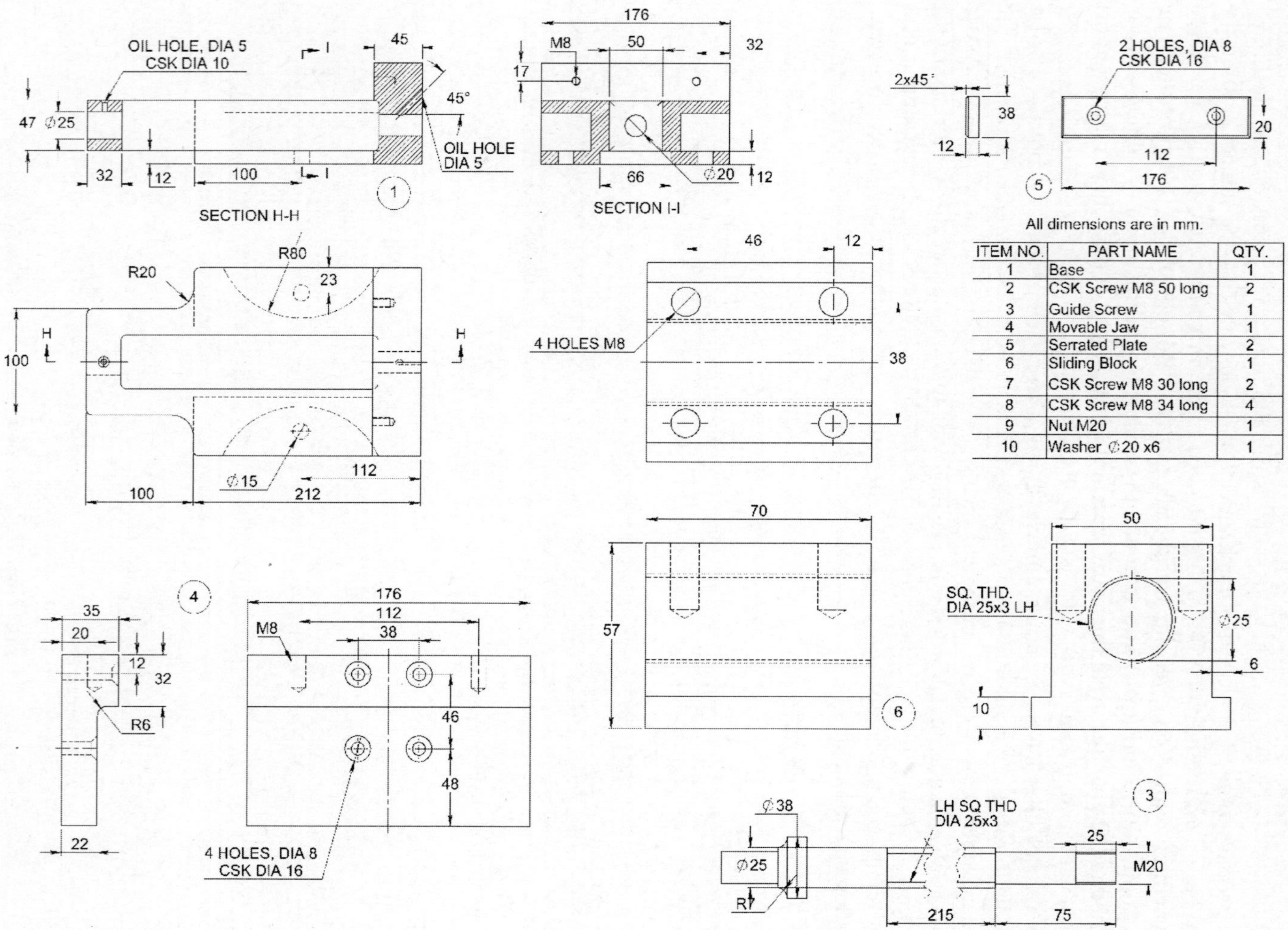

All dimensions are in mm.

ITEM NO.	PART NAME	QTY.
1	Base	1
2	CSK Screw M8 50 long	2
3	Guide Screw	1
4	Movable Jaw	1
5	Serrated Plate	2
6	Sliding Block	1
7	CSK Screw M8 30 long	2
8	CSK Screw M8 34 long	4
9	Nut M20	1
10	Washer ϕ20 x6	1

Fig. 7.26: Detailed views of machine vice

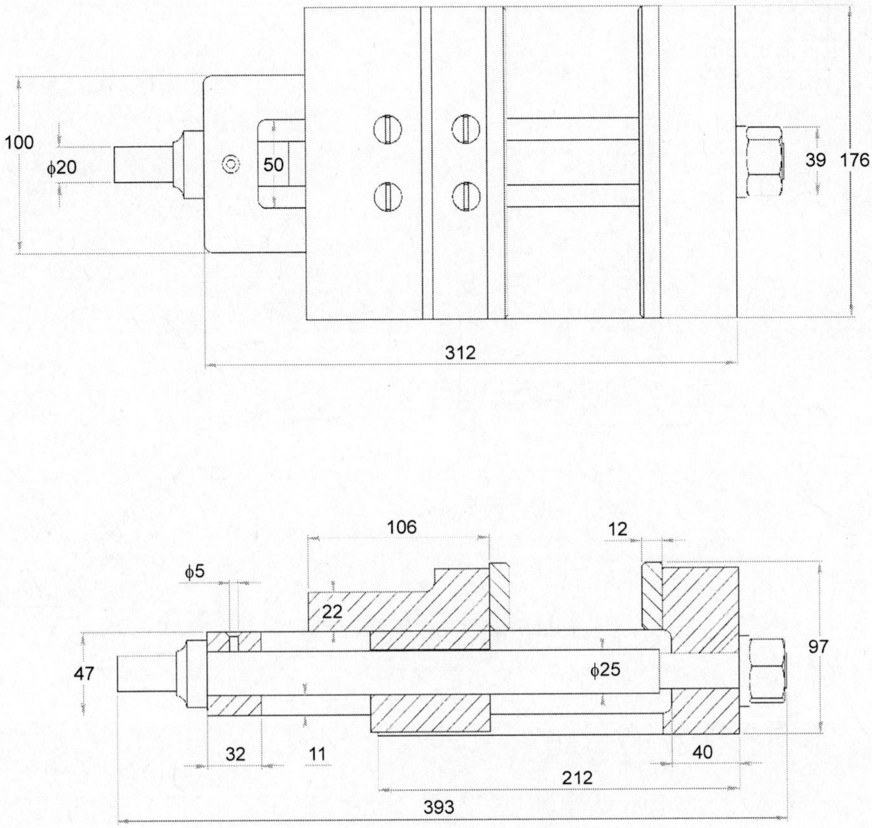

Fig. 7.27: Assembly drawing of machine vice

7.10 CONNECTING ROD

A connecting rod is a popular choice as the coupler link in the 4-bar mechanism of an internal combustion engine – though it is not limited to use in engines only. Also called a 'con rod', it is a sturdy coupler that connects the crank to the piston slider link and helps convert rotary motion into linear reciprocatory motion by acting as an intermediate. While doing so, the connecting rod itself undergoes a complex motion that is described as neither pure rotational nor pure translational. The connecting rod's smaller end is attached to the piston head using a wrist pin or a gudgeon pin and the bigger end is connected to the crankshaft using a crank pin lined by split plain bearings. With these pin joints, a connecting rod can vary its angle with the piston's axis and the force components acting on the rod hence vary through the cycle.

The connecting rod is usually made up of steel and aluminium alloys and is responsible for transmitting the compressive and tensile forces exerted by the piston head to the crank while under operation. They are also designed to have a low inertial mass and retain enough strength to resist buckling and fracture under the sheer forces of an engine.

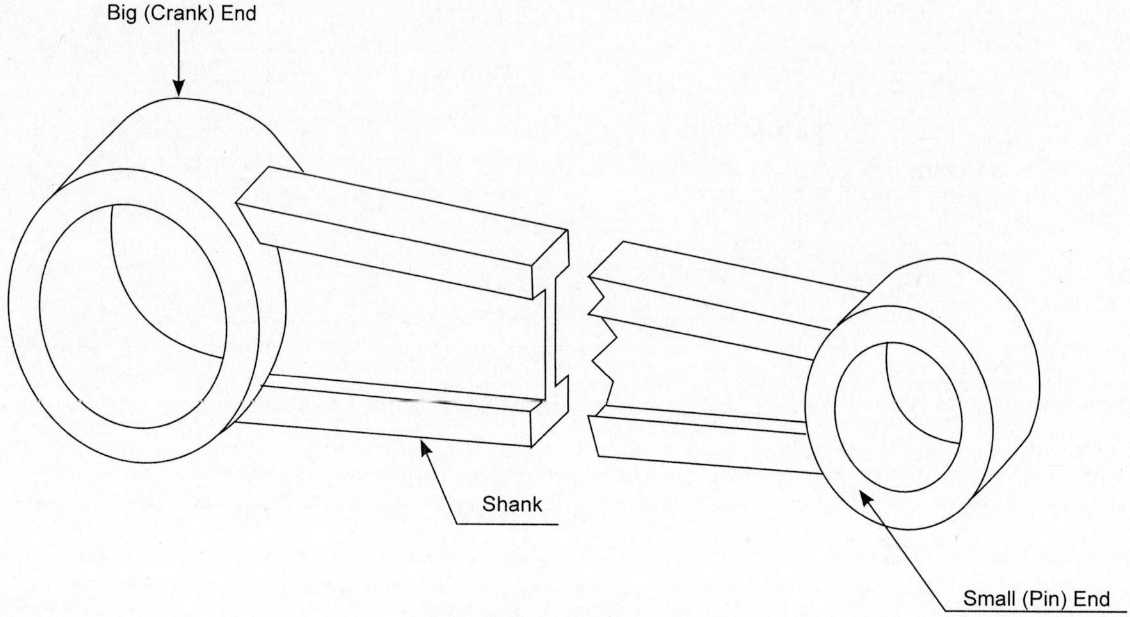

Fig. 7.28: Connecting rod

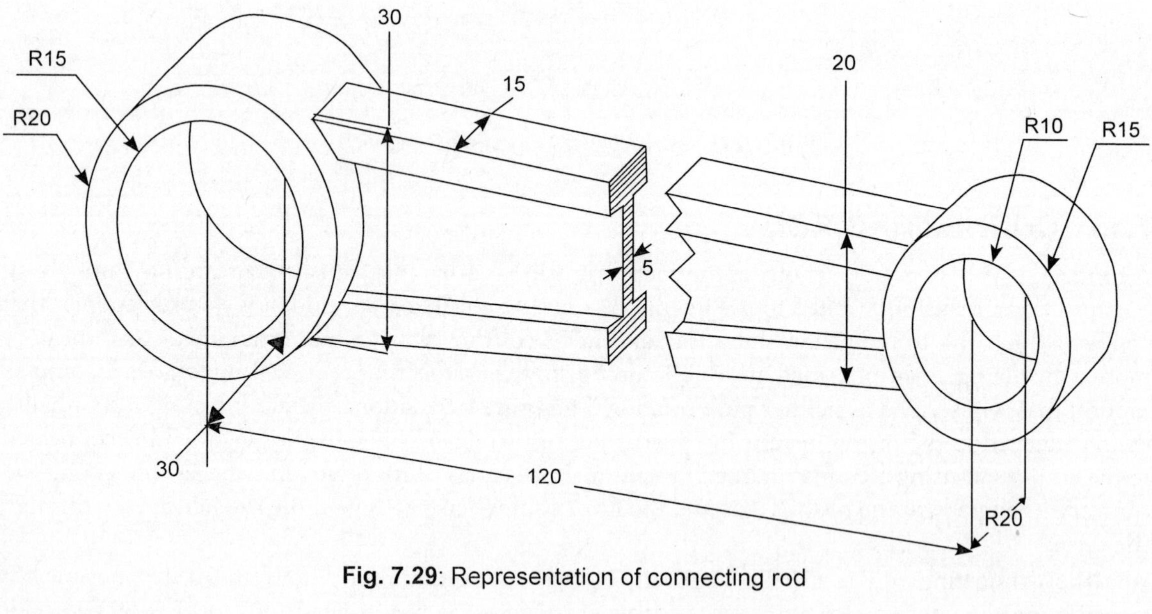

Fig. 7.29: Representation of connecting rod

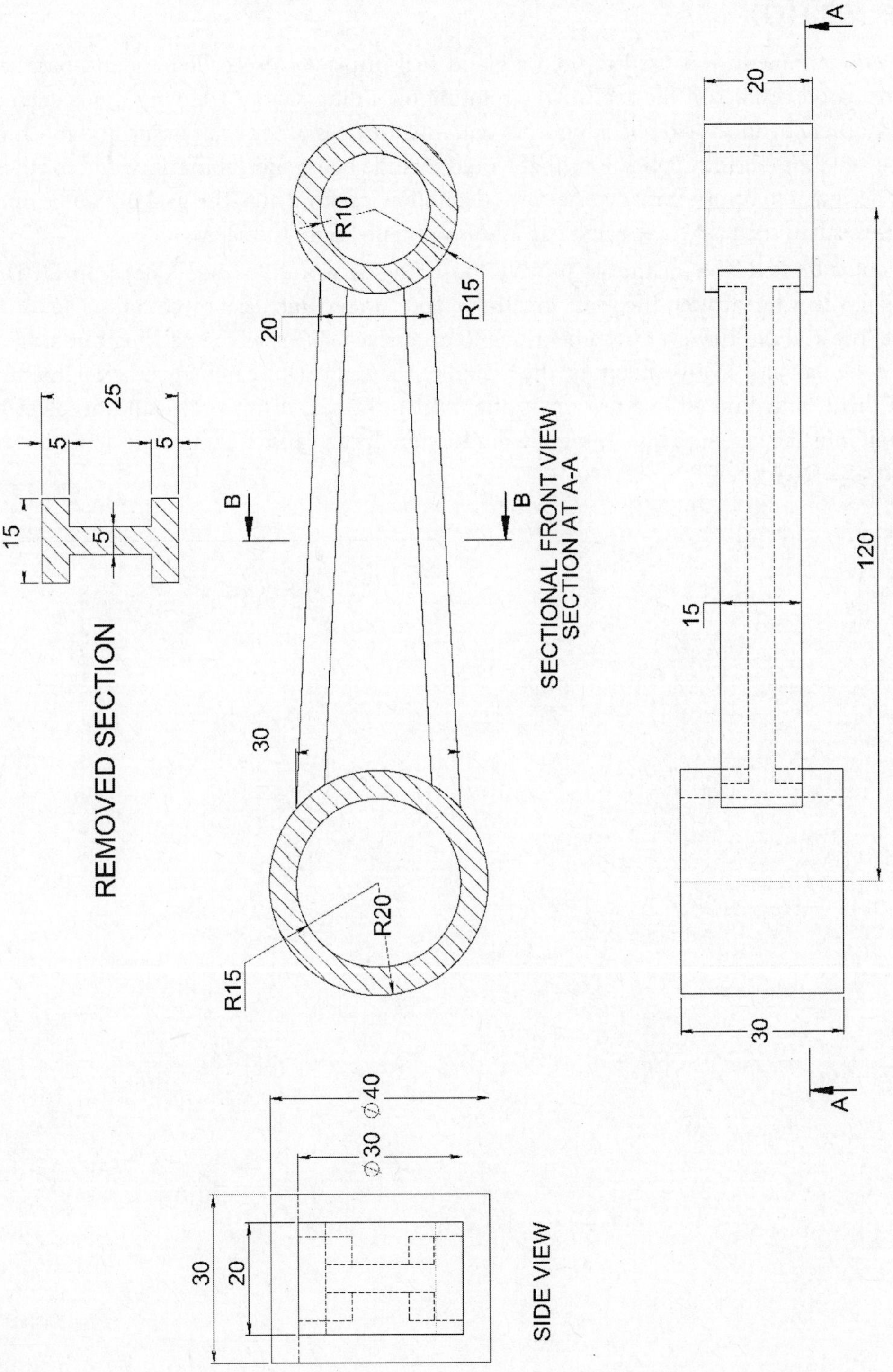

Fig. 7.30: Detailed views of connecting rod

7.11 STUFFING BOX

A stuffing box is a containment unit used to hold a gland seal in place. Also called a gland package, the housing ensures good contact of the seal with a rotating or sliding shaft – trapping fluids such as water, oil, steam, or other chemicals from leaking from and into crevices of the shaft lining and machine parts. It is a critical need for marine applications that would include boats and submarines and for these specific cases, it is designed to have a threaded sleeve and a hollow packing nut. The packing nut is lined with a lubricant and sealant such as wax-treated flax and is tightened onto the sleeve.

The packing nut is backed by a locknut to prevent loosening by vibrations and water impact. This allows the packing flax to wrap around the shaft, creating a good water-tight seal by compressive forces exerted by the nut. The seal can however become prone to excessive leakage as the packing nut loosens up with time when the flax packing wears out by the action of the shaft. This can be fixed by tightening the nut again, but this is only limited to a few times after which the seal disintegrates and needs to be replaced. Precaution must be taken, as tightening the nut too much can cause high force indentation on the shaft and create excessive wear.

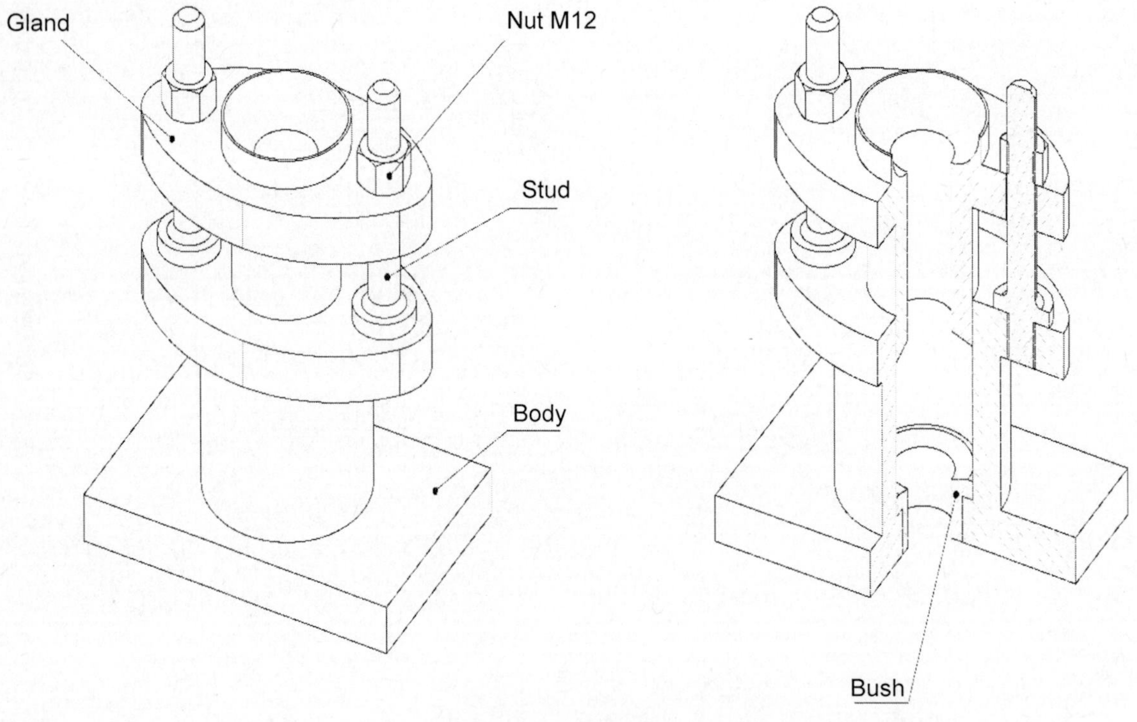

Fig. 7.31: Stuffing box

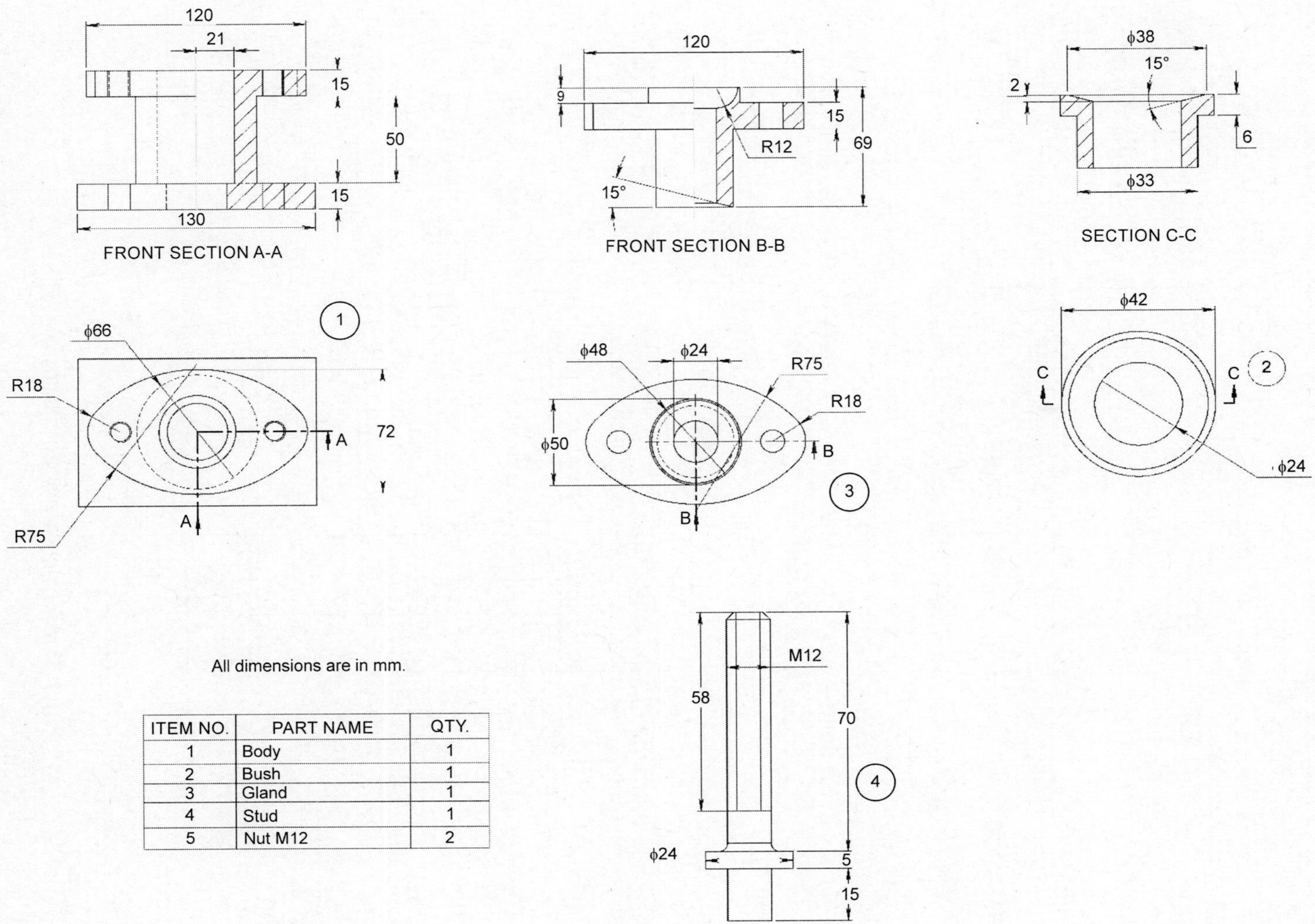

All dimensions are in mm.

ITEM NO.	PART NAME	QTY.
1	Body	1
2	Bush	1
3	Gland	1
4	Stud	1
5	Nut M12	2

Fig. 7.32: Detailed views of stuffing box

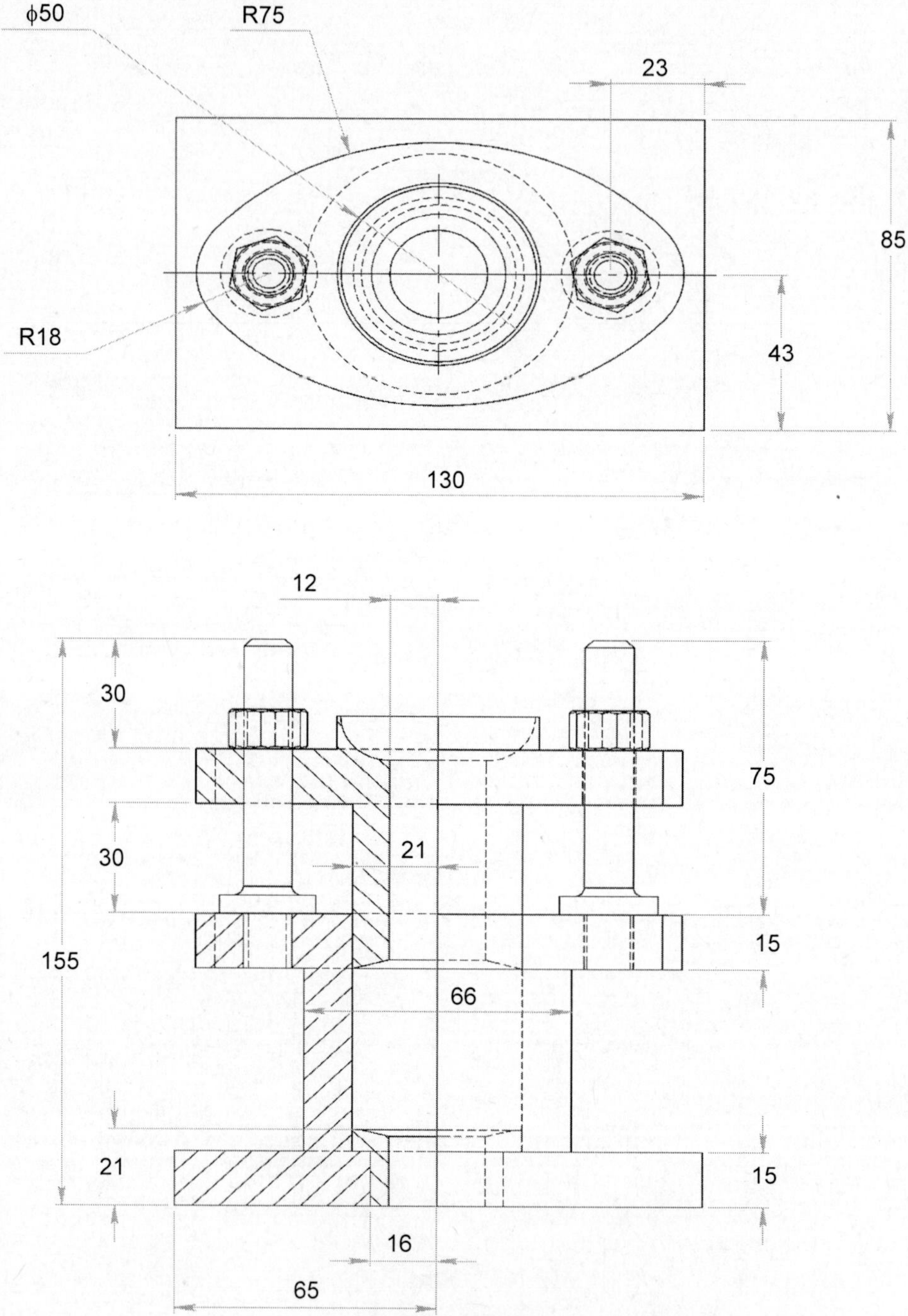

Fig. 7.33: Assembly drawing of stuffing box

7.12 SCREW JACK

A screw jack is a simple load raising/lowering device designed to have a very high mechanical advantage. Mechanical advantage is termed as the ratio of load raised/lowered to the actual human effort required at the input and screw jacks achieve that by spreading the force over the threaded wedge of the screw and raising only small increments with time. A screw jack assembly is composed of a heavy and durable vertical screw, placed into a tapped cylindrical body and connected to a handle shaft via mating bevel or worm gear sets. By rotating the handle, the vertical screw slowly moves up or down depending on the direction of the input rotation and is capable of lifting very heavy loads. Although it takes a lot of time and the cost of efficiency to raise the load, one can lift tons with minimal effort by using the right screw jack.

Screw jacks are being replaced by hydraulic jacks in the modern era due to ease of use, better automation and the capability of lifting higher loads. However, a screw jack can be made self-locking and therefore the main screw will not fall under a compressive load even if the effort is removed. This feature is absent in modern hydraulic jacks, which need constant input line pressure to hold loads. Screw jacks have popular applications in lifting cars and heavy industrial loads.

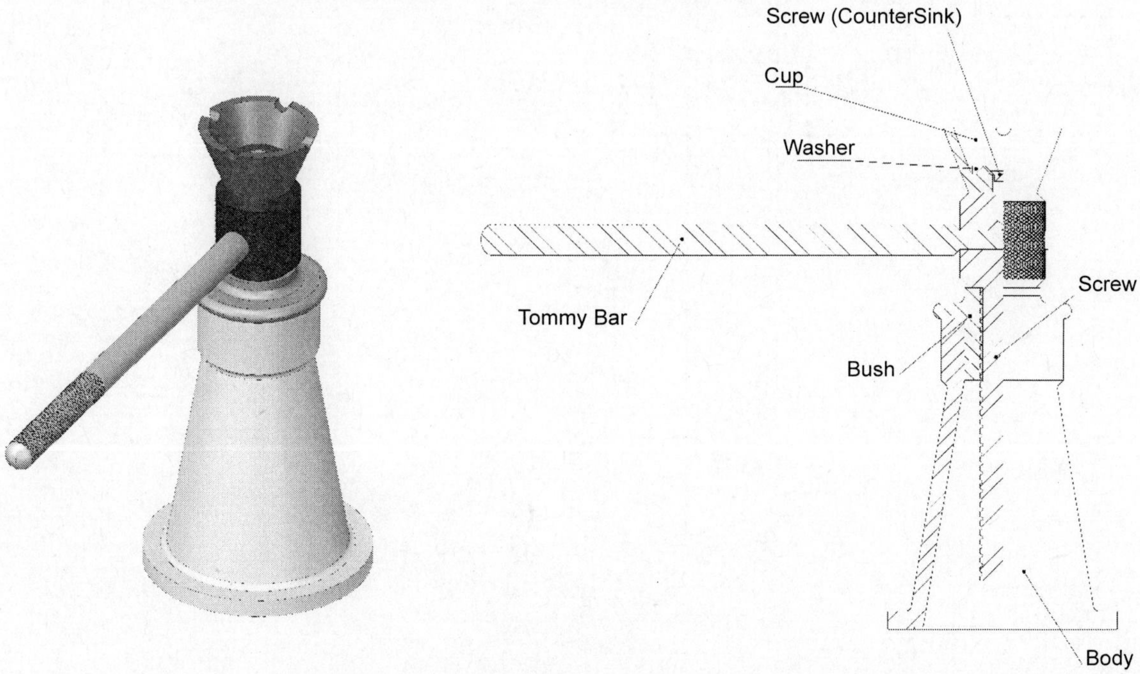

Fig. 7.34: Three-dimensional model and sectional view of screw jack

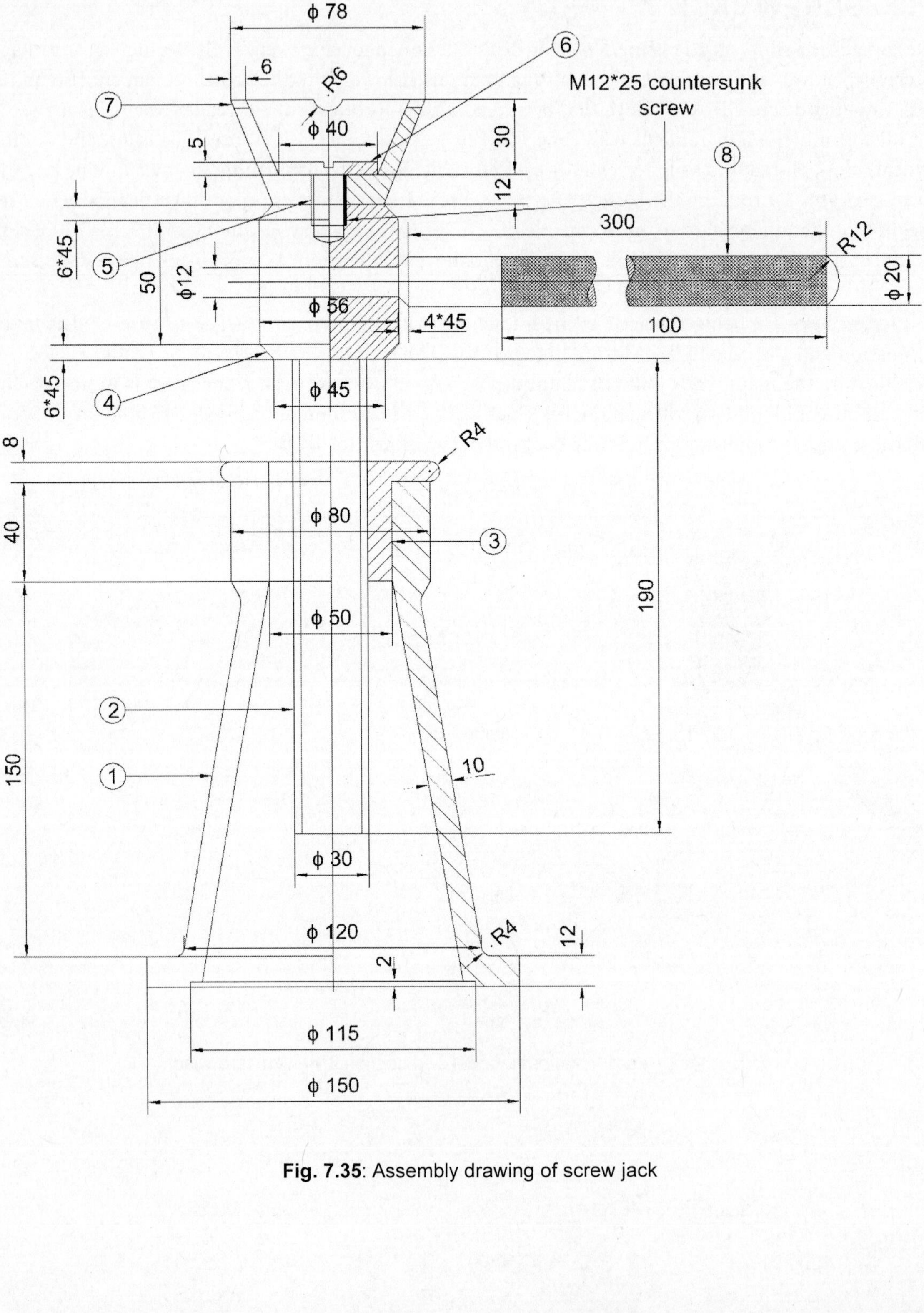

Fig. 7.35: Assembly drawing of screw jack

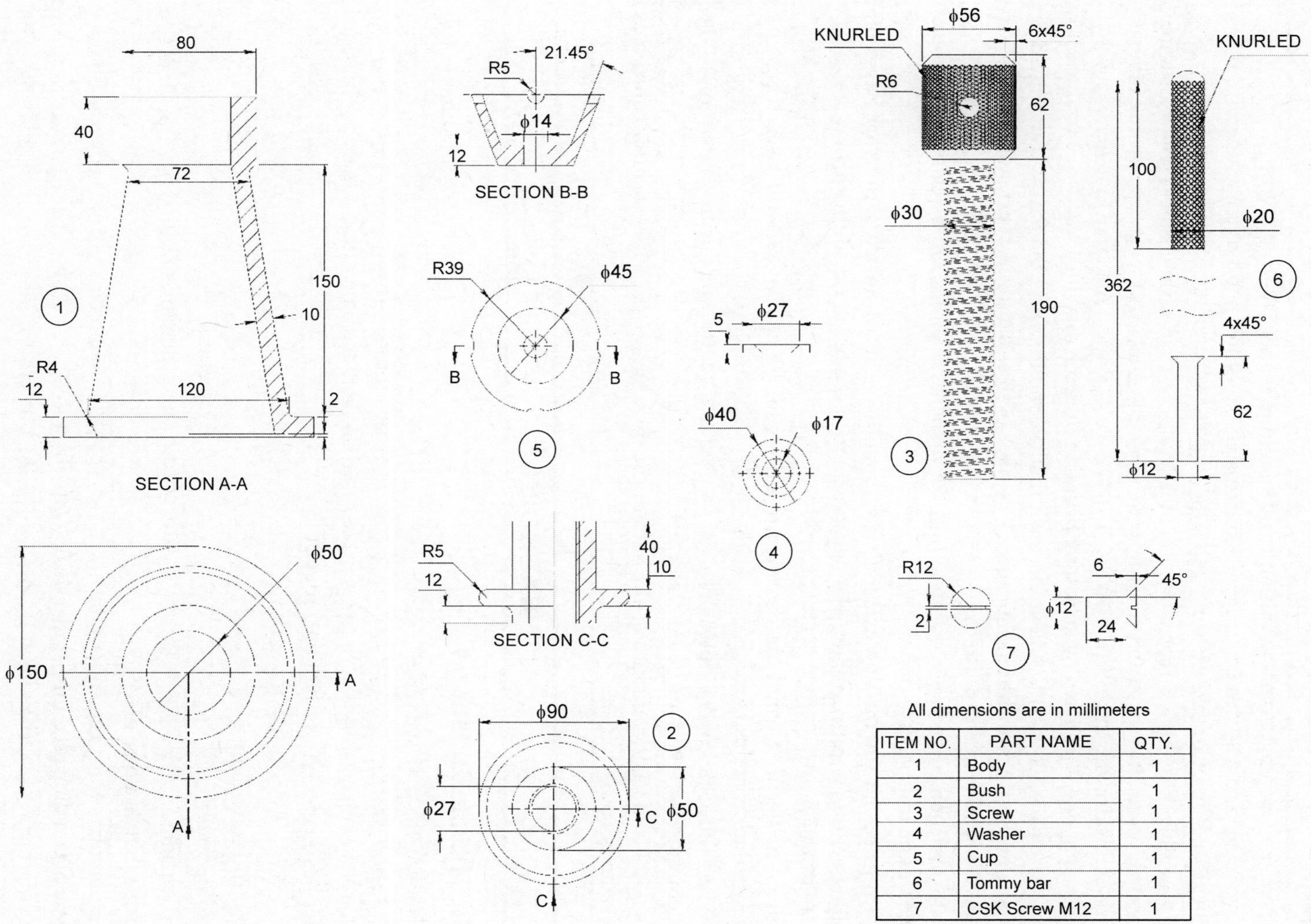

All dimensions are in millimeters

ITEM NO.	PART NAME	QTY.
1	Body	1
2	Bush	1
3	Screw	1
4	Washer	1
5	Cup	1
6	Tommy bar	1
7	CSK Screw M12	1

Fig. 7.36: Detailed views of screw jack

UNIT SUMMARY

Based on the contents covered briefly in this unit you should remember that:

- A working drawing provides details and instructions for building machines or other structures. Working drawings can be classified into detail drawings and assembly drawings.
- An assembly drawing is a representation of a product in its finished state. Design assembly drawings, installation assembly drawings and assembly drawings for catalogues are examples of assembly drawing types. An item list (also known as a bill of materials) is a detailed list of every component depicted on an assembly design or detail drawing.
- To secure parts like gears, pulleys etc. such that the motion from the part is transmitted to the shaft or vice versa without slippage, a key made of steel is utilized. It extends into another groove in a hub and lies partially in a groove in a shaft.
- Shafts are coupled or joined via couplings. Permanent couplings can be classified as solid, flexible and universal.
- Bearings permit smooth, low-friction movement between two surfaces. Bearings based on rolling action are called rolling-element bearings; those based on sliding action are called plain bearings.
- Journal or sleeve bearings are designed to carry radial loads whereas thrust bearings are designed to carry axial loads.
- Devices called jigs and fixtures are used to hold and locate work or to guide tools while machining operations are being performed.
- A jig holds the work and locates the path of the tool. Jigs are used for machining purposes or assembly purposes.

EXERCISES

A. Multiple Choice Questions

7.1 The function of key is to join

(a) two shafts
(b) two rotors
(c) a shaft and a rotor
(d) a shaft and a bearing

7.2 A cotter has a taper on width of the order of

(a) 1:5 (b) 1:10 (c) 1:20 (d) 1:30

7.3 A sleeve and cotter joint uses

(a) one cotter
(b) two cotters
(c) two gibs
(d) one cotter and one gib

7.4 A knuckle joint is a type of

(a) pin joint (b) cotter joint (c) permanent joint (d) spherical joint

7.5 The type of coupling used for parallel shafts with offset is

(a) marine (b) oldham (c) universal (d) rigid flange

7.6 A knuckle joint is also called as

(a) coupling (b) bearing (c) forked pin joint (d) socket pin joint

7.7 The portion of a shaft inside the sleeve bearing is called as
(a) journal (b) coupling (c) flange (d) cotter

7.8 Which bearing can support loads acting along the axis of the shaft?
(a) radial (b) thrust (c) longitudinal (d) transverse

7.9 A Plummer block supports
(a) shafts (b) blocks
(c) bearings (d) vertical shafts only

7.10 A footstep bearing is used for
(a) radial load (b) axial load (c) variable load (d) transverse load

7.11 Eccentric is mounted on
(a) piston (b) connecting rod (c) camshaft (d) crankshaft

7.12 Eccentric converts rotary motion to
(a) power (b) torque
(c) energy (d) reciprocating motion

7.13 The piston is connected to connecting rod at
(a) shank (b) big end (c) small end (d) edge of shank

7.14 The crankshaft is connected to connecting rod at
(a) shank (b) big end (c) small end (d) edge of shank

7.15 The part that connects the small end and the big end of the connecting rod is known as
(a) connector (b) shank (c) cap (d) bridge

7.16 The rod that transmits reciprocating motion to rotary motion is known as
(a) camshaft (b) crankshaft (c) piston (d) connecting rod

7.17 Cap is attached to which end of the connecting rod?
(a) wrist pin (b) shank (c) big end (d) small end

7.18 The prime objective of the stuffing box is
(a) Reducing the resistance of fluid flow
(b) Absorbing the contraction/expansion of pipelines
(c) Facilitating smooth opening and closing of a valve
(d) Prevention of fluid leakage around moving parts

7.19 The material used in the construction of stuffing box is
(a) Aluminium (b) Cast iron (c) Titanium (d) Chromium

7.20 Which thread is used in the screw Jack?
(a) square (b) trapezoidal (c) buttress (d) acme

Answers to Multiple-Choice Questions
7.1 (c), 7.2 (d), 7.3 (b), 7.4 (a), 7.5 (b), 7.6 (c), 7.7 (a), 7.8 (b), 7.9 (c), 7.10 (b), 7.11 (d), 7.12 (d), 7.13 (c), 7.14 (b), 7.15 (b), 7.16 (d), 7.17 (c), 7.18 (d), 7.19 (b), 7.20 (a)

B. Subjective Questions

7.1 Describe the purpose of a cotter joint.

7.2 Contrast different types of cotter joints.

7.3 Analyze why slots are made slightly bigger than the width of the cotter.

7.4 Examine the difference between a spigot cotter joint and a sleeve cotter joint.

7.5 Describe the utility of a knuckle joint.

7.6 Contrast the construction and functioning of the cotter and pin joint.

7.7 Articulate the advantages of protected type flange coupling over other types.

7.8 Explaining the construction and working of universal coupling.

7.9 Differentiate between Universal coupling and Constant Velocity joint.

7.10 Differentiate between Plummer block and pedestal support bearings.

7.11 Explain the utility of a footstep bearing.

7.12 Write the need for a hollow base of bearing supports and non-circular holes.

7.13 Describe how an eccentric works and where is it used in steam engines.

7.14 Discuss the functions of the small and big ends of a connecting rod.

7.15 Identify and illustrate the type of thread used in screw jack and lathe lead screw.

KNOW MORE

Cloud-based CAD is a growing trend that has significantly increased access to CAD in the modern day. SolidWorks, Autodesk and Onshape are some of the companies to initiate the offering of Cloud-based CAD in 2010, 2013 and 2015 respectively. Since then, this development has contributed to a wide-ranging shift in the CAD sector. Cloud-based CAD allows designers in different locations to collaborate on the same CAD model in real-time.

Due to the internet and cloud-based CAD, CAD software has become widely used in a variety of fields. In its early days, CAD was mostly employed by manufacturing and aerospace companies. These days, though, it is unusual to come across a sector that does not employ its use.

The prevalence of CAD's widespread adoption has grown steadily over the years. Design, simulation and production processes for almost all modern products rely on CAD software. Since CAD is so crucial, it has also broken through to the job market. Many job openings in the design industry today specifically want CAD experience or knowledge. Proficiency in AutoCAD, SolidWorks, Revit etc. is a highly prized commodity in the job market.

With *Augmented Reality*, CAD models may be seen in the physical world. Existing augmented reality software makes use of mobile devices like smartphones and tablets, recording the environment through a camera and superimposing a digital effect to examine the model. Successful augmented reality applications in the past include the sensational *Pokémon Go* game. Augmented reality's various uses in the CAD industry include providing designers with a preview of their models from all possible perspectives before they go to production. Apps developed by companies like Augment give designers a new way to interact with and preview 3D models built in SketchUp, Revit, SolidWorks and other CAD programmes.

Augmented and virtual reality-driven CAD is an exciting development with potentially far-reaching implications for the design business. It is possible that in the future before a product is even created, designers may be able to provide clients with augmented reality product renderings to examine in their own environments.

Some CAD businesses have already attained cloud-based CAD and many more are planning to do so soon. If this trend continues, users may never again need to install software patches. There would be no need to update or download cloud-based CAD because it would be accessible in your browser. Problems associated with data management, for example, may become irrelevant. With cloud-based CAD software, you can avoid saving several versions of a design or worrying about overwritten data.

The ability of CAD software to anticipate a designer's next action and respond appropriately is an exciting prospect for the future of computer-aided design. It is called *Generative Design*. A future when designers may collaborate with computers to arrive at a flawless design is not far off. *Dream Catcher*, which Autodesk is currently developing, is an artificial intelligence-aided tool that can generate hundreds of unique designs in hours. CAD has developed and shifted significantly through the years and this trend will only continue.

REFERENCES AND SUGGESTED READINGS

1. Pohit, G.; Ghosh, G. Machine Drawing with AutoCAD, 1st ed.; Pearson Education India: New Delhi, India, 2007.
2. Jensen, C.H.; Helsel, J.D.; Short, D. Engineering Drawing and Design, 7th ed.; McGraw Hill Higher Education: Maidenhead, England, 2007.
3. Singh, A. Machine Design Includes AutoCAD, 2nd ed.; Tata McGraw Hill Education Private Limited: New Delhi, India, 2018.
4. Patil, H.G.; Pilli, S.C.; Malagi, R.R.; Patil, M.S. Design of Machine Elements; I K International Publishing House: New Delhi, India, 2020.

Dynamic QR Code for Further Reading

Sleeve & Cotter Joint

Spigot & Cotter Joint

Knuckle Joint

Protected Type Flanged Coupling

Universal Coupling

Plummer Block

Foot Step Bearing

Simple Eccentric

Machine Vice

Connecting Rod

Stuffing Box

Screw Jack

CO AND PO ATTAINMENT TABLE

Course outcomes (COs) for this course can be mapped with the programme outcomes (POs) after the completion of the course and a correlation can be made for the attainment of POs to analyze the gap. After proper analysis of the gap in the attainment of POs necessary measures can be taken to overcome the gaps.

Table for CO and PO attainment

Course Outcomes	**ATTAINMENT OF PROGRAMME OUTCOMES** (1 – Weak Correlation; 2 – Medium correlation; 3 – Strong Correlation)						
	PO-1	**PO-2**	**PO-3**	**PO-4**	**PO-5**	**PO-6**	**PO-7**
CO-1							
CO-2							
CO-3							
CO-4							
CO-5							

The data filled in the above table can be used for gap analysis.

INDEX

Absolute Coordinate, 19
Aligned Dimension,
Angular Dimension,
Arc Command, 39
Array Command, 62
AutoCAD 2022, 4
Block Definitions, Creating,
Cartesian Coordinate System, 21
Chamfer Command, 61
Circle Command, 35
Command Dimensioning, 79
Components, AutoCAD Software Window,
Computer Aided Drafting, 2
Construction Line Command, 33
Coordinate System in AutoCAD, 19
Copy Command, 54
Cotter Joint with Sleeve, 178
Dimension Editing, 91
Dimension Types, 82
Dimensioning in AutoCAD, 77
Draw Menu, 29
Drawing, Setting Up New,
Dynmode, 13
Ellipse Command, 47
Extend Command, 59
File Features, 9
Fillet Command, 60
Freezing Layers in Viewports,
Grid, Creating,
Hatch Tool, 105
Isodraft, 15
Isometric Drawing,
Jogged Dimension,
Knuckle Joint, 184
Layer Override in Viewports,
Layers Command, 117
Linear Dimension,
Line Command, 31
Measure Command, 52
Mirror Command, 56
Modify Menu, 53
Move Command, 53
Object Snap, 15
Offset Command, 64
Orthomode, 14
Plotting, 159
Point Command, 30
Polar Coordinate, 20
Polyline Command, 45
Pre-requisites, 29
Quick Access Toolbar, 6
Radius Dimension,
Ray Command, 32
Rectangle Command, 44
Relative Coordinate, 20
Rotate Command, 56